Cambridge IGCSE® & O Level
Complete
Chemistry
Fourth Edition

RoseMarie Gallagher
Paul Ingram
Livien Khor
Nerissa Puntawe
Suriyani Rahamat

OXFORD
UNIVERSITY PRESS

Great Clarendon Street, Oxford, OX2 6DP, United Kingdom

Oxford University Press is a department of the University of Oxford. It furthers the University's objective of excellence in research, scholarship, and education by publishing worldwide. Oxford is a registered trade mark of Oxford University Press in the UK and in certain other countries

© Oxford University Press 2021

The moral rights of the authors have been asserted

First published in 2021

All rights reserved. No part of this publication may be reproduced, stored in a retrieval system, or transmitted, in any form or by any means, without the prior permission in writing of Oxford University Press, or as expressly permitted by law, by licence or under terms agreed with the appropriate reprographics rights organization. Enquiries concerning reproduction outside the scope of the above should be sent to the Rights Department, Oxford University Press, at the address above.

You must not circulate this work in any other form and you must impose this same condition on any acquirer

British Library Cataloguing in Publication Data
Data available

978-1-38-200585-2 (standard)

10 9 8 7 6 5 4 3 2 1

978-1-38-200584-5 (enhanced)

10 9 8 7 6 5 4 3 2 1

Paper used in the production of this book is a natural, recyclable product made from wood grown in sustainable forests. The manufacturing process conforms to the environmental regulations of the country of origin.

Printed in Italy by L.E.G.O. SpA

Acknowledgements

IGCSE® is the registered trademark of Cambridge International Examinations. All answers have been written by the authors. In examination, the way marks are awarded may be different.

The publisher and authors would like to thank the following for permission to use photographs and other copyright material:

Cover: Sputnik/Science Photo Library.

Photos: p2(t): Radius Images/Alamy Stock Photo; p2(bl): Vovan Dvolitko/Shutterstock; p2(br):Suhendri/Bigstock; p3(l): RoseMarie Gallagher; p3(r): RoseMarie Gallagher; p4(l): Dreamsquare/Shutterstock; p4(m): mahirart/Shutterstock; p4(r):AndreaAstes/iStockPhoto; p5(t): Pegasus/Visuals Unlimited Inc./Science Photo Library; p5(m): Prostock-studio/Alamy Stock Photo; p5(bl): Stu Shaw/Shutterstock; p5(br): Scott Mattock/Shutterstock; p6(t): KC Slagle/Shutterstock; p6(m): Alexandr Vlassyuk/Shutterstock; p6(b): KC Slagle/Shutterstock; p10: Bbtree/Alamy Stock Photo; p11(t): kurtjurgen/Shutterstock; p11(b): Mira Drozdowski/Shutterstock; p14(tl): RoseMarie Gallagher; p14(tm): Dr Alexandra Waine; p14(tr): Dr Alexandra Waine; p15: Dr Alexandra Waine; p14(bl): Cavan/Alamy Stock Photo; p14(br):Andri Tambunan/Corbis; p16: Iofoto/Shutterstock; p18: Photograph by Paul Ehrenfest, copyright status unknown. Coloured by Science Photo Library; p19(t): Dr Alexandra Waine; p19(b): Georgeclerk/iStock Unreleased/Getty Images; p20(t): Krakenimages.com/Shutterstock; p20(b): Hans engbers/Shutterstock; p21: Martyn F. Chillmaid/Science Photo Library; p22(t): Robert-Horvat/Shutterstock; p22(m): Vkilikov/Shutterstock; p22(br):Stephen Kiers/Shutterstock; p22(bl): Science Photo Library; p23(t): The Print Collector/Alamy Stock Photo; p23(m): Mark Yuill/ Shutterstock; p23(b): IBM/Science Photo Library; p26(l): Shutterstock; p26(m): Shutterstock; p26(r): Shutterstock; p27: Tlorna/Shutterstock; p28(tl): Andrew Lambert Photography/Science Photo Library; p28(tm): Joerg Beuge/Shutterstock; p28(m): Photocritical/Shutterstock; p28(b): Lubos Chlubny/Shutterstock; p29: Turtle Rock Scientific/Science Photo Library; p30: Giphotostock/Science Photo Library; p31: Sebalos/iStockPhoto; p32: KMNPhoto/Shutterstock; p33: RoseMarie Gallagher; p34(m): Shutterstock; p34(b): Ian Miles-Flashpoint Pictures/Alamy Stock Photo; p35(t):RoseMarie Gallagher; p35(m): RoseMarie Gallagher; p35(b): RoseMarie Gallagher; p36(a): RoseMarie Gallagher; p36(b): RoseMarie Gallagher; p36(d): RoseMarie Gallagher; p36(c): RoseMarie Gallagher; p37(a): RoseMarie Gallagher; p37(b): RoseMarie Gallagher; p37(c): RoseMarie Gallagher; p38(tm): Shutterstock; p38(tr): Daniel Taeger/Shutterstock; p38(bm): Stefan Glebowski/Shutterstock; p38(br): IP Galanternik D.U./iStockPhoto; p39(t): Igor Nikushin/Shutterstock; p39(b): Stefan Redel/Shutterstock; p40(t): Dmitry Kalinovsky/ Shutterstock; p40(m):Shutterstock; p41(t): RoseMarie Gallagher; p41(b): RoseMarie Gallagher; p42(t): James Steidl/Shutterstock; p42(bm): Shutterstock; p42(br):Zbynek Burival/Shutterstock; p43(t): TonyV3112/Shutterstock; p43(b): NigelSpiers/Shutterstock; p46: Napawadee Thaisonthi/Shutterstock; p47:Paul Rapson/Science Photo Library; p48(t): NASA; p48(b): NASA; p49(t): Dr Alexandra Waine; p49(m): RoseMarie Gallagher; p49(b):GIphotostock/Science Photo Library; p50: PNWL/Alamy Stock Photo; p51: RoseMarie Gallagher; p52(t): RoseMarie Gallagher; p52(b): Jean-Loup Charmet/Science Photo Library; p53(t): Chanishka Colombage/Shutterstock; p53(b): Morozov Anatoly/Shutterstock; p56(l): Dr Alexandra Waine; p56(m): Dr Alexandra Waine; p56(r): Dr Alexandra Waine; p58: RoseMarie Gallagher; p59(t): RoseMarie Gallagher; p59(b): Dr Alexandra Waine; p60(t): Nastco/iStockPhoto; p60(b): RoseMarie Gallagher; p61: Vulcan.wr.usgs.gov; p64: RoseMarie Gallagher; p65: Ron Kloberdanz/Shutterstock; p66: Emptyclouds/iStockPhoto; p67(t): Pramote Polyamate/Alamy Stock Photo; p67(b): Raksha Shelare/Shutterstock; p68(t): Thor Jorgen Udvang/Shutterstock; p68(b): Chepko/iStockPhoto; p69: T-lorien/E+/Getty Images; p72(t): RoseMarie Gallagher; p72(bl): Jordache/Shutterstock; p72(br): Claudio Arnese/iStockPhoto; p73(t): Ulga/Shutterstock; p73(b): vasa/Alamy Stock Photo; p74: Redfrisbee/Shutterstock; p75: Charles D. Winters/Science Photo Library; p76(t): RoseMarie Gallagher; p76(b): Dr Alexandra Waine; p77(t): RoseMarie Gallagher; p77(b): RoseMarie Gallagher; p78: Andrew Lambert Photography/Science Photo Library; p79(t): Andrew Lambert Photography/Science Photo Library; p79(m): Yuri Shevtsov/Shutterstock; p79(bl): Surya Nair/Alamy Stock Photo; p79(br): Lewis Houghton/Science Photo Library; p82(tl): RoseMarie Gallagher; p82(mr): Muellek Josef/Shutterstock; p83(tl): RoseMarie Gallagher; p83(tm): RoseMarie Gallagher; p83(tr): RoseMarie Gallagher; p89(tl): Mlle Sonyah/Alamy Stock Photo; p89(tr): Harshit Srivastava S3/Shutterstock; p92(tl): RoseMarie Gallagher; p92(tm): Charles D. Winters/Science Photo Library; p92(tr): RoseMarie Gallagher; p93(tl): RoseMarie Gallagher; p93(tm): Sciencephotos/Alamy Stock Photo; p93(tr): RoseMarie Gallagher; p94: Vera Larina/Shutterstock; p95: Jan Kaliciak/Shutterstock; p96(bl): Icosha/Shutterstock; p96(br): Charles D. Winters/Science Photo Library; p98: Yoshikazu Tsuno/AFP/Getty Images; p99: Paceman/Shutterstock; p102(tl): RoseMarie Gallagher; p102(tm): Hafizzuddin/Shutterstock; p102(tr): Charles Schug/iStockPhoto; p102(bl): Best Images/Shutterstock; p102(bm): Siriporn-88/Shutterstock; p102(br): 06photo/Shutterstock; p103(tl): RoseMarie Gallagher; p103(tm): RoseMarie Gallagher; p103(tr): RoseMarie Gallagher; p104: Patryk Kosmider/Shutterstock; p105:RoseMarie Gallagher; p106: RoseMarie Gallagher; p107(t): Varela/iStockPhoto; p107(b): cglade/iStockPhoto; p108(l): RoseMarie Gallagher; p108(r): RoseMarie Gallagher; p109(t): PA Images/Alamy Stock Photo; p109(b): DmyTo/Shutterstock; p112: PRILL/Shutterstock; p113(t): James L. Amos/The Image Bank /Getty Images; p113(b): Jasminko Ibrakovic/Alamy Stock Photo; p114(tr): Knorre/Shutterstock; p114(mr):Grasko/Shutterstock; p114(bl): Volker Steger/Science Photo Library; p114(br): RoseMarie Gallagher; p115(t): Shannon Matteson/Shutterstock; p115(m): Jinga80/Fotolia; p115(b): NOAA Photo Library; p118(tl): RoseMarie Gallagher; p118(tm): RoseMarie Gallagher; p118(tr): AlixWaine: Dr Alexandra Waine; p118(bl): Dr Alexandra Waine; p118(bm): Dr Alexandra Waine; p118(br): Dr Alexandra Waine; p119: Peter Albrektsen/Shutterstock; p122: Mosista/Shutterstock; p123: Dinodia Photos/Alamy Stock Photo; p124: Alexey Rezvykh/Alamy Stock Photo; p125(m): GIphotostock/Science Photo Library; p125(b): Southtownboy Studio/Shutterstock; p128(tl): Dr Alexandra Waine; p128(tm): Zoart Studio/Shutterstock; p128(tr): RoseMarie Gallagher; p128(m): RoseMarie Gallagher; p128(bl): Viktor_LA/Shutterstock; p128(br): Self-taught/Shutterstock; p129(tr): SolStock/iStockPhoto; p129(ml): Martyn F. Chillmaid/Science Photo Library; p129(m): Dr Alexandra Waine; p129(mr): Zoart Studio/Shutterstock; p130: RoseMarie Gallagher; p131(t): Andrew Lambert Photography/Science Photo Library; p131(b):Robcruse/iStockPhoto; p132(t): Dr Alexandra Waine; p132(m): RoseMarie Gallagher; p132(b): RoseMarie Gallagher; p133(t): Natee Photo/Shutterstock; p133(b): WAYHOME studio/Shutterstock; p134: Dr Alexandra Waine; p135: David R. Frazier Photolibrary, Inc./Alamy Stock Photo; p137: RoseMarie Gallagher; p139: Dr Alexandra Waine; p140: RoseMarie Gallagher; p141(t): Science Photo Library; p141(m): liliya Vantsura/Shutterstock; p141(b): Ihor Matsiievskyi/Shutterstock; p142: Trevor Clifford Photography/Science Photo Library; p143(tl): Turtle Rock Scientific/Science Photo Library; p143(tm): Science Photo Library; p143(mr): Goir/Shutterstock; p147: TungCheung/Shutterstock; p148(t):RoseMarie Gallagher; p148(b): Trevor Clifford Photography/Science Photo Library; p149(t): Martyn F. Chillmaid/Science Photo Library; p149(b):RoseMarie Gallagher; p150(t): Andrew Lambert Photography/Science Photo Library; p150(b): Dlewis33/iStockPhoto; p151: Andrew Lambert Photography/Science Photo Library; p152: Turtle Rock Scientific/Science Source/Science Photo Library; p153: Foto-Ruhrgebiet/Shutterstock; p154(tl): RoseMarie Gallagher; p154(tm): RoseMarie Gallagher; p154(tr): RoseMarie Gallagher; p154(b): Javarman/Shutterstock; p155(l):hbaumann/Shutterstock; p155(r): Dennis Hallinan/Alamy Stock Photo; p156(t): March Cattle/Shutterstock; p156(b): Africa Studio/Shutterstock; p157(tl): Time & Life Pictures/Getty Images; p157(m): Dmitry Yashkin/Shutterstock; p157(b): G. Victoria/Shutterstock; p161: Maarten Zeehandelaar/Shutterstock; p163: GIphotostock/Science Photo Library; p165: Charles D. Winters/Science Photo Library; p166(t): Martin Anderson/Shutterstock; p166(b): Sergey Peterman/Shutterstock; p167(t): Dean Conger/Corbis/Getty Images; p167(bl): Simon Turner/Alamy Stock Photo; p167(br): Kapuska/Shutterstock; p168(tr): Piotr Tomicki/Shutterstock; p169(t): Robert Buchanan Taylor/ Shutterstock; p168(bl):VT750/Shutterstock; p168(br): VDWimages/Shutterstock; p169(bl): John Panella/Shutterstock; p169(br): Ian Cartwright/LGPL/Alamy Stock Photo; p172(t): Thorsten Rust/Shutterstock; p172(m): YegoroV/Shutterstock; p172(bl): Alejandro Lafuente Lopez/Shutterstock; p172(bm):Telia/Shutterstock; p172(br): Albert Russ/Shutterstock; p173: Paul Fleet/Shutterstock; p174(tr): Crown Copyright/Health & Safety Laboratory/Science Photo Library; p174(mr): Jiang Zhongyan/Shutterstock; p174(br): Sergey Zavalnyuk/Dreamstime; p175:Lost_in_the_Midwest/Shutterstock; p176(tl): Bill Bachman/Alamy Stock Photo; p176(tm): electra/Shuterstock; p176(tr): Ngataringa/iStockPhoto; p176(bl): Andrey Rudakov/Bloomberg/Getty Images; p176(bm): Pavel L Photo and Video/Shutterstock; p176(br): Martyn Jandula/Shutterstock; p177(t): Maximilian Stock Ltd/Science Photo Library; p177(b): boyphare/Shutterstock; p178(tl): Alexey Rezvykh/Shutterstock; p178(tm):kaband/Shutterstock; p178(tr): John van Hasselt/Sygma/Getty Images; p178(bl): RoseMarie Gallagher; p178(bm): Novinit/Shutterstock; p178(br):Kaushik Ranjan Haldar/Shutterstock; p179(tl): Bokeh Blur Background/Shutterstock; p179(tr): imageBROKER/Alamy Stock Photo; p179(br): dodotone/Shutterstock; p180(tl): RoseMarie Gallagher; p180(tm): Anton Foltin/ Shutterstock; p180(tr): Svetlana Lukienko/Shutterstock; p180(m): motive56/Shutterstock; p180(bl): W_NAMKET/Shutterstock; p180(br): Gatot Adri/Shutterstock; p183(tl): Erhan Dayi/Shutterstock; p183(tr):Cobalt/iStockPhoto; p183(m): Eoghan McNally/Shutterstock; p183(b): GeorgiosArt/iStockPhoto; p186(t): S.E.A. Photo/Alamy Stock Photo; p186(b): lidia hayaty/Shutterstock; p187(t): Spring Images/Alamy Stock Photo; p187(b): Weerayuth Kanchanacharoen/Shutterstock; p188: Sakarin Sawasdinaka/Shutterstock; p189: Greentellect Studio/Shutterstock; p190: Martyn F. Chillmaid/Science Photo Library; p192(tl): P Wei/iStockPhoto; p192(tm): Aggphotographer/Bigstock; p192(tr): Adam Hart Davis/Science Photo Library; p192(br): ~Userc0373230_9/iStockPhoto; p192(bl):RoseMarie Gallagher; p193(t): Vitaliy Kyrychuk/Shutterstock; p193(b): HeikeKampe/iStockPhoto; p194(bl): NASA; p194(br): Elisei Shafer/Shutterstock; p195(t): Eraxion/iStockPhoto; p195(b): Viridis/iStockPhoto; p196(t): RoseMarie Gallagher; p196(b): Ian Bracegirdle/Shutterstock; p197: Anticiclo/Shutterstock; p198: Ramniklal Modi/Shutterstock; p199(t): Tanja Esser/Shutterstock; p199(bl): Marc Van Vuren/Shutterstock; p199(br): Tongra239/Shutterstock; p200(l): Jenson/Shutterstock; p200(r): Pla2na/Shutterstock; p201: Ashley Cooper/Alamy Stock Photo; p204(tl): Paul Rapson/Science Photo Library; p204(tm): tonton/Shutterstock; p204(tr): Hywit Dimyadi/Shutterstock; p205(tl):Lya_Cattel/iStockPhoto; p205(tm): RoseMarie Gallagher; p205(tr): TY Lim/Shutterstock; p205(b): Panya7/Shutterstock; p206: ymgerman/Shutterstock; p207(t): Peter Gudella/Shutterstock; p207(b): Kozmoat98/iStockPhoto; p208(t): Ulga/Shutterstock; p208(b): Leszek Czerwonka/Shutterstock; p210: Dimas Ardian/Bloomberg/Getty Images; p211: Abdul Razak Latif/Shutterstock; p212(t): MirAgreb/iStockPhoto; p212(b): agchinook/Shutterstock; p213(t): littlenySTOCK/Shutterstock; p213(bl): Avigator Fortuner/Shutterstock; p213(br): Gold Picture/Shutterstock; p214(t): Loekiepix/Shutterstock; p214(b): Ekkamai Chaikanta/Shutterstock; p215(t): Trevor Clifford Photography/Science Photo Library; p215(bl): dpa picture alliance/Alamy Stock Photo; p215(br): Mr.Sompong Kantotong/Shutterstock; p216: Navinpeep/Moment/Getty Images; p217(t): PJiiiJane/Shutterstock; p217(b): Chutharat Kamkhuntee/Shutterstock; p218(t): RoseMarie Gallagher; p218(b): Turtle Rock Scientific/Science Photo Library; p222(tl): RoseMarie Gallagher; p222(tm): RoseMarie Gallagher; p222(tr): NiceProspects-Prime/Alamy Stock Photo; p222(b): Ollyy/Shutterstock; p223(tl): Highviews/iStockPhoto; p223(tm): Fizkes/Shutterstock; p223(tr): Gabi Moisa/Shutterstock; p223(b): quangmooo/Shutterstock; p223(b):Chris Hellyar/Shutterstock; p224(t): Pressmaster/Shutterstock; p224(b): Andriscam/Shutterstock; p225: Dinodia Photos/Alamy Stock Photo; p226:Charles D. Winters/Science Photo Library; p227: RoseMarie Gallagher; p228(t): Oksix/Shutterstock; p228(b): RoseMarie Gallagher; p229(t): Alexey U/Shutterstock; p229(bl): Imaginechina Limited/Alamy Stock Photo; p229(br): Bluestocking/iStockPhoto; p230(t): Dorling Kindersley ltd/Alamy Stock Photo; p230(m): kakteen/Shutterstock; p230(bl): jaiman taip/Shutterstock; p230(br): Paulo Oliveira/Alamy Stock Photo; p231(t):Mr.anaked/Shutterstock; p231(b): Roman Mikhailiuk/Shutterstock; p232(t): photka/Shutterstock; p232(bl): Imagebroker/Alamy Stock Photo; p232(br): Albert Karimov/Alamy Stock Photo; p233(t): Nito100/iStock/Getty Images Plus/Getty Images; p233(m): Tribalium/Shutterstock; p233(b):whity2j/Shutterstock; p234(t): Maryna Olyak/Shutterstock; p234(b): Bennyartist/Shutterstock; p235(tl): Alissa Eckert, MSMI, Dan Higgins, MAMS; p235(tr): Felix Dlangamandla/Beeld/Gallo Images/Getty Images; p235(m): AGITA LEIMANE/Shutterstock; p235(b): Igor Dutina/Shutterstock; p238(tr): Artsiom P/Shutterstock; p238(tl): Dr Alexandra Waine; p238(tm): Dr Alexandra Waine; p238(ml): Dr Alexandra Waine; p238(m): Dr Alexandra Waine; p238(mr): Dr Alexandra Waine; p238(bl): Dr Alexandra Waine; p238(bm): Dr Alexandra Waine; p238(br): Dr Alexandra Waine; p240(t): Shutterstock; p240(b): Shutterstock; p241: Art Directors & TRIP/Alamy Stock Photo; p242: Imagestock/iStockPhoto; p243(t):Wloven/iStockPhoto; p243(m): RoseMarie Gallagher; p243(b): RoseMarie Gallagher; p244(tr): aluxum/iStokphoto; p244(bl): Daniel Balakov/iStokphoto; p244(br): Art Directors & TRIP/Alamy Stock Photo; p245: Rin Boonprasan/Shutterstock; p246: Andrew Lambert Photography/Science Photo Library; p250(mr): JPC-PROD/Shutterstock; p250(bl): RoseMarie Gallagher; p250(br): RoseMarie Gallagher; p251:Photodsotiroff/iStokphoto; p252(tl): LorenRodgers/Shutterstock; p252(tr): Natasa Adzic/Shutterstock; p253(mr): Mtr/Shutterstock; p253(bl): PhilAugustavo/iStockPhoto; p253(br): Kim Christensen/Shutterstock; p256: Tonkid/Shutterstock; p257(tl): Ktsdesign/Shutterstock; p257(tm): SDI Productions/iStockPhoto; p257(tr): Martyn F. Chillmaid/Science Photo Library; p257(ml): Golfcphoto/iStockPhoto; p257(m): Steven Collins/Shutterstock; p257(mr): Arieliona/Shutterstock; p257(br): Science History Images/Alamy Stock Photo; p259: RoseMarie Gallagher; p262(t):Andrew Lambert Photography/Science Photo Library; p262(b): Science Photo Library; p264: Science Photo Library; p265(t): Yeko Photo Studio/Shutterstock; p265(b): Andrew Lambert Photography/Science Photo Library; p285: Sujitra Chaowdee/Shutterstock.

Artwork by Q2A Media Services Pvt. Ltd.

Every effort has been made to contact copyright holders of material reproduced in this book. Any omissions will be rectified in subsequent printings if notice is given to the publisher.

Introduction

If you are taking IGCSE chemistry, this book is for you. It has been written to help you understand chemistry – and do well in your exams. It covers the syllabus fully.

Finding your way around the book
The contents list on page xiv shows how the book is organised. The chapters are divided into two-page units. And note the IGCSE practice questions and the reference section at the back of the book.

Using the book
These points will help you use this book effectively.

The Core syllabus content If you are following the Core syllabus, you can ignore any material with a red line beside it.	**The Extended syllabus content** For this, you need all the main material, including the extended material marked with a red line.	**Chapter checkups** The last unit in each chapter has a revision checklist, and exam-level questions on the chapter content.
Extra material Some units, *with no numbers*, set chemistry in a wider context. You do not need these for your exams!	**The IGCSE practice questions** The twenty-four pages of questions at the end of the book are exam-level, with mark schemes.	**The reference section** The glossary has the vocabulary you need. The Periodic Table and A_r values will be useful too.

Answering questions
Answering questions is a great way to get to grips with a topic. This book has hundreds of questions: at the end of each unit and each chapter, as well as the IGCSE practice questions at the back of the book.

Answers are given at the back of the book, except for the IGCSE practice questions. Those answers can be accessed online at   www.oxfordsecondary.com/complete-igcse-science

Preparing for the exam
It is important to know what is on your syllabus! The **syllabus grid** on pages iv–xii shows the syllabus content, to help you. Page xiii outlines the different papers in the exam, and the types of questions to expect.

Online material
The **Enhanced Online Book** is a separate purchase. It supports this book, to help you build and practise the knowledge and skills for success in IGCSE chemistry. It provides a wealth of additional resources:

- a worksheet and interactive quiz for every unit
- *On Your Marks* activities to help you achieve your best
- glossary quizzes to test your understanding of chemical terminology
- additional practice papers with mark schemes.

And finally …
Chemistry is a big exciting subject, of great importance in today's world. We hope this book will help you to understand chemistry … and enjoy it.

RoseMarie Gallagher and Paul Ingram

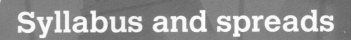
Syllabus and spreads

This grid lists the topics on the IGCSE chemistry syllabus, and where to find them in the textbook. It will help you with revision.
Red text shows Extended syllabus content.

Topic in the IGCSE chemistry syllabus	Unit in this book
1 STATES OF MATTER	
1.1 Solids, liquids and gases	
1 State the distinguishing properties of solids, liquids and gases	1.2
2 Describe the structures of solids, liquids and gases in terms of particle separation, arrangement and motion	1.3
3 Describe changes of state in terms of melting, boiling, evaporating, freezing and condensing	1.2
4 Describe the effects of temperature and pressure on the volume of a gas	1.5
5 Explain changes of state in terms of kinetic particle theory, including the interpretation of heating and cooling curves	1.4
6 Explain, in terms of kinetic particle theory, the effects of temperature and pressure on the volume of a gas	1.5
1.2 Diffusion	
1 Describe and explain diffusion in terms of kinetic particle theory	1.1
2 Describe and explain the effect of relative molecular mass on the rate of diffusion of gases	1.2
2 ATOMS, ELEMENTS AND COMPOUNDS	
2.1 Elements, compounds and mixtures	
1 Describe the differences between elements, compounds and mixtures	2.1, 3.1
2.2 Atomic structure and the Periodic Table	
1 Describe the structure of the atom as a central nucleus containing neutrons and protons surrounded by electrons in shells	2.2
2 State the relative charges and relative masses of a proton, a neutron and an electron	2.2
3 Define proton number / atomic number as the number of protons in the nucleus of an atom	2.2
4 Define mass number / nucleon number as the total number of protons and neutrons in the nucleus of an atom	2.2
5 Determine the electronic configuration of elements and their ions with proton number 1 to 20, e.g. 2,8,3	2.3
6 State that: (a) Group VIII noble gases have a full outer shell (b) the number of outer shell electrons is equal to the group number in Groups I to VII (c) the number of occupied electron shells is equal to the period number	2.3
2.3 Isotopes	
1 Define isotopes as different atoms of the same element with the same number of protons but different numbers of neutrons	2.4
2 Interpret and use symbols for atoms, e.g. $^{12}_{6}C$, and ions, e.g. $^{35}_{17}Cl^-$	2.4
3 State that isotopes of an element have the same chemical properties because they have the same number of outer shell electrons	2.4
4 Calculate the relative atomic mass of an element from the relative masses and abundances of its isotopes	2.4
2.4 Ions and ionic bonds	
1 Describe the formation of positive ions, known as cations, and negative ions, known as anions	3.2
2 State that an ionic bond is a strong electrostatic attraction between oppositely charged ions	3.3
3 Describe the formation of ionic bonds between elements from Group I and Group VII (including dot-and-cross diagrams)	3.3
4 Describe the properties of ionic compounds: (a) high melting points and boiling points (b) good electrical conductivity when aqueous or molten and poor when solid	3.5, 3.6

5	Describe the giant lattice structure of ionic compounds as a regular arrangement of alternating positive and negative ions	3.7
6	Describe the formation of ionic bonds between ions of metallic and non-metallic elements, including dot-and-cross diagrams	3.3
7	Explain in terms of structure and bonding the properties of ionic compounds (see 4)	3.7
2.5 Simple molecules and covalent bonds		
1	State that a covalent bond is formed when a pair of electrons is shared leading to noble gas electronic configurations	3.5
2	Describe the formation of covalent bonds in simple molecules including H_2, Cl_2, H_2O, CH_4, NH_3 and HCl, using dot-and-cross diagrams	3.5
3	Describe in terms of structure and bonding the properties of simple molecular compounds: (a) low melting points and boiling points (b) poor electrical conductivity	3.7
4	Describe the formation of covalent bonds in simple molecules, including: CH_3OH, C_2H_4, O_2, CO_2 and N_2	3.5, 3.6
5	Explain in terms of structure and bonding: (a) the low melting points and boiling points of simple molecular compounds (b) their poor electical conductivity	3.7
2.6 Giant covalent structures		
1	Describe the giant covalent structures of graphite and diamond	3.8
2	Relate the structures and bonding of graphite and diamond to their uses	3.8
3	Describe the giant covalent structure of silicon(IV) oxide, SiO_2	3.9
4	Describe the similarity in properties between diamond and silicon(IV) oxide, related to their structures	3.9
2.7 Metallic bonding		
1	Describe metallic bonding as the electrostatic attraction between the positive ions in a giant metallic lattice and a 'sea' of delocalised electrons	3.9
2	Explain in terms of structure and bonding the properties of metals: (a) good electrical conductivity (b) malleability and ductility	3.9
3 STOICHIOMETRY		
3.1 Formulae		
1	State the formulae of the elements and compounds named in the syllabus	throughout
2	Define the molecular formula of a compound as the number and type of different atoms in one molecule	4.1
3	Deduce the formula of a simple compound from the relative numbers of atoms present in a model or diagram	4.1
4	Construct word equations and symbol equations to show how reactants form products, including state symbols	4.2
5	Define the empirical formula of a compound as the simplest whole number ratio of the different atoms or ions in a compound	5.5
6	Deduce the formula of an ionic compound from a model or diagram, or from the charges on the ions	4.1
7	Construct symbol equations with state symbols, including ionic equations	throughout
8	Deduce the symbol equation with state symbols for a chemical reaction, given relevant information	throughout
3.2 Relative masses of atoms and molecules		
1	Describe relative atomic mass, A_r.	2.4, 4.3
2	Define relative molecular mass and relative formula mass M_r as the sum of the relative atomic masses	4.3
3	Calculate reacting masses in simple proportions	4.4
3.3 The mole and the Avogadro constant		
1	State that concentration can be measured in g/dm^3 or mol/dm^3	5.4
2	Define the mole (mol) and Avogadro constant	5.1
3	Use this relationship in calculations: amount of substance (mol) = mass (g) ÷ molar mass (g/mol)	5.1
4	Use the molar gas volume, taken as 24 dm^3 at room temperature and pressure, r.t.p., in calculations involving gases	5.3
5	Calculate reacting masses, limiting reactants, gas volumes at r.t.p., volumes and concentrations of solutions, in correct units	5.2, 5.3, 5.4
6	Use experimental data from a titration to calculate the moles of solute, or the concentration or volume of a solution	5.7, 11.8

7	Calculate empirical formulae and molecular formulae	5.5, 5.6
8	Calculate percentage yield, percentage composition by mass and percentage purity	5.7

4 ELECTROCHEMISTRY

4.1 Electrolysis

1	Define electrolysis as the decomposition of an ionic compound, by the passage of an electric current	7.1
2	Identify in simple electrolytic cells: (a) the anode (b) the cathode (c) the electrolyte	7.2
3	Identify the products at the electrodes and describe observations made during the electrolyis of: (a) molten lead(II) bromide (b) concentrated aqueous sodium chloride (c) dilute sulfuric acid using inert electrodes	7.2
4	State that metals or hydrogen are formed at the cathode and that non-metals (other than hydrogen) are formed at the anode	7.2
5	Predict the identity of the products at each electrode for the electrolysis of a binary compound in the molten state	7.2
6	State that metal objects are electroplated to improve their appearance and resistance to corrosion	7.4
7	Describe how metals are electroplated	7.4
8	Describe the transfer of charge during electrolysis to include: (a) the movement of electrons in the external circuit (b) the loss or gain of electrons at the electrodes (c) the movement of ions in the electolyte	7.3
9	Identify the products formed, and describe the observations made, during the electrolysis of aqueous copper(II) sulfate when using inert electrodes and copper electrodes	7.4
10	Predict the products at each electrode for the electrolysis of a halide compound in dilute or concentrated aqueous solution	7.3
11	Construct ionic half-equations for reactions at the anode (to show oxidation) and at the cathode (to show reduction)	7.3

4.2 Hydrogen – oxygen fuel cells

1	State that a hydrogen – oxygen fuel cell uses hydrogen and oxygen to produce electricity, with water as the only chemical product	8.4
2	Describe the advantages and disadvantages of using hydrogen – oxygen fuel cells in comparison with gasoline / petrol engines in vehicles	8.4

5 CHEMICAL ENERGETICS

5.1 Exothermic and endothermic reactions

1	State that an exothermic reaction transfers thermal energy to the surroundings leading to an increase in the temperature of the surroundings	8.1
2	State that an endothermic reaction takes in thermal energy from the surroundings leading to a decrease in the temperature of the surroundings	8.1
3	Interpret reaction pathway diagrams showing exothermic and endothermic reactions	8.2
4	State that the transfer of thermal energy during a reaction is called the enthalpy change, ΔH, of the reaction. ΔH is negative for exothermic reactions and positive for endothermic reactions	8.2
5	Define activation energy, E_a, as the minimum energy that colliding particles must have to react	8.2
6	Draw and fully label reaction pathway diagrams for exothermic and endothermic reactions	8.2
7	Explain the enthalpy change of a reaction in terms of bond breaking and bond making	8.3
8	Calculate the enthalpy change of a reaction using bond energies	8.3

6 CHEMICAL REACTIONS

6.1 Physical and chemical changes

1	Identify physical and chemical changes, and describe the differences between them	3.1

6.2 Rate of reaction

1	Describe the effect on the rate of reaction of changes in: (a) concentration of solutions, (b) gas pressure, (c) surface area of solids, (d) temperature, and of adding or removing a catalyst, including enzymes	9.3, 9.4
2	State that a catalyst increases the rate of a reaction and is unchanged at the end of a reaction	9.6

3	Describe practical methods for investigating the rate of a reaction	9.2, 9.3, 9.4
4	Interpret data, including graphs, from rate of reaction experiments	9.1
5	Describe collision theory in terms of: (a) number of particles (b) frequency of collisions (c) kinetic energy (d) activation energy, E_a	9.5
6	Use collision theory to explain the effect on the rate reaction of changes in: (a) concentration of solutions, (b) gas pressure, (c) surface area of solids, (d) temperature, and of adding or removing a catalyst, including enzymes	9.5
7	State that a catalyst decreases the activation energy E_a of a reaction	9.6
8	Evaluate practical methods for investigating the rate of a reaction	9.2, 9.3, 9.4
6.3 Reversible reactions and equilibrium		
1	State that some chemical reactions are reversible as shown by the symbol $\rightleftharpoons$	10.1
2	Describe how changing the conditions can change the direction of a reversible reaction, using the effect of heat and water on copper(II) sulfate and colbalt(II) chloride as examples	10.1
3	State that a reversible reaction in a closed system is at equilibrium when: (a) the rate of the forward reaction is equal to the rate of the reverse reaction (b) the concentrations of reactants and products are no longer changing	10.1
4	Predict and explain, for a reversible reactions, how the position of equilibrium is affected by: (a) changing temperature, (b) changing pressure, (c) changing concentration, (d) using a catalyst	10.2
5	State the symbol equation for the production of ammonia in the Haber process	10.2
6	State the sources of the hydrogen (methane) and nitrogen (air) in the Haber process	10.3
7	State the typical conditions in the Haber process	10.3
8	State the symbol equation for the conversion of sulfur dioxide to sulfur trioxide in the Contact process	10.4
9	State the sources of the sulfur dioxide (burning sulfur or roasting sulfide ores) and oxygen (air) in the Contact process	10.4
10	State the typical conditions for the conversion of sulfur dioxide to sulfur trioxide in the Contact process	10.4
11	Explain why the typical conditions stated are used in the Haber Process and in the Contact process, including safety considerations and economics	10.3, 10.4
6.4 Redox		
1	Use a Roman numeral to indicate the oxidation number of an element in a compound	6.3
2	Define redox reactions as involving simultaneous oxidation and reduction	6.1
3	Define oxidation as gain of oxygen and reduction as loss of oxygen	6.1
4	Identify redox reactions as reactions involving gain and loss of oxygen	6.1
5	Identify oxidation and reduction in redox reactions	6.1
6	Define oxidation in terms of: (a) loss of electrons (b) an increase in oxidation number	6.2
7	Define reduction in terms of: (a) gain of electrons (b) a decrease in oxidation number	6.2
8	Identify redox reactions as reactions involving gain and loss of electrons	6.2
9	Identify redox reactions by changes in oxidation number (using the rules for oxidation number)	6.3
10	Identify redox reactions by the colour changes involved when using acidified aqueous potassium manganate(VII) or aqueous potassium iodide	6.4
11	Define an oxidising agent as a substance that oxidises another substance and is itself reduced	6.4
12	Define a reducing agent as a substance that reduces another substance and is itself oxidised	6.4
13	Identify oxidising agents and reducing agents in redox reactions	6.4
7 ACIDS, BASES AND SALTS		
7.1 The characteristic properties of acids and bases		
1	Describe the characteristic properties of acids in terms of their reactions with: (a) metals (b) bases (c) carbonates	11.3
2	Describe acids in terms of their effect on: (a) litmus (b) thymolphthalein (c) methyl orange	11.1
3	State that bases are oxides or hydroxides of metals and that alkalis are soluble bases	11.1
4	Describe the characteristic properties of bases in terms of their reactions with: (a) acids (b) ammonium salts	11.3

5	Describe alkalis in terms of their effect on: (a) litmus (b) thymolphthalein (c) methyl orange	11.1
6	State that aqueous solutions of acids contain H⁺ ions and aqueous solutions of alkalis contain OH⁻ ions	11.2
7	Describe how to compare hydrogen ion concentration, neutrality, in terms of colour and pH using universal indicator paper	11.2
8	Describe the neutralisation reaction between an acid and an alkali to produce water: $H^+(aq) + OH^-(aq) \longrightarrow H_2O(l)$	11.3
9	Define acids as proton donors and bases as proton acceptors	11.4
10	Define a strong acid as an acid that is completely dissociated in aqueous solution and a weak acid as an acid that is partially dissociated in aqueous solution	11.2
11	State that hydrochloric acid is a strong acid, as shown by the symbol equation: $HCl(aq) \longrightarrow H^+(aq) + Cl^-(aq)$	11.2
12	State that ethanoic acid is a weak acid, as shown by the symbol equation: $CH_3COOH(aq) \longrightarrow H^+(aq) + CH_3COO^-(aq)$	11.2

7.2 Oxides

1	Classify oxides as acidic or basic, related to metallic and non-metallic character	11.5
2	Describe amphoteric oxides as oxides that react with acids and with bases to produce a salt and water	11.5
3	Classify Al_2O_3 and ZnO as amphoteric oxides	11.5

7.3 Preparation of salts

1	Describe the preparation, separation and purification of soluble salts, by reaction of an acid with: (a) an alkali by titration (b) excess metal (c) excess insoluble base (d) excess insoluble carbonate	11.6
2	Describe the general solubility rules for salts	11.7
3	Understand the difference between a hydrated substance and an anhydrous substance	10.1, 11.7
4	Describe the preparation of insoluble salts by precipitation	11.7
5	Define the term water of crystallisation as the water molecules present in hydrated crystals	11.7

8 THE PERIODIC TABLE

8.1 Arrangement of elements

1	Describe the Periodic Table as an arrangement of elements in periods and groups and in order of increasing proton number	12.1
2	Describe the change from metallic to non-metallic character across a period	12.1, 12.4
3	Describe the relationship between group number and the charge of the ions formed from elements in that group	12.4
4	Explain similarities in the chemical properties of elements in the same group in terms of their electronic configuration	12.2, 12.3
5	Explain how the position of an element in the Periodic Table can be used to predict its properties	12.1
6	Identify trends in groups, given information about the elements	12.4

8.2 Group I properties

1	Describe the Group I alkali metals, lithium, sodium and potassium, as relatively soft metals with these general trends down the group: (a) decreasing melting point (b) increasing density (c) increasing reactivity	12.2
2	Predict the properties of other elements in Group I, given information about the elements	12.2

8.3 Group VII properties

1	Describe the Group VII halogens, chlorine, bromine and iodine, as diatomic non-metals with these general trends down the group: (a) increasing density (b) decreasing reactivity	12.3
2	State the appearance of the halogens at r.t.p.	12.3
3	Describe and explain the displacement reactions of halogens with other halide ions	12.3
4	Predict the properties of other elements in Group VII, given information about the elements	12.3

8.4 Transition elements

1	Describe the properties of transition elements: (a) high densities (b) high melting points (c) coloured compounds (d) use as catalysts	12.5
2	Describe transition elements as having ions with variable oxidation numbers, including iron(II) and iron(III)	12.5

8.5 Noble gases

1	Describe the Group VIII noble gases as unreactive, monatomic gases and explain this in terms of electronic configuration	2.3, 12.1

	9 METALS	
	9.1 Properties of metals	
1	Compare the general physical properties of metals and non-metals	12.5, 13.1
2	Describe the general chemical properties of metals reacting with: (a) dilute acids (b) cold water and steam (c) oxygen	13.2
	9.2 Uses of metals	
1	Describe the uses of metals in terms of their physical properties	13.1, 14.4
	9.3 Alloys and their properties	
1	Describe an alloy as a mixture of a metal with other elements, including: (a) brass as a mixture of copper and zinc (b) stainless steel as a mixture of iron and other elements such as chromium, nickel and carbon	14.5
2	State that alloys can be harder and stronger than the pure metals and are more useful	14.5
3	Describe the uses of alloys in terms of their physical properties (including use of stainless steel in cutlery)	14.5
4	Identify representations of alloys from diagrams of structure	14.5
5	Explain in terms of structure how alloys can be harder and stronger than the pure metals	14.5
	9.4 Reactivity series	
1	State the order of the reactivity series	13.4
2	Describe the reactions, if any, of metals with: (a) cold water (b) steam (c) dilute hydrochloric acid and explain these reactions in terms of the position of the metals in the reactivity series	13.2
3	Deduce an order of reactivity from a given set of experimental results	13.2
4	Describe the relative reactivities of metals in terms of their tendency to form positive ions in displacement reactions	13.3
5	Explain the apparent unreactivity of aluminium in terms of its oxide layer	13.4
	9.5 Corrosion of metals	
1	State the conditions required for the rusting of iron and steel to form hydrated iron(III) oxide	13.5
2	State some common barrier methods, including painting, greasing and coating with plastic	13.5
3	Describe how barrier methods prevent rusting by excluding oxygen or water	13.5
4	Describe the use of zinc in galvanising as an example of a barrier method and sacrificial protection	13.5
5	Explain sacrificial protection in terms of the reactivity series and in terms of electron loss	13.5
	9.6 Extraction of metals	
1	Describe the ease in obtaining metals from their ores, related to the position of the metal in the reactivity series	14.1
2	Describe the extraction of iron from hematite in the blast furnace	14.2
3	State that the main ore of aluminium is bauxite and that aluminium is extracted by electrolysis	14.3
4	State the symbol equations for the extraction of iron from hematite	14.2
5	Describe the extraction of aluminium from purified bauxite / aluminium oxide, including: (a) the role of cryolite (b) why the carbon anodes need to be regularly replaced (c) the reactions at thr electrodes, including ionic equations	14.3
	10 CHEMISTRY OF THE ENVIRONMENT	
	10.1 Water	
1	Describe chemical tests for the presence of water using anhydrous cobalt(II) chloride and anhydrous copper(II) sulfate	10.1, 15.3
2	Describe how to test for the purity of water using melting point and boiling point	15.3
3	Explain that distilled water is used in practical chemistry rather than tap water because it contains fewer chemical impurities	15.3
4	State that water from natural sources may contain substances, including: (a) dissolved oxygen (b) metal compounds (c) plastics (d) sewage (e) harmful microbes (f) nitrates from fertilisers (g) phosphates from fertilisers and detergents	15.2
5	State that some of the substances in 4 are beneficial, and give examples	15.2
6	State that some of the substances in 4 are potentially harmful, and give examples	15.2
7	Describe the treatment of the domestic water supply	15.3
	10.2 Fertilisers	
1	State that ammonium salts and nitrates are used as fertilisers	15.2
2	Describe the use of NPK fertilisers to provide the elements nitrogen, phosphorus and potassium for improved plant growth	15.2

	10.2 Air quality and climate	
1	State the composition of clean, dry air	15.4
2	State some of the sources of each of these air pollutants: (a) carbon dioxide (b) carbon monoxide and particulates (c) methane (d) oxides of nitrogen (d) sulfur dioxide	15.5
3	State some of the adverse effect of the air pollutants in 2, including climate change and acid rain	15.5, 15.6
4	State and explain some of the strategies to reduce the effects of: (a) climate change (b) acid rain	15.5, 15.7
5	Describe photosynthesis as the reaction between carbon dioxide and water to produce glucose and oxygen in the presence of chlorophyll and using energy from light	15.7
6	State the word equation for photosynthesis, carbon dioxide + water $\longrightarrow$ glucose + oxygen	15.7
7	Describe, in terms of transfer of thermal energy, how the greenhouse gases carbon dioxide and methane cause global warming	15.6
8	Explain how oxides of nitrogen form in car engines and describe their removal by catalytic converters	15.5
9	State the symbol equation for photosynthesis, $6CO_2 + 6H_2O \longrightarrow C_6H_{12}O_6 + 6O_2$	15.7

11 ORGANIC CHEMISTRY

	11.1 Formulae, functional groups and terminology	
1	Draw and interpret the displayed formula of a molecule to show all the atoms and all the bonds	16.3
2	Write and interpret general formulae of compounds in the same homologous series: (a) alkanes (b) alkenes (c) alcohols (d) carboxylic acids	16.3
3	Identify a functional group as an atom or group of atoms that determine the chemical properties of a homologous series	16.3
4	State that a homologous series is a family of similar compounds with similar chemical properties, due to the presence of the same functional group	16.3
5	State that a saturated compound has molecules in which all carbon–carbon bonds are single bonds	16.4
6	State that an unsaturated compound has molecules in which one or more carbon–carbon bonds are not single bonds	16.6
7	State that a structural formula is an unambiguous description of the way the atoms in a molecule are arranged	16.3
8	Define structural isomers as compounds with the same molecular formula, but different structural formulae	16.4
9	Describe the general characteristics of a homologous series	16.3
	11.2 Naming organic compounds	
1	Name and draw the displayed formulae for compounds stated in sections 11.4–11.7 and the products of their reactions	11.4–11.7
2	State the type of compound present, given a chemical name ending in -ane, -ene, -ol, or -oic acid or from a molecular formula or displayed formula	11.4–11.7
3	Name and draw the structural and displayed formulae of unbranched: (a) alkanes (b) alkenes (c) alcohols (d) carboxylic acids containing up to four carbon atoms per molecule	11.4–11.7
4	Name and draw the displayed formulae of the unbranched esters which can be made from unbranched alcohols and carboxylic acids, each containing up to four carbon atoms	11.7
	11.3 Fuels	
1	Name the fossil fuels: coal, natural gas and petroleum	16.1
2	Name methane as the main constituent of natural gas	16.1
3	State that hydrocarbons are compounds that contain hydrogen and carbon only	16.1
4	State that petroleum is a mixture of hydrocarbons	16.1
5	Describe the separation of petroleum into useful fractions by fractional distillation	16.2
6	Describe how these properties of fractions obtained from petroleum change from the bottom to the top of the fractionating column: (a) chain length (b) volatility (c) boilng points (d) viscosity	16.2
7	Name the fractions obtained from petroleum and their uses	16.2
	11.4 Alkanes	
1	State that the bonding in alkanes is single covalent and that alkanes are saturated hydrocarbons	16.4
2	Describe the properties of alkanes as being generally unreactive, except in terms of combustion and substitution by chlorine	16.4

3	State that in a substitution reaction one atom or group of atoms is replaced by another atom or group of atoms	16.4
4	Describe the substitution reaction of alkanes with chlorine as a photochemical reaction, with ultraviolet light providing the activation energy E_a, and draw the structural or displayed products for subsitution of one hydrogen atom	16.4
11.5 Alkenes		
1	State that the bonding in alkenes includes a double carbon–carbon covalent bond and that alkenes are unsaturated hydrocarbons	16.6
2	Describe the manufacture of alkenes and hydrogen by the cracking of larger alkane molecules using high temperature and a catalyst	16.5
3	Describe the reasons for the cracking of larger alkane molecules	16.5
4	Describe the test to distinguish between saturated and unsaturated hydrocarbons by their reaction with aqueous bromine	16.6
5	State that in an addition reaction only one product is formed	16.6
6	Describe the addition reactions of alkenes with: (a) bromine (b) hydrogen (c) steam and draw the structural or displayed formulae of the products	16.6
11.6 Alcohols		
1	Describe the manufacture of ethanol by: (a) the fermentation of aqueous glucose (b) the catalytic addition of steam to ethene	16.6
2	Describe the combustion of ethanol	16.7
3	State the uses of ethanol as: (a) a solvent (b) a fuel	16.7
4	Describe the advantages and disadvantages of the manufacture of ethanol by: (a) fermentation (b) catalytic addition of steam to ethene	16.7
11.7 Carboxylic acids		
1	Describe the reaction of ethanoic acid with: (a) metals (b) bases (c) carbonates including names and formulae of the salts produced	16.8
2	Describe the formation of ethanoic acid by the oxidation of ethanol: (a) with acidified aqueous potassium manganate(VII) (b) by bacterial oxidation during vinegar production	16.8
3	Describe the reaction of a carboxylic acid with an alcohol using an acid catalyst to form an ester	16.8
11.8 Polymers		
1	Define polymers as large molecules built up from many smaller molecules called monomers	17.1
2	Describe the formation of poly(ethene) as an example of addition polymerisation using ethene monomers	17.2
3	State that plastics are made from polymers	17.4
4	Describe how the properties of plastics have implications for their disposal	17.5
5	Describe the environmental challenges caused by plastics: (a) in landfill sites (b) in the oceans (c) when burned	17.5, 17.6
6	Identify the repeat units and / or linkages in addition polymers and in condensation polymers	17.2, 17.3
7	Deduce the structure or repeat unit of an addition polymer from a given alkene and vice versa	17.2
8	Deduce the structure or repeat unit of a condensation polymer from given monomers and vice versa, for: (a) polyamides (b) polyesters:	17.3
9	Describe the differences between addition and condensation polymerisation	17.3
10	Describe and draw the structure of: (a) nylon, a polyamide (b) PET, a polyester	17.3
11	State that PET can be converted back into monomers and re-polymerised	17.3
12	Describe proteins as natural polyamides and that they are formed from amino acid monomers with the general structure where R represents different types of side chain	17.7

13	Describe and draw the structure of proteins as: ―N―▨―C―N―☐―C―N―☐―C― (with H, O, H, O substituents)	17.7

12 EXPERIMENTAL TECHNIQUES AND CHEMICAL ANALYSIS

12.1 Experimental design

1	Name appropriate apparatus for the measurement of time, temperature, mass and volume	18.1, 19.2, 19.3
2	Suggest advantages and disadvantages of experimental methods and apparatus	18.3, 18.4
3	Explain each term: (a) solvent (b) solute (c) solution (d) saturated solution (e) residue (f) filtrate	18.2

12.2 Acid–base titrations

1	Describe an acid–base titration using a burette and volumetric pipette with a suitable indicator	11.6, 19.2
2	Describe how to identify the end-point of a titration using an indicator	11.6, 19.2

12.3 Chromatography

1	Describe how paper chromatography is used to separate mixtures of soluble coloured substances, using a suitable solvent	18.5
2	Interpret simple chromatograms to identify: (a) unknown substances (b) pure and impure substances	18.5
3	Describe how paper chromatography is used to separate mixtures of soluble colourless substances	18.6
4	State and use the equation: $R_f = \dfrac{\text{distance travelled by substance}}{\text{distance travelled by solvent}}$	18.6

12.4 Separation and purification

1	Describe and explain methods of separation and purification using: (a) a suitable solvent (b) filtration (c) crystallisation (d) simple distillation (e) fractional distillation	18.1, 18.3, 18.4
2	Suggest suitable separation and purification techniques, given information about the substances involved	18.3, 18.4
3	Identify substances and assess their purity using melting point and boiling point information	18.7

12.5 Identification of ions and gases

1	Describe tests to identify these anions: (a) carbonate (b) chloride, bromide, iodide (c) nitrate (d) sulfate (e) sulfite	19.3
2	Describe tests to identify these aqueous cations: (a) aluminium (b) ammonium (c) calcium (d) chromium(III) (e) copper(II) (f) iron(II) (g) iron(III) (h) zinc	19.4
3	Describe tests to identify these gases: (a) ammonia (b) carbon dioxide (c) chlorine (d) hydrogen (e) oxygen (f) sulfur dioxide	19.3
4	Describe the use of a flame test to identify these cations: (a) lithium (b) sodium (c) potassium (d) calcium (e) barium (f) copper(II)	19.4

The syllabus and your exams

The syllabus shows content to be tested in exams. But how does it relate to the types of questions you will be asked, and the papers you will take? Let's see.

What types of questions will you be asked?

There are different types of questions, to meet three different Assessment Objectives (or AO for short). This table explains what they mean in the exam.

Assessment Objective	This objective covers ...	It leads to these kinds of exam question ...	Share of total marks in the exam
AO1	knowledge with understanding	questions which mainly test recall and understanding of what you have learned	about 50%
AO2	information handling and problem solving	questions which ask you to use what you have learned in unfamiliar situations; you may be asked to examine data in graphs or tables, or to carry out calculations	about 30%
AO3	experimental skills and investigations	questions on practical work, which are tested in the practical paper; the skills you develop in practising these questions may help you with questions on your theory paper too	about 20%

So it is not enough to be able to answer questions that require recall with understanding (AO1). The other two types of questions are important too.

Which exam papers test these objectives?

If you are studying the Core content you will take ...

Paper 1: Multiple choice, 45 minutes
40 multiple-choice questions, 40 marks (30% of the total marks)

Paper 3: Theory (Core), 1 hour 15 minutes
short-answer and structured questions, 80 marks (50% of the total marks)

Both papers test **AO1** and **AO2**.

If you are studying the Extended content you will take ...

Paper 2: Multiple choice, 45 minutes
40 multiple-choice questions, 40 marks (30% of the total marks)

Paper 4: Theory (Extended), 75 minutes
short-answer and structured questions, 80 marks (50% of the total marks)

Both papers test **AO1** and **AO2**.

... and you will also take **one** practical paper ...

Paper 5, Practical test, 1 hour 15 minutes
This paper will include experimental work in the lab; 40 marks (20% of the total marks)
OR
Paper 6, Alternative to practical, 75 minutes
This paper will not include work in the lab; 40 marks (20% of the total marks)
Both papers test **AO3**.

The end-of-chapter questions throughout this book, and the IGCSE practice questions at the end of the book, include questions to test all three objectives, AO1, AO2 and AO3. Your teacher will help you to identify and work through the three types. Practice makes perfect!

Contents

1 States of matter

1.1	Everything is made of particles	2
1.2	Solids, liquids, and gases	4
1.3	The particles in solids, liquids, and gases	6
1.4	Heating and cooling curves	8
1.5	A closer look at gases	10
	Checkup on Chapter 1	12

2 Atoms and elements

2.1	Meet the elements	14
2.2	More about atoms	16
2.3	How electrons are arranged	18
2.4	Isotopes and A_r	20
	How our model of the atom developed	22
	Checkup on Chapter 2	24

3 Atoms combining

3.1	Compounds, mixtures, and chemical change	26
3.2	Why do atoms form bonds?	28
3.3	The ionic bond	30
3.4	More about ions	32
3.5	The covalent bond	34
3.6	Covalent compounds	36
3.7	Comparing ionic and covalent compounds	38
3.8	Giant covalent structures	40
3.9	The bonding in metals	42
	Checkup on Chapter 3	44

4 Reacting masses, and chemical equations

4.1	The names and formulae of compounds	46
4.2	Equations for chemical reactions	48
4.3	The masses of atoms, molecules, and ions	50
4.4	Calculations about mass and %	52
	Checkup on Chapter 4	54

5 Using moles

5.1	The mole	56
5.2	Calculations from equations	58
5.3	Reactions involving gases	60
5.4	The concentration of a solution	62
5.5	Finding the empirical formula	64
5.6	From empirical to final formula	66
5.7	Finding % yield and % purity	68
	Checkup on Chapter 5	70

6 Redox reactions

6.1	Oxidation and reduction	72
6.2	Redox and electron transfer	74
6.3	Redox and oxidation numbers	76
6.4	Oxidising and reducing agents	78
	Checkup on Chapter 6	80

7 Electricity and chemical change

7.1	Conductors and non-conductors	82
7.2	The principles of electrolysis	84
7.3	The reactions at the electrodes	86
7.4	Electroplating	88
	Checkup on Chapter 7	90

8 Energy changes in reactions

8.1	Energy changes in reactions	92
8.2	A closer look at energy changes	94
8.3	Calculating enthalpy changes	96
8.4	The hydrogen-oxygen fuel cell	98
	Checkup on Chapter 8	100

9 The rate of reaction

9.1	Introducing reaction rates	102
9.2	Measuring the rate of a reaction	104
9.3	Changing the rate (part I)	106
9.4	Changing the rate (part II)	108
9.5	Explaining rate changes	110
9.6	Catalysts	112
	More about enzymes	114
	Checkup on Chapter 9	116

10 Reversible reactions and equilibrium

10.1	Reversible reactions	118
10.2	Shifting the equilibrium	120
10.3	The Haber process	122
10.4	The Contact process	124
	Checkup on Chapter 10	126

11 Acids, bases, and salts

11.1	Acids and bases	128
11.2	A closer look at acids and alkalis	130
11.3	The reactions of acids and bases	132

11.4	A closer look at neutralisation	134
11.5	Oxides	136
11.6	Making salts (part I)	138
11.7	Making salts (part II)	140
11.8	Finding concentration by titration	142
	Checkup on Chapter 11	144

12 The Periodic Table

12.1	The Periodic Table: an overview	146
12.2	Group I: the alkali metals	148
12.3	Group VII: the halogens	150
12.4	More about the trends	152
12.5	The transition elements	154
	How the Periodic Table developed	156
	Checkup on Chapter 12	158

13 The behaviour of metals

13.1	Comparing metals and non-metals	160
13.2	Comparing metals for reactivity	162
13.3	Metals in competition	164
13.4	The reactivity series	166
13.5	The rusting of iron	168
	Checkup on Chapter 13	170

14 Extracting and using metals

14.1	Metal ores and metal extraction	172
14.2	Extracting iron	174
14.3	Extracting aluminium	176
14.4	Making use of metals	178
14.5	Alloys	180
	Metals, civilisation, and you	182
	Checkup on Chapter 14	184

15 Chemistry of the environment

15.1	Our environment and us	186
15.2	What is in river water?	188
15.3	Our water supply	190
15.4	Fertilisers	192
15.5	Air, the gas mixture we live in	194
15.6	Air pollution from fossil fuels	196
15.7	Two greenhouse gases	198
15.8	Tackling climate change	200
	Checkup on Chapter 15	202

16 Organic chemistry

16.1	Petroleum: a fossil fuel	204
16.2	Refining petroleum	206
16.3	Four families of organic compounds	208
16.4	The alkanes	210
16.5	Cracking alkanes	212
16.6	The alkenes	214
16.7	The alcohols	216
16.8	The carboxylic acids	218
	Checkup on Chapter 16	220

17 Polymers

17.1	Introducing polymers	222
17.2	Addition polymerisation	224
17.3	Condensation polymerisation	226
17.4	Plastics	228
17.5	The plastics problem	230
17.6	Tackling the plastics problem	232
17.7	Proteins	234
	Checkup on Chapter 17	236

18 Separation and purification

18.1	Making a substance in the lab	238
18.2	Solutions and solubility	240
18.3	Separating a solid from a liquid	242
18.4	Separating by distillation	244
18.5	Paper chromatography (part I)	246
18.6	Paper chromatography (part II)	248
18.7	Checking purity	250
	The chromatography detectives	252
	Checkup on Chapter 18	254

19 Experiments and tests in the lab

19.1	The scientific method, and you	256
19.2	Writing up an experiment	258
19.3	Preparing, and testing, gases	260
19.4	Testing for cations	262
19.5	Testing for anions	264
	Checkup on Chapter 19	266

Answers for Chapters 1 – 19	268

IGCSE practice questions for the exam

IGCSE practice questions for Paper 3	287
IGCSE practice questions for Paper 4	295
IGCSE practice questions for Paper 6	303

Reference

Glossary	311
The Periodic Table and atomic masses	316
Index	318

1.1 Everything is made of particles

Objectives: understand that everything is made of particles; give evidence for particles, from both outside and inside the lab; name the three types of particle

Particles: the big idea in chemistry

All chemistry is based on one big idea: that everything is made of very tiny pieces called **particles**.

Most single particles are far too small to see. When you look at a jar of a chemical in the lab, you are looking at a collection of countless trillions of tiny particles. Think about that!

And when you observe a chemical reaction – for example, an acid dissolving a piece of metal – you are seeing the result of trillions of tiny particles reacting together. This is one of the things that makes chemistry so exciting.

It took a while for the big idea to develop ...

For centuries, people had guessed that water and air were made of tiny particles. But they could not see the particles, so could not prove it.

Then in 1827, a botanist called Robert Brown was studying pollen from a flower, in water, under a microscope. He noticed that the pollen granules jiggled around. They were not alive – so why were they moving?

78 years later, Albert Einstein came up with the answer. The granules were moving because they were being struck by tiny invisible particles of water. In fact their movement proved that water is made of tiny particles.

Today, everyone accepts that things are made of particles.

▲ All made of particles!

The evidence around you

Most particles are so small that we cannot see them directly, even with the most powerful microscope. But there is evidence for them all around you. Think about these four examples.

Outside the lab

1 In sunlit rooms, you sometimes see dust dancing in the air. It dances because the dust specks are being bombarded by tiny particles in the air, too small to see. Just like those pollen granules in water.

2 Cooking smells spread. The 'smells' are due to particles which spread because they are bombarded by the particles in air. This is an example of diffusion. (See pages 3 and 11.) Some end up in your nose!

STATES OF MATTER

In the lab

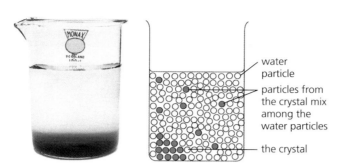

3 Place a crystal of purple potassium manganate(VII) in a beaker of water. The colour spreads through the water, because particles leave the solid crystal – it **dissolves** – and spread among the water particles.

 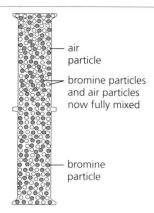

4 Place an open gas jar of air upside down on an open gas jar containing a few drops of red-brown bromine. The colour spreads upwards because the particles of bromine vapour mix among the gas particles in the air.

Diffusion

Look again at the two examples above. The particles mix and spread by colliding with other particles, and bouncing off in all directions. This mixing process is called **diffusion**. It takes place in liquids and gases.

The result is that particles spread from where they are more concentrated, until all the particles are evenly mixed. Look at the drawings on the right.

So what are these tiny particles?

- The smallest particles, that we cannot break down further in chemical reactions, are called **atoms**.
- In some substances, the particles are just single atoms. For example, air contains single atoms of argon.
- In many substances, the particles consist of two or more atoms joined together. These particles are called **molecules**. Water and bromine exist as molecules. Air is mostly nitrogen and oxygen molecules.
- In other substances the particles are atoms or groups of atoms that carry a charge. These particles are called **ions**. In example 3 above, the particles in potassium manganate(VII) are ions.

You'll learn more about all these particles in Chapters 2 and 3.

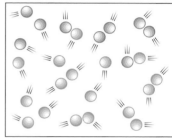

The moving particles collide, then bounce apart in all directions.

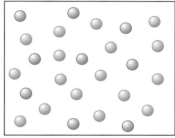

In the end they are all mixed up. (But they keep on moving.)

▲ The process of diffusion.

'Seeing' atoms

- We can't see atoms through any microscopes that use light.
- But images of them can be taken using **electron microscopes**.
- These use beams of electrons, not light. (We meet electrons later!)

1 Dust specks appear to dance in air. Why do they move?

2 Example 3 above proves that potassium manganate(VII) is made of particles. Explain why.

3 Bromine vapour is heavier than air. Even so, it spreads upwards, in example 4 above. Why?

4 a Describe *diffusion*.
  b Use the idea of diffusion to explain how the smell of perfume travels across a room.

5 a List the three types of particle you will meet in chemistry.
  b What is the difference between an *atom* and a *molecule*?

1.2 Solids, liquids, and gases

Objectives: distinguish between solids, liquids, and gases; describe the change of state during melting, boiling, evaporation, freezing, and condensing

What's the difference?
It is easy to tell the difference between a solid, a liquid, and a gas:

A solid has a fixed shape and a fixed volume. It does not flow. It cannot be compressed into a smaller space. Are you sitting on a solid?

A liquid flows easily. It has a fixed volume, but its shape changes. It takes the shape of the container you pour it into.

A gas does not have a fixed volume or shape. It spreads out to fill its container. It is much lighter than the same volume of solid or liquid.

Water: solid, liquid and gas
Water can be a solid (ice), a liquid (water), and a gas (steam). Its state can be changed by heating or cooling:

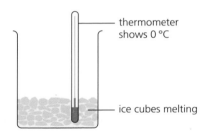
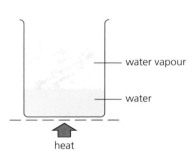
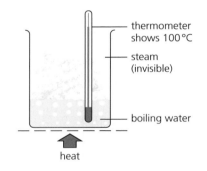

1 Ice changes to water when it is warmed. This change is called **melting**. The thermometer shows 0 °C until all the ice has melted. So 0 °C is called its **melting point**.

2 When the water is heated, its temperature rises. As it warms up, some of it changes from liquid to gas, which goes into the air. This is called **evaporation**.

3 Now the water is **boiling**. It evaporates fast. The gas it forms is called **steam**. The thermometer shows 100 °C while the water boils off. 100 °C is its **boiling point**.

And when steam is cooled, the opposite changes take place:

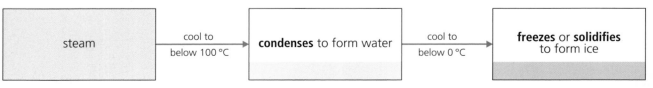

You can see that:
- condensing is the opposite of evaporating
- freezing is the opposite of melting
- the freezing point of water is the same as the melting point of ice, 0 °C.

STATES OF MATTER

Other things can change state too

It's not just water. All substances can exist as solid, liquid, and gas. Even iron and diamond can melt and boil!

Compare the melting and boiling points in this table:

Substance	Melting point / °C	Boiling point / °C
oxygen	−219	−183
mercury	−39	357
ethanol	−15	78
water	0	100
sodium	98	890
table salt (sodium chloride)	801	1465
iron	1540	2900
diamond	3550	4832

▲ Molten iron being poured out at an iron works. Hot – over 1540 °C!

Why are they all different? It's because their particles are different. You will find out more about all these substances during your course.

The state at room temperature

What usually matters most to us is the state of a substance at room temperature. (Room temperature is taken to be 20 °C.)

Look again at the figures in the table. At 20 °C, oxygen is a gas. That's lucky, since we cannot do without it, and we breathe it in.

Iron has a very high melting point, and is hard and strong at room temperature. So we use it for building structures. Mercury is a metal too – but it is liquid at room temperature, so no good for building anything.

▲ Water – liquid at room temperature. How does that benefit us?

▲ Enjoying water in its solid state, in Antarctica.

▲ Once upon a time, we depended on steam for rail transport.

1. Give two properties of:
   a a solid   b a liquid   c a gas
2. What do we call the substance that forms when water:
   a freezes?   b evaporates at its boiling point?
3. Give a word which means the opposite of:
   a evaporate   b freeze
4. Look at the substances in the table above.
   a Which of them has the lowest melting point?
   b Which has the highest boiling point?
   c Which will boil off first on heating: iron or table salt?
   d Which are liquids at room temperature?
   e Why are the melting and boiling points so different?

1.3 The particles in solids, liquids, and gases

Objectives: describe how the particles are arranged, and move, in solids, liquids, and gases; explain how and why their state changes with temperature

How the particles are arranged, and move

Water can change from solid to liquid to gas. Its *particles* do not change. They are the same in each state. But their *arrangement*, and *movements*, change. The same is true for all substances. Look:

State		How the particles are arranged, and move	Diagram of particles
Solid 		The particles in a solid are arranged in a regular pattern or **lattice**. Strong forces hold them together. So they cannot leave their positions. The only movements they make are tiny vibrations to and fro.	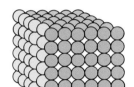
Liquid 		The particles in a liquid can move about and slide past each other. They are still close together, but not in a lattice. The forces that hold them together are weaker than in a solid.	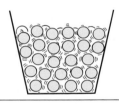
Gas 		The particles in a gas are far apart, and they move about very quickly. There are almost no forces holding them together. They collide with each other and bounce off in all directions.	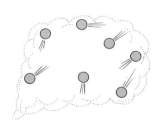

Changing state

So why do substances change state when you heat them? It is because the particles take in heat energy – and this changes how they move.

Melting When a solid is heated, its particles get more energy and vibrate more. This makes the solid **expand**. At the melting point, the particles vibrate so much that they break away from their positions. The solid turns liquid.

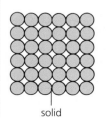 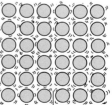 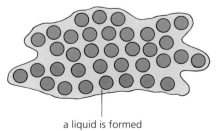

solid   heat energy → the vibrations get larger   heat energy at melting point → a liquid is formed

Boiling When a liquid is heated, its particles get more energy and move faster. They collide more often, and bounce further apart. This makes the liquid expand. At the boiling point, the particles get enough energy to overcome the forces between them. They break away to form a gas. Look:

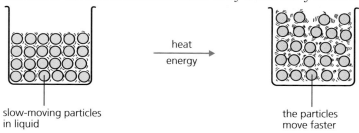
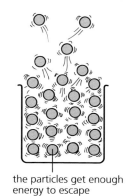

Evaporating Even well below the boiling point, some particles in a liquid have enough energy to escape and form a gas.

Evaporation takes place *at the surface of the liquid*, and over a range of temperatures. It is a slower process than boiling. (Boiling takes place throughout the liquid, and at a specific temperature.)

How much heat is needed?

The amount of heat needed to melt or boil a substance is different for every substance. That's because the particles in each substance are different, with different forces of attraction between them.

The stronger the forces, the more heat energy is needed to overcome them. So the higher the melting and boiling points will be.

Reversing the changes

You can reverse those changes again by cooling. As a gas cools, its particles lose energy and move less quickly. When they collide, they do not have enough energy to bounce away. So they stay close, and form a liquid. On further cooling, the liquid turns to a solid. Look:

> **The kinetic particle theory**
> Look at the key ideas you have met.
> - A substance can be a solid, a liquid, or a gas, and change from one state to another.
> - It has different characteristics in each state. (For example, solids do not flow.)
> - The differences are due to the way its particles are arranged, and move, in each state.
>
> Together, these ideas make up the **kinetic particle theory**. (*Kinetic* means *about motion*.)

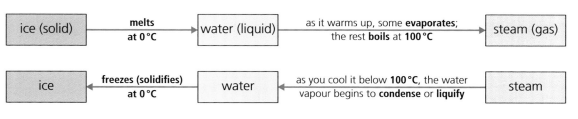

1. Draw diagrams to show the arrangement of particles in a solid, a liquid, and a gas. Add notes to each diagram.
2. Using the idea of particles, explain why:
   a. you cannot pour solids   b. you can pour liquids
3. Draw a diagram to show what happens to the particles, when a liquid cools to a solid.
4. Why do puddles of rain dry in the sunshine?
5. State *three* differences between evaporation and boiling.
6. Oxygen is the gas we breathe in. It can be separated from the air. It freezes at −219°C and boils at −183°C.
   a. In which state is oxygen, at: i 0°C? ii −200°C?
   b. How would you turn oxygen gas into solid oxygen?

1.4 Heating and cooling curves

Objectives: describe the heating and cooling curves for water; interpret heating and cooling curves in terms of changes of state, and particle behaviour

The heating curve for water

Suppose you heat up some ice at a steady rate.

You might expect the temperature to keep rising as the ice warms up and turns to water, and then from water to steam. But this does not happen.

Instead, the temperature remains constant for a time in two places, even though you keep on heating. Look at the two horizontal lines on graph ①.

This graph is called a **heating curve**.

A heating curve shows how the temperature of a substance changes as you heat it up.

Note the axes labels. The *time* is a measure of how much heat energy has been added, on heating at a steady rate.

As you can see from the graph:

Temperatures remain constant while water changes state.
These temperatures are its melting and boiling points.

Explaining the heating curve

The shape of the heating curve can be explained by the way the particles behave. Read the notes on this larger copy of the graph. Start at the bottom.

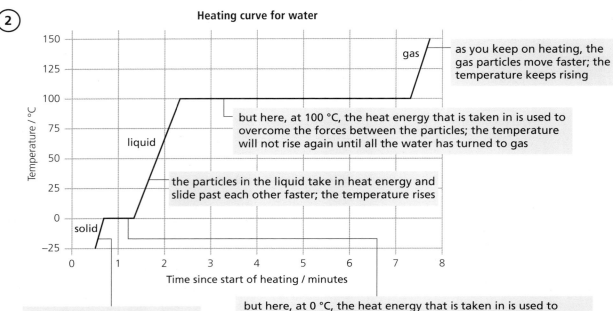

The cooling curve for water

The graph below is a **cooling curve** for water.

A cooling curve shows how the temperature of a substance changes as you cool it down.

As you can see, this curve is *the mirror image* of the heating curve. The temperature remains constant at 100 °C and 0 °C, as the water changes state. This time, read the notes on the graph from the top down.

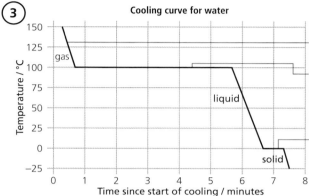

- as the gas cools, the particles move less quickly, with less energy; they don't bounce so far apart when they collide
- by 100 °C, the particles in the gas do not have enough energy to bounce apart when they collide; the forces of attraction between them take over; the gas condenses to a liquid; this releases heat, so even though you keep on cooling, the temperature does not fall again until all the gas has liquified
- by 0 °C the particles in the liquid move so slowly that stronger forces of attraction take over and the lattice starts to form – the water freezes (solidifies) to ice; this process also releases heat, so the temperature does not fall again until all the water has frozen

Not just water ...

You can draw heating and cooling curves for any substance. They are different for each substance. That's because they depend on the particles and the forces between them. The heating curve on the right is for iron.

What if a substance is not pure?

The graphs here are for **pure** water and iron. A pure substance has only one type of particle. And it has sharp melting and boiling points, which you can find from the heating curve.

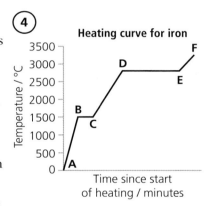

But if other types of particle are mixed in, they affect the forces between particles. Changes of state will now occur over a range of temperatures, not sharply. So lines like BC and DE on graph ④ will be tilted, not flat.

This means that melting and boiling points can be used to check whether a substance is pure. You can find out more on page 250.

1. What is a *heating curve*?
2. Explain how you can find the melting and boiling point of a substance from a heating curve like graph ①.
3. Look at graph ②. The temperature stays steady at 0 °C for over half a minute, even though heating continues. Explain why, using the term *particles* in your answer.
4. Which takes more energy: melting ice, or boiling water? Give evidence to support your answer.
5. Look at cooling curve ③, for water.
   a. About how long does the temperature stay at 100 °C?
   b. Explain why the temperature stays steady for this time, even though you keep on cooling the container.
6. Look at graph ④, for iron.
   a. What is happening between points D and E?
   b. Between which points is iron turning into a liquid?
   c. Give approximate melting and boiling points for iron.

1.5 A closer look at gases

Objectives: describe how gas volume changes with temperature and pressure; explain why the rate of diffusion of a gas depends on the mass of its particles

What is gas pressure?

When you blow up a balloon, you fill it with air particles. As they move about, they collide with the sides of the balloon, and exert **pressure** on it. This pressure keeps the balloon inflated. In the same way, all gases exert pressure on the walls of their containers.

When you change the temperature of a gas

Look at the container of gas below. The piston can move freely up or down, *until the pressure is the same inside and outside the container*.

▲ The more air in, the higher the pressure.

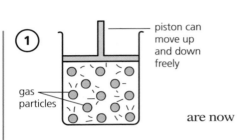

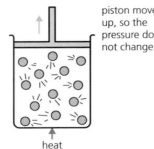

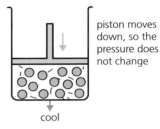

The pressure is the same inside and outside the container. But if you heat the gas, its particles will move faster, and collide with the walls …

… more often, and with more force. So the piston moves up, as shown here. Now the gas takes up more space: its **volume** has increased.

But if you cool it, the particles slow down, and strike the walls less often, with less force. The piston moves down. The gas volume decreases.

The same is true for all gases, at constant pressure:

When you heat a gas, its volume increases.
When you cool a gas, its volume decreases.

When you change the pressure of a gas

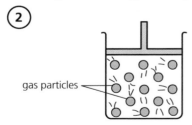

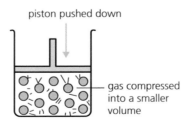

Suppose you *push* the piston down. The gas pressure will rise, because the particles are now in a smaller space, so they will …

… strike the walls more often. As you can see, increasing the gas pressure in this way means that the volume of the gas decreases.

But if you now release the piston, it will move up again. So the pressure of the gas falls again – and its volume increases.

The same is true for all gases, at constant temperature:

An increase in pressure means a decrease in volume, for a gas.
A decrease in pressure means an increase in volume, for a gas.

In fact, if you increase the pressure on a gas enough, you can push its particles so close together that a gas turns into a liquid.

STATES OF MATTER

The rate of diffusion of gases
On page 3 you saw that gases **diffuse** because the particles collide with other particles, and bounce off in all directions.

But two gases at the same temperature will not diffuse at the same rate. The rate depends on the mass of their particles.

An experiment to compare rates of diffusion
The particles in hydrogen chloride gas are twice as heavy as those in ammonia gas. So which gas do you think will diffuse faster? Let's see:

- Cotton wool soaked in ammonia solution is put into one end of a long tube (at A below). It gives off ammonia gas.
- *At the same time*, cotton wool soaked in hydrochloric acid is put into the other end of the tube (at B). It gives off valid hydrogen chloride gas.
- Gases diffuse along the tube. White smoke forms where they meet:

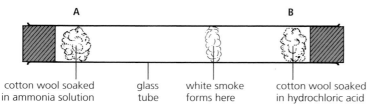

The white smoke forms closer to B. So the ammonia particles have travelled further than the hydrogen chloride particles – which means they have travelled *faster*.

The lower the mass of its particles, the faster a gas will diffuse.

This makes sense when you think about it. When particles collide and bounce away, the lighter particles will bounce further and faster.

The particles in the two gases above are molecules. The mass of a molecule is called its **relative molecular mass**. So we can also say:
The lower its relative molecular mass, the faster a gas will diffuse.

Note that everything in the experiment above needs to be at the same temperature (room temperature) for a valid test. That is because, as you know, the temperature also affects how fast gas particles move.

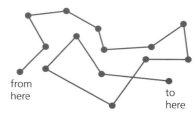

▲ The typical random path of a gas particle during diffusion. After each collision, it bounces off in a different direction. But overall, it travels to where the gas is less concentrated.

▲ Imagine the marble on the right is the heavier one. Which marble will bounce further and faster, when they collide?

▲ The particles that carry scent spread by diffusion. No need for help from you!

1. What causes the *pressure* in a gas?
2. What happens to the volume of a gas:
   a as you heat the gas?   b as you cool the gas?
3. How does the volume of the gas change if the pressure is:
   a increased?   b reduced?
4. In the diagrams labelled ①, the piston can move up and down freely. Using the idea of particles, explain why:
   a the volume of the gas increases as the gas is heated
   b the volume of the gas decreases as the gas is cooled
5. Look at the diagrams labelled ②. You can force the gas into a smaller volume by pushing the piston down. Using the idea of particles, explain why its pressure rises.
6. In the diffusion experiment above, there is air in the tube.
   a Explain how the ammonia particles move from A to B.
   b Explain why the hydrogen chloride particles don't move as far as the ammonia particles.
7. Of all gases, hydrogen diffuses fastest at any given temperature. What can you deduce from this?

Checkup on Chapter 1

Revision checklist

Core syllabus content
Make sure you can …
- [ ] state that everything is made of particles, and name three types of particles
- [ ] explain why this shows that particles exist:
  - the movement of dust specks in air
  - the purple colour spreading when a crystal of potassium manganate(VII) is placed in water
- [ ] describe diffusion, and use the idea of particles to explain why it occurs
- [ ] name the three states of matter, and give the properties of each state
- [ ] explain that heating a substance will bring about a change of state
- [ ] describe the change of state that occurs during: melting   boiling   evaporating   condensing
- [ ] define these terms:
  melting point   *boiling point*   *freezing point*
- [ ] describe how the particles are arranged, and move, in each state of matter
- [ ] explain why a gas exerts a pressure
- [ ] describe how this affects the volume of a gas:
  - changing the temperature
  - changing the pressure

Extended syllabus content
Make sure you can also …
- [ ] *explain* the changes of state that occur when a substance is heated, or cooled, in terms of particles and their movement
- [ ] define these terms: *heating curve   cooling curve*
- [ ] identify these on heating and cooling curves:
  changes of state   melting / freezing points
  boiling / condensing points
- [ ] interpret heating and cooling curves in terms of particles and their movement
- [ ] explain, in terms of particles, the effect of temperature and pressure on the volume of a gas
- [ ] outline an experiment to show that a gas will diffuse faster than another gas with heavier particles
- [ ] explain the effect of relative molecular mass on the rate of diffusion of gases

Questions

Core syllabus content

1 Give the correct name for each change of state:
   a solid ⟶ liquid   **b** liquid ⟶ solid
   c gas ⟶ liquid   **d** liquid ⟶ gas

2 A large crystal of potassium manganate(VII) was placed in the bottom of a beaker of cold water, and left for several hours.

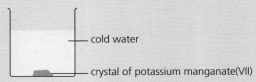

   a Describe what would be seen:
      i after five minutes   **ii** after several hours
   b Explain your answers using the idea of particles.
   c Name the two processes that took place during the experiment.

3 Use the idea of particles to explain why:
   a solids have a definite shape
   b liquids fill the bottom of a container
   c you can't store gases in open containers
   d you can't squeeze a sealed plastic syringe that is completely full of water
   e a balloon expands as you blow into it.

4 Below is a heating curve for a pure substance. It shows how the temperature rises over time, when the substance is heated until it melts, then boils.

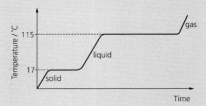

   a What is the melting point of the substance?
   b What happens to the temperature while the substance changes state?
   c The graph shows that the substance takes longer to boil than to melt. Suggest a reason for this.
   d How can you tell that the substance is not water?
   e Sketch a rough heating curve for pure water.

5 A **cooling curve** is the opposite of a heating curve. It shows how the temperature of a substance changes with time, as it is cooled from a gas to a solid. Here is the cooling curve for one substance:

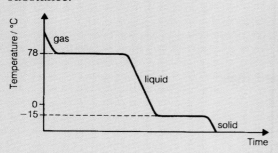

  a What is the state of the substance at room temperature (20 °C)?
  b Use the list of melting and boiling points on page 5 to identify the substance.
  c Sketch a cooling curve for pure water.

6 Using the idea of particles, explain each of these.
  a When two solids are placed on top of each other, they do not mix.
  b Heating a gas in a closed container will increase its pressure.
  c Poisonous gases from a factory chimney can affect a large area.
  d In a darkened cinema, dust particles appear to dance in the beam of light.

7 a Pick out an example of diffusion.
   i   a helium-filled balloon floating in the air
   ii   an ice lollipop melting
   iii a creased shirt losing its creases when you iron it with a hot iron
   iv   a balloon bursting when you stick a pin in it
   v   a blue crystal forming a blue solution, when it is left sitting in a glass of water
   vi   brushing paint onto a wall from a paint tin
   vii the yolk hardening, when you boil an egg
  b For your choice in **a**, draw a diagram showing the particles before and after diffusion.
  c Explain why this diffusion occurs.

Extended syllabus content

8 This is about the cooling curve in question **5**.
  a Using the kinetic particle theory, describe and explain the changes in *arrangement* of particles that occurs at 78 °C.
  b Explain how the *movement* of particles is changing in the part of the curve labelled:
     i   gas   ii   liquid

9 You can measure the rate of diffusion of a gas using this apparatus. The gas enters through the thin tube:

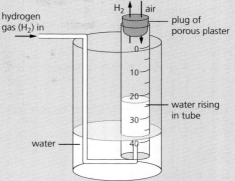

The measuring tube is sealed at the top with a plug of porous plaster. Air and other gases can diffuse in and out through the tiny holes in the plug.

The water rises in the measuring tube if the chosen gas diffuses out through the plug faster than air diffuses in. Air is mainly nitrogen and oxygen.

  a When you use hydrogen gas, the water rises in the measuring tube. Why?
  b What does this tell you about the rate of diffusion of hydrogen, compared with the gases in air?
  c Explain your answer to **b**. Use the term *mass*!
  d The molecules in carbon dioxide are heavier than those in nitrogen and oxygen. Predict what will happen to the water in the measuring tube, when you use carbon dioxide. Explain your answer.

10

Gas	Formula	Relative atomic or molecular mass
methane	CH_4	16
helium	He	4
oxygen	O_2	32
nitrogen	N_2	28
chlorine	Cl_2	71

Look at the table above.
  a Which two gases will mix fastest? Explain.
  b Which gas will take least time to escape from a gas syringe?
  c Would you expect chlorine to diffuse more slowly than the gases in air? Explain.
  d An unknown gas diffuses faster than nitrogen, but more slowly than methane. What you can deduce about its relative molecular mass?

2.1 Meet the elements

Objectives: define *atom, element, compound, mixture*; understand the Periodic Table as a way to arrange the elements and show patterns; point out its groups and periods

First, atoms ...
In Chapter 1 you met the idea that everything is made of particles. One type of particle is the **atom**.

Atoms are the smallest particles of matter that we cannot break down further by chemical means.

The elements
Look at these substances:

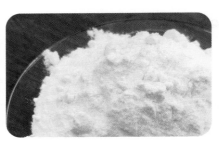

Sodium is made *only* of sodium atoms. So it is called an **element**.

Carbon is made only of carbon atoms. So it is an element.

Sulfur is made only of sulfur atoms. It is an element too.

An element contains only one kind of atom.

The elements above are different because their atoms are different.

How many elements are there?
By 2020, there were 118 known elements. 94 of them occur naturally on Earth. Scientists made the rest in the lab. Many of these synthetic elements last just seconds before breaking down into other elements. (That is why they are not found in nature.)

Symbols for the elements
Every element has a symbol, as shorthand. The symbol for sodium is Na, from its Latin name *natrium*. The symbol for carbon is C. Sulfur is S. Some elements are named after the people who discovered them.

▲ This painting shows Hennig Brand, who discovered the element phosphorus in 1669. It glows in the dark!

◄ Collecting the element sulfur from a volcano crater in Indonesia. It is used as an ingredient in many cosmetics.

ATOMS AND ELEMENTS

The Periodic Table

Period	I	II												III	IV	V	VI	VII	VIII
1						1 H hydrogen 1													2 He helium 4
2	3 Li lithium 7	4 Be beryllium 9												5 B boron 11	6 C carbon 12	7 N nitrogen 14	8 O oxygen 16	9 F fluorine 19	10 Ne neon 20
3	11 Na sodium 23	12 Mg magnesium 24				the transition elements								13 Al aluminium 27	14 Si silicon 28	15 P phosphorus 31	16 S sulfur 32	17 Cl chlorine 35.	18 Ar argon 40
4	19 K potassium 39	20 Ca calcium 40	21 Sc scandium 45	22 Ti titanium 48	23 V vanadium 51	24 Cr chromium 52	25 Mn manganese 55	26 Fe iron 56	27 Co cobalt 59	28 Ni nickel 59	29 Cu copper 64	30 Zn zinc 65	31 Ga gallium 70	32 Ge germanium 73	33 As arsenic 75	34 Se selenium 79	35 Br bromine 80	36 Kr krypton 84
5	37 Rb rubidium 85	38 Sr strontium 88	39 Y yttrium 89	40 Zr zirconium 91	41 Nb niobium 93	42 Mo molybdenum 96	43 Tc technetium –	44 Ru ruthenium 101	45 Rh rhodium 103	46 Pd palladium 106	47 Ag silver 108	48 Cd cadmium 112	49 In indium 115	50 Sn tin 117	51 Sb antimony 122	52 Te tellurium 128	53 I iodine 127	54 Xe xenon 131
6	55 Cs caesium 133	56 Ba barium 137	57–71 lanthanoids	72 Hf hafnium 178	73 Ta tantalum 181	74 W tungsten 184	75 Re rhenium 186	76 Os osmium 190	77 Ir iridium 192	78 Pt platinum 195	79 Au gold 197	80 Hg mercury 201	81 Tl thallium 204	82 Pb lead 207	83 Bi bismuth 209	84 Po polonium –	85 At astatine –	86 Rn radon –
7	87 Fr francium –	88 Ra radium –	89–103 actinoids	104 Rf rutherfordium –	105 Db dubnium –	106 Sg seaborgium –	107 Bh bohrium –	108 Hs hassium –	109 Mt meitnerium –	110 Ds darmstadtium –	111 Rg roentgenium –	112 Cn copernicium –	113 Nh nihonium –	114 Fl flerovium –	115 Mc moscovium –	116 Lv livermorium –	117 Ts tennessine –	118 Og oganesson –

	57 La lanthanum 139	58 Ce cerium 140	59 Pr praseodymium 141	60 Nd neodymium 144	61 Pm promethium –	62 Sm samarium 150	63 Eu europium 152	64 Gd gadolinium 157	65 Tb terbium 159	66 Dy dysprosium 163	67 Ho holmium 165	68 Er erbium 167	69 Tm thulium 169	70 Yb ytterbium 173	71 Lu lutetium 175
actinoids	89 Ac actinium –	90 Th thorium 232	91 Pa protactinium 231	92 U uranium 238	93 Np neptunium –	94 Pu plutonium –	95 Am americium –	96 Cm curium –	97 Bk berkelium –	98 Cf californium –	99 Es einsteinium –	100 Fm fermium –	101 Md mendelevium –	102 No nobelium –	103 Lr lawrencium –

The table above is called the **Periodic Table**. It is very important in chemistry. It is a way to arrange the elements to show their patterns.

- It gives the names and symbols for the elements. (Look at the key.)
- The column and row it is in gives us lots of clues about an element. For example, look at the columns numbered I, II, …, VIII. The elements in each of them form a **group** with similar properties.
- The rows numbered 1 – 7 are **periods**. Period 1 has just two elements.
- Look at the zig-zag red line. It separates **metals** from **non-metals**, with the non-metals to the right of the line, except for hydrogen.

Compounds and mixtures

So, you have met the elements. Everything else you meet in chemistry is either a **compound** or a **mixture**.

A compound is made of atoms of different elements bonded together.

A mixture contains different elements or compounds that are *not* bonded together. You can usually separate them quite easily.

Key for Periodic Table

atomic number
symbol
name
relative atomic mass

▲ **A** contains water, a compound. Its particles are *molecules*, made of hydrogen and oxygen atoms bonded together. A chemical reaction would be needed to break the bonds.
B contains a mixture of water and carbon. You could easily filter the carbon out.

Q

1. What is: **a** an atom? **b** an element?
2. How many elements occur naturally?
3. **a** What is the Periodic Table?
   b Give names and symbols for three elements in Group I.
4. Which element in the Periodic Table has this symbol?
   a Ca **b** Mg **c** N (Hint: look in periods 2 – 4.)
5. Which elements are there more of: metals, or non-metals?
6. From the Periodic Table, name:
   a three metals **b** three non-metals
   that you expect to behave in a similar way.
7. Water is *not* an element. Explain why. (Photo!)
8. Define: **a** a compound **b** a mixture

2.2 More about atoms

Objectives: give the names and properties of the subatomic particles; describe how they are arranged; define *proton number* and *nucleon number*

Protons, neutrons, and electrons

We cannot break down an atom by chemical means. But in fact an atom is made of even smaller particles. It consists of a **nucleus**, surrounded by **electrons**. The nucleus is a cluster of **protons** and **neutrons**.

These subatomic particles are very light. So their masses must be given in **atomic mass units** rather than grams. Protons and electrons also have an **electric charge**. Look:

Subatomic particle	Mass /atomic mass units (amu)	Charge
proton ●	1	positive charge (1+)
neutron ●	1	none
electron •	taken as 0	negative charge (1−)

In fact the mass of an electron is 0.0005 amu, but we take it as zero.

▲ The nucleus is very tiny compared with the rest of the atom. If the atom were the size of a football stadium, the nucleus would be the size of a pea!

How the particles are arranged

The sodium atom is a good one to start with. It has **11** protons, **11** electrons, and **12** neutrons. They are arranged like this:

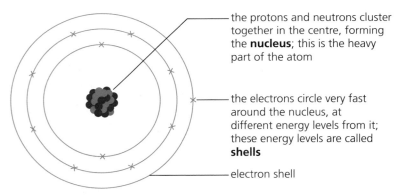

- the protons and neutrons cluster together in the centre, forming the **nucleus**; this is the heavy part of the atom
- the electrons circle very fast around the nucleus, at different energy levels from it; these energy levels are called **shells**
- electron shell

> **Subatomic!**
> Protons, neutrons, and electrons are called **subatomic particles** because they are smaller than the atom, and within it.

Proton number (or atomic number)

A sodium atom has 11 protons. This can be used to identify it, since *only* a sodium atom has 11 protons. Every other atom has a different number. **You can identify an atom by the number of protons it has**.

The number of protons in the nucleus of an atom is called its proton number (or atomic number).
The proton number for sodium is 11.

How many electrons?

The sodium atom also has 11 electrons. So it has an equal number of protons and electrons. The same is true for every type of atom:
Every atom has an equal number of protons and electrons. So atoms have no overall charge.

The panel on the right shows how the charges cancel for the sodium atom.

> **The charge on a sodium atom:**
> ●●●● 11 protons
> ●●●● Each has a
> ●●●   charge of 1+
>   Total charge 11+
> ×××× 11 electrons
> ×××× Each has a
> ×××   charge of 1−
>   Total charge 11−
> Adding the charges: 11+
>   11−
>   ———
>   0
> The answer is zero. The atom has no overall charge.

16

ATOMS AND ELEMENTS

Nucleon number (or mass number)
Protons and neutrons form the nucleus, so they are called **nucleons**. They give the atom its mass.

The total number of protons and neutrons in the nucleus of an atom is called its nucleon number, or mass number.

The nucleon number for the sodium atom is 23. (11 + 12 = 23)

So sodium can be described in a short way like this: $^{23}_{11}$Na.

The upper number is the nucleon number. The lower one is the proton number. So you can tell that the atom has 12 neutrons. (23 − 11 = 12)

> **Try it yourself!**
> You can describe an atom of any element in a short way, like this:
>
> nucleon number
>   symbol for element
> proton number
>
> For example: $^{16}_{8}$O

The atoms of the first 20 elements
Look back at the Periodic Table on page 15. The small number above each symbol is the proton number. As you can see, the elements are in order of increasing proton number. Here are the first 20 elements:

Element	Symbol	Proton number	Electrons	Neutrons	Nucleon number (protons + neutrons)
hydrogen	H	1	1	0	1
helium	He	2	2	2	4
lithium	Li	3	3	4	7
beryllium	Be	4	4	5	9
boron	B	5	5	6	11
carbon	C	6	6	6	12
nitrogen	N	7	7	7	14
oxygen	O	8	8	8	16
fluorine	F	9	9	10	19
neon	Ne	10	10	10	20
sodium	Na	11	11	12	23
magnesium	Mg	12	12	12	24
aluminium	Al	13	13	14	27
silicon	Si	14	14	14	28
phosphorus	P	15	15	16	31
sulfur	S	16	16	16	32
chlorine	Cl	17	17	18	35
argon	Ar	18	18	22	40
potassium	K	19	19	20	39
calcium	Ca	20	20	20	40

So the numbers of protons and electrons increase by 1 at a time – and are always equal, for an atom. Is there a pattern for the number of neutrons?

1. Name the subatomic particles that make up the atom.
2. Which subatomic particle has:
   a a positive charge? b no charge? c almost no mass?
3. An atom has 9 protons. Which element is it?
4. Why do atoms have no overall charge?
5. Define: a proton number   b nucleon number
6. Give another term that can be used in place of:
   a proton number   b nucleon number
7. The Periodic Table gives the elements in order of …?
8. Identify each atom, and say how many protons, electrons, and neutrons it has:
   $^{12}_{6}$C $^{16}_{8}$O $^{24}_{12}$Mg $^{27}_{13}$Al $^{64}_{29}$Cu

2.3 How electrons are arranged

Objectives: describe the arrangement of electrons in atoms; explain its link to the patterns in the Periodic Table; explain the unreactivity of the Group VIII elements

Electron shells

Electrons are arranged in **shells** around the nucleus.
The first shell, closest to the nucleus, is the lowest energy level.
The further a shell is from the nucleus, the higher the energy level.

Each shell can hold only a certain number of electrons. These are the rules:

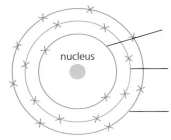

- The first shell can hold only 2 electrons. It fills first.
- The second shell can hold 8 electrons. It fills next.
- The third shell can hold only 8 electrons, *when it is the outer shell.* (The next 2 electrons go into a fourth shell. But after that the third shell can take 10 more electrons.)

This arrangement of electrons is called the **electronic configuration** of the atom. For the atom above, it can be described in a short way as **2,8,8**.

Electronic configuration for the first 20 elements

The electronic configuration for the atoms of the first 20 elements of the Periodic Table is shown below.

The number of electrons increases by 1 each time. (It is the same as the proton number.) The shells fill according to the rules above.

▲ Famous scientists: Neils Bohr (left) with Albert Einstein, around 1925. Neils Bohr put forward the idea that electrons occupy specific energy levels around the nucleus. We call these levels **shells**.

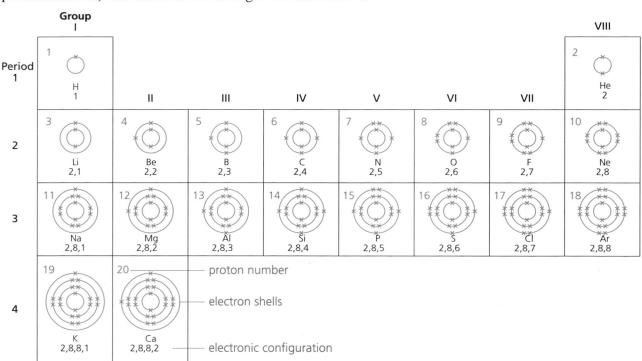

ATOMS AND ELEMENTS

Patterns in the Periodic Table

Electronic configuration is linked closely to the Periodic Table (page 15):

- The **period number** tells you how many electron shells are occupied.
- The **group number** is the same as the number of outer shell electrons, for Groups I to VII. So the atoms of Group I elements have 1 outer shell electron. The atoms of Group II elements have 2, and so on.
- The number of outer shell electrons dictates how an element reacts. That is why the elements in a group have similar reactions.

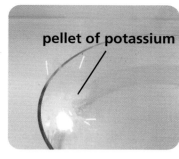

▲ Potassium (Group I) reacts violently with water. Hydrogen gas forms, and it may catch fire. All the Group I metals react in a similar way, because all their atoms have 1 outer shell electron. (The reaction with water is explosive for some of them.)

Group VIII, a special group

The elements in Group VIII have a special electronic configuration: **their atoms have a full outer shell of electrons.**

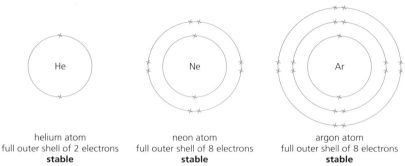

helium atom
full outer shell of 2 electrons
stable

neon atom
full outer shell of 8 electrons
stable

argon atom
full outer shell of 8 electrons
stable

A full outer shell make an atom stable. And this in turn has a very important result: it means the Group VIII elements are **unreactive**.

As you will see later, atoms of other elements react in order to obtain a full outer shell. But the Group VIII elements do not react because their atoms have a full outer shell already.

After calcium …

Calcium is the 20th element in the Periodic Table. After calcium, the electron shells fill in a more complex order. Let's look at rubidium, Rb. It is the 37th element in the table, so its proton number is 37. It is in Group I, Period 5.

- Group I tells you there is 1 electron in the outer shell.
- Period 5 tells you there are five shells.
- The proton number is 37, so there are also 37 electrons.
- The third shell holds 18 electrons, when full.

So the electronic configuration for rubidium is: 2,8,18,8,1.

▲ Helium from Group VIII is a light gas. Since it is unreactive, it is ideal for filling balloons. The other Group VIII elements are also gases. They are often called the **noble gases**, because they are unreactive.

1 One element has atoms with 13 electrons.
   a Draw a diagram to show how its electrons are arranged.
   b Give its electronic configuration in this form: 2,…
   c Name the element.
2 The electronic configuration for boron is 2,3. What is it for:
   a lithium?   b magnesium?   c hydrogen?
3 An element has 5 outer shell electrons. Which group is it in?
4 How many electron shells do the elements of Period 3 have?
5 The atoms of the Group VIII elements have a full outer shell of electrons. How does this affect these elements?
6 The element krypton, Kr, is in Group VIII, Period 4.
   a Which of these describes its electronic configuration?
      2,8,18,8,3   2,8,18,8   2,8,18,6 .
   b What can you say about the chemical reactivity of krypton?

2.4 Isotopes and A_r

Objectives: explain what isotopes are; understand that carbon-12 is taken as the standard atom; define *relative atomic mass* (A_r); know how to calculate A_r

How to identify an atom: a reminder
Only sodium atoms have 11 protons.
You can identify an atom by the number of protons in it.

Isotopes
All carbon atoms have 6 protons. So their proton number is 6. But not all carbon atoms are identical. Some have more *neutrons* than others.

 6 protons
6 electrons
6 neutrons

Most carbon atoms are like this, with **6** neutrons. That makes **12** nucleons (protons + neutrons) in total. So it is called **carbon-12**.

 6 protons
6 electrons
7 neutrons

But about one in every hundred carbon atoms is like this, with **7** neutrons. It has **13** nucleons in total. So it is called **carbon-13**.

 6 protons
6 electrons
8 neutrons

A very tiny number of carbon atoms are like this, with **8** neutrons. It has **14** nucleons in total. So it is called **carbon-14**.

The three atoms above are **isotopes** of carbon.
Isotopes are atoms of the same element that have the same number of protons, but different numbers of neutrons.

Most elements have isotopes that occur naturally. For example, oxygen has three, iron has four, magnesium has three, and chlorine has two.

How do isotopes react?
All the isotopes of an element have the same number of electrons. It is only the number of neutrons that differs.

This means they also have the same number of outer shell electrons.

As you saw in the last unit, the number of outer shell electrons dictates how atoms react. So all the isotopes of an element will behave in the same way, in a chemical reaction. We say they have the same **chemical properties**.

For example, all three isotopes of carbon react with oxygen, forming molecules of the gas carbon dioxide.

▲ Breathing in the three isotopes of oxygen. By far the most abundant of them is oxygen-16.

Unstable isotopes
The natural isotopes you meet in this unit are stable. But some elements have isotopes with an unstable nucleus, that breaks down by splitting.
- The breakdown is a **nuclear reaction**, not a chemical one.
- The breakdown gives out a great deal of energy, including radiation. So these isotopes are called **radioisotopes**.
- The radioisotope uranium-235 is used in nuclear power stations. The energy it gives out is used to boil water, to drive a steam turbine. This generates electricity.

ATOMS AND ELEMENTS

Remember that shorthand!
The shorthand way to describe any atom is shown again on the right. Using this method, the natural isotopes of oxygen and magnesium are:

$^{16}_{8}O$ $\quad$ $^{17}_{8}O$ $\quad$ $^{18}_{8}O$ $\quad$ and $\quad$ $^{24}_{12}Mg$ $\quad$ $^{25}_{12}Mg$ $\quad$ $^{26}_{12}Mg$

This is a quick and easy way to show isotopes.

Shorthand for an atom
You can describe any atom like this:

nucleon number
$\qquad$ **symbol for element**
proton number

Carbon-12 as our standard atom
As you already know, an atom is so light that it cannot be measured in grams. So scientists came up with the idea of **atomic mass units**.

They chose the carbon-12 atom as the standard. They assigned to it a mass of 12 units. That meant each of its nucleons had a mass of 1 unit. (In other words, one-twelfth of the total mass.)

Then they used the mass of a nucleon to work out the masses of other atoms. (They took the mass of an electron as zero.)

So the masses of all atoms are *relative* to the carbon-12 atom.

The relative atomic mass of an element, A_r
Isotopes have different masses. So when working out the relative atomic mass for an element, you must take its isotopes into account. An *average* value is calculated.

The relative atomic mass of an element, A_r, is the average mass of its isotopes, compared to 1/12 the mass of an atom of carbon-12.

How to calculate A_r
The A_r for an element is calculated like this:

A_r = (atomic mass of first isotope × its % abundance) +
$\qquad$ (atomic mass of second isotope × its % abundance) … and so on.

Example
Chlorine has two natural isotopes: $^{35}_{17}Cl$ and $^{37}_{17}Cl$.

75% of the atoms in chlorine have a mass of 35, and 25% a mass of 37.
So A_r for chlorine = 75% × 35 + 25% × 37
$\qquad\qquad\qquad\quad = \frac{75}{100} \times 35 + \frac{25}{100} \times 37$ (changing % to fractions)
$\qquad\qquad\qquad\quad = 26.25 + 9.25$
$\qquad\qquad\qquad\quad = 35.5$

▲ A gas jar of chlorine. It contains the two isotopes of chlorine.

1 a What are *isotopes*?
$\quad$ **b** Name the three isotopes of carbon, and write the shorthand symbols for them.
2 The three isotopes of oxygen can be written as:
$\quad$ **a** oxygen-16 $\quad$ **b** oxygen-17 $\quad$ **c** oxygen-18
$\quad$ State the number of protons and neutrons in each isotope.
3 Hydrogen, the first element in the Periodic Table, has three naturally occuring isotopes. They have 0, 1 and 2 neutrons. Write the shorthand symbols for all three.
4 a What does A_r stand for?
$\quad$ **b** Which atom is used as the standard for working out A_r?
$\quad$ **c** What does the r in A_r tell you?
5 Write out the definition of A_r, and make sure you understand each part of it.
6 a An imaginary element Z has two isotopes. 30% of its atoms have a mass of 45. The rest have a mass of 48. Calculate its A_r.
$\quad$ **b** Explain why both isotopes of Z react in the same way.

How our model of the atom developed

Discovering the atom, by indirect evidence
Today, we have powerful electron microscopes that can capture images of atoms. But long before such microscopes existed, we had learned a great deal about atoms, even though we could not see them. It was amazing. How did we do it? Through **indirect evidence**.

It began with the Ancient Greeks
In Ancient Greece (around 750 BCE–150 BCE), the philosophers thought hard about the world around them. Is water continuous matter? Is air just empty space? If you crush a stone to dust, then keep on crushing the dust, what will you end up with?

A philosopher called Democritus came up with an answer. He said that everything is made of tiny particles that cannot be divided. He called them **atoms**. (In Greek, *atomos* means *cannot be cut*.) They came in different shapes and colours, and had different tastes. When mixed in different amounts, they give us … everything.

So over 2000 years ago, the idea of the atom had begun to take shape.

▲ Indirect evidence … of what?

On to the alchemists
The Greek philosophers did a lot of thinking – but no experiments. The **alchemists** were different. They experimented day and night. They were searching for the **elixir of life**, which would keep us young, and for a way to turn metals like iron into gold. So we would all be young, and rich!

From about 600 CE, the practice of alchemy spread to many countries, including Persia (Iran), India, China, Greece, France, and Britain.

The alchemists did not find the elixir of life, or make gold. They did not develop ideas about the atom. But they carried out a great many reactions, and knew how much of each reactant to use. This helped chemists later.

▲ Democritus shown on an old Greek banknote. He gave us the term *atom*.

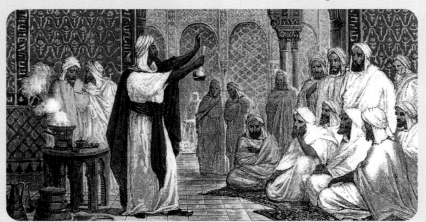

▲ The Persian alchemist Geber, who lived around 721 – 815 BCE, is often called 'the father of chemistry'.

▲ An alchemist's lab book. Is yours like this? When you purify things in the lab by crystallisation and distillation, you are using techniques that the alchemists developed many hundreds of years ago.

ATOMS AND ELEMENTS

Make way for us chemists ...
By around 1600 CE, the alchemists gave way to the **chemists** – like you. They carried out experiments in a very systematic way. They collected a great deal of indirect evidence about particles of matter. For example:

- A gas can be compressed into a smaller space. So it must be made of particles with empty space between them.
- Copper(II) carbonate *always* contains 5.3 parts of copper to 1 part of carbon to 4 parts of oxygen, by mass. So copper, carbon, and oxygen must be made of particles, which combine in a fixed ratio.
- Pollen granules jiggle around in water, even though they are not alive. It must be because they are being struck by water particles.

And so, little by little, chemists proved that there were indeed tiny particles that could not be divided further by chemical means. The particles were named **atoms**, in tribute to Democritus.

And then came the shock!
By 200 years ago, the existence of atoms had been finally accepted. But then came the shock. There were even smaller particles inside atoms – the **subatomic particles**! These were discovered by **physicists**.

Electrons
In 1897, the English physicist J.J. Thomson was investigating mystery rays that glowed inside an empty glass tube, when it was plugged into an electric circuit. He had discovered **electrons**.

Protons and neutrons
A year earlier, a French physicist called Becquerel was working with crystals of a uranium salt. He found they glowed in the dark, and left images on a photographic plate. He had discovered **radioactivity**.

The English physicist Ernest Rutherford investigated radioactivity. He found it produced **alpha particles**: 7000 times heavier than electrons, with a positive charge. He could speed them up and shoot them like tiny bullets! (We now know they consist of two protons and two neutrons.)

In 1911, Rutherford shot alpha particles at some gold foil. Most passed right through the foil. But some bounced back, as if they had collided with something! He deduced that an atom is mostly empty space, but with something small and dense at the centre. It was the **nucleus**. He assumed it was made of particles of positive charge, and called them **protons**.

Those electron shells
If the nucleus is positive, why don't the negative electrons rush straight into it? In 1913 the Danish scientist Niels Bohr came up with the theory of **electron shells**. It fitted all the experiments.

At last, the neutrons
In 1930, two German physicists, Bothe and Becker, shot alpha particles at beryllium – and knocked a stream of new particles from it. These had the same mass as protons, but no charge. They were named **neutrons**.

So finally, around 2300 years after the death of Democritus, our model of the atom was complete. It's the model you study in chemistry.

▲ J.J. Thomson at work in his lab. In 1897 he discovered the electron.

▲ The Polish scientist Marie Curie also researched radioactivity. She was awarded the Nobel Prize twice – in physics and chemistry. She worked in France.

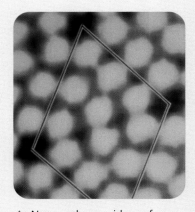

▲ Now we have evidence for atoms from microscopes that can 'see' them with the help of electrons. This image shows silicon atoms. (The lines were added to mark a repeating 'unit' of atoms, and false colour was added too.) The atoms above are about 75 million times their real size!

Checkup on Chapter 2

Revision checklist

Core syllabus content

Make sure you can …
- explain what each term means, with examples:
  atom   element   compound   mixture
- explain what the Periodic Table is
- describe an atom as having a central nucleus surrounded by electrons in shells
- name the subatomic particles in the nucleus
- state the mass and charge of:
  a proton   an electron   a neutron
- explain why an atom has no charge
- define these terms:
  proton number (atomic number)
  nucleon number (mass number)
- name the first 20 elements of the Periodic Table in order of proton number, and give their symbols
- describe the order in which electrons fill the electron shells, and the number in each shell (for a maximum of 20 electrons)
- sketch the electronic configuration for an atom, given its proton number
- show the electronic configuration for an atom in this form: 2,8,…
- describe how these are connected:
  – the outer shell electrons of an element, and its group number in the Periodic Table
  – the number of electron shells in its atoms, and its period number in the Periodic Table
- explain why the Group VIII elements are unreactive
- explain what an isotope is, and give examples
- state the number of protons, neutrons and electrons in an atom, from a description like this: $^{23}_{11}$Na
- define *relative atomic mass (A_r)*

Extended syllabus content

Make sure you can also …
- explain why all the isotopes of an element have the same chemical properties
- give, and use, the formula for calculating the relative atomic mass of an element with different isotopes

Questions

Core syllabus content

1. Magnesium is the eighth most abundant element in Earth's crust. It is a typical metallic element.
   a i Give the symbol for magnesium.
      ii About how many natural elements are there? Choose:
         about 10   about 100   about 1000
   b i Name the table that displays the elements.
      ii Where is magnesium found in this table?
      iii Name another element in the same group as magnesium.
   c Magnesium forms compounds when it reacts with non-metallic elements.
      i Name a non-metallic element.
      ii How do compounds differ from elements?
      iii How do compounds differ from mixtures?

2. Statements **A – J** below are about the particles that make up the atom. For each statement write:
   p if it describes the **proton**
   e if it describes the **electron**
   n if it describes the **neutron**

   A the positively-charged particle
   B found with the proton, in the nucleus
   C can occur in different numbers, in atoms of the same element
   D held in shells around the nucleus
   E has a negative charge
   F the particle with negligible mass
   G to find how many there are, subtract the proton number from the nucleon number
   H has no charge
   I has the same mass as a neutron
   J dictates the position of the element in the Periodic Table

3. The atoms of an element can be represented by a set of three letters, $^{y}_{z}X$ as shown on the right.
   a What does this letter in the set stand for?
      i X   ii y   iii z
   b How many neutrons do these atoms have?
      i $^{107}_{47}$Ag   ii $^{63}_{29}$Cu   iii $^{1}_{1}$H   iv $^{20}_{10}$Ne   v $^{238}_{92}$U
   c Bromine atoms have 45 neutrons and 35 protons. Describe them using the method in **b**.

4 These are six elements: aluminum (Al), boron (B), nitrogen (N), oxygen (O), phosphorus (P), and sulfur (S). They are shown in the Periodic Table on page 15.
For each element write down:
a i whether it is a metal or a non-metal
ii its period number in the Periodic Table
iii its group number in the Periodic Table
iv its proton number (atomic number)
v the number of electrons in an atom of the element
vi the electronic configuration of its atoms
vii the number of outer shell electrons in an atom of the element
b Which of the six elements would you expect to have similar properties? Explain your answer.

5 Boron has two types of atom, shown below.

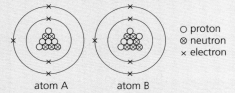

○ proton
⊗ neutron
× electron

atom A   atom B

a What is different about these two atoms?
b What name is given to atoms like these?
c Describe each atom in shorthand form, as in **3**.
d What is the nucleon number of atom A?
e Is atom B heavier, or lighter, than atom A?
f i Give the electronic configuration for A and B.
ii Comment on your answer for **i**.

6 The two metals sodium (proton number 11) and magnesium (proton number 12) are found next to each other in the Periodic Table.
a Say whether this is the same, or different, for their atoms:
i the number of electron shells
ii the number of outer shell electrons
The relative atomic mass of sodium is 23.0. The relative atomic mass of magnesium is 24.3.
b Which of the two elements may exist naturally as a single isotope? Explain your answer.

7 Strontium is in Group II, period 5 of the Periodic Table. Its proton number is 38. Give the missing numbers, for the electrons in an atom of strontium:

a ..... electrons in total
b ..... electron shells
c ..... electrons in the outer shell

8 This diagram shows how the electrons are arranged, in the atoms of an element we will call X.

a i Give the electronic configuration for the atoms of element X.
ii What is special about this arrangement?
b In which group of the Periodic Table is X?
c Identify element X. (See the Periodic Table on page 15.)
d Name another element with the same number of outer shell electrons in its atoms.

Extended syllabus content

9 The structures of the three natural isotopes of hydrogen are shown below:

hydrogen deuterium tritium

a Copy and complete this key for the diagram:
× represents ....
○ represents ....
⊗ represents ....
b Give the mass numbers of the three isotopes.
c Copy and complete this statement:
Isotopes of an element contain the same number of .... and .... but different numbers of ....
d The relative atomic mass of hydrogen is 1.008. Which of the three isotopes is present in the highest proportion? Explain your answer.

10 Gallium has two natural isotopes, gallium-69 and gallium-71. Its proton number is 31.
a Give the number of neutrons in an atom of:
i gallium-69   ii gallium-71
b If gallium contained 50% of each isotope, what would its relative atomic mass be?
c The A_r of gallium is in fact 69.723. Show that this value is consistent with a mixture of approximately 60% gallium-69 and 40% gallium-71.

11 Question **5** shows the natural isotopes of boron. Boron contains 80.1% of atom A. Calculate:
a the percentage of atom B in boron
b the relative atomic mass of boron

3.1 Compounds, mixtures, and chemical change

Objectives: define *element*, *compound*, and *mixture*; give the signs of a chemical change; understand the difference between chemical and physical change

Elements: a reminder
An **element** contains only one kind of atom. For example, the element sodium contains only sodium atoms.

Compounds
A compound is made of atoms of different elements, bonded together.

The compound is described by a **formula**, made from the symbols of the atoms in it. (The plural of formula is **formulae**.)

There are millions of compounds. This table shows three common ones.

Name of compound	Elements in it	How the atoms are joined	Formula of compound
water	hydrogen and oxygen		H_2O
carbon dioxide	carbon and oxygen		CO_2
ethanol	carbon, hydrogen, and oxygen		C_2H_5OH

Water has two hydrogen atoms joined or bonded to an oxygen atom. So its formula is H_2O. Note where the 2 is written. Now check the formulae for carbon dioxide and ethanol. Do they match the drawings?

Compounds and mixtures: the difference
A mixture contains different substances that are *not* bonded together.
So you can usually separate the substances quite easily. For example:

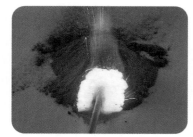

This is a **mixture** of iron powder and sulfur. You could separate them by dissolving the sulfur in methylbenzene (a solvent), and filtering the iron off.

But if you heat the end of a metal rod in a Bunsen burner, and push it into the mixture, the mixture starts to glow brightly. A **chemical change** is taking place.

The result is a black **compound** called iron(II) sulfide. It is made of iron and sulfur atoms bonded together. Its formula is FeS. It will not dissolve in methylbenzene.

ATOMS COMBINING

The signs of a chemical change

When you heat a mixture of iron and sulfur, a chemical change takes place. The iron and sulfur atoms **bond together** to form a compound. You can tell when a chemical change has taken place, by these three signs:

1. **One or more new chemical substances are formed.**
   You can describe the change by a **word equation** like this:
   iron + sulfur ⟶ iron(II) sulfide
   The + means *reacts with*, and the ⟶ means *to form*.
   The new substances usually look different from the starting substances. For example, sulfur is yellow, but iron(II) sulfide is black.

2. **Energy is taken in or given out, during the reaction.**
   Energy was needed to start off the reaction between iron and sulfur, in the form of heat from the hot metal rod. But the reaction gave out heat once it began – the mixture glowed brightly.

3. **The change is usually difficult to reverse.**
   You would need to carry out several reactions to get the iron and sulfur back from iron sulfide. (But it can be done!)

A chemical change is usually called *a chemical reaction*.

▲ Burning gas, to fry eggs. Are chemical changes taking place?

It is different from physical change

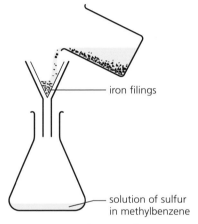

 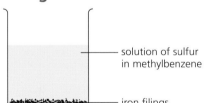

When you mix iron powder with sulfur, that is a **physical change**. No new substance has formed. If you then dissolve the sulfur …

… in methylbenzene, that is also a physical change. The solvent could be removed again by distilling it. (Danger! It is highly flammable.)

Now separate the iron by filtering. That is a physical change. You can reverse it by putting the iron back into the flask again.

No new chemical substances are formed in these changes.

If no new chemical substance is formed, a change is a physical change.

Unlike chemical changes, a physical change is usually easy to reverse.

1. Hydrogen is an *element*. Water is a *compound*. Explain the difference between them.
2. Give the formulae for: **a** water **b** carbon dioxide
3. Write the formulae for these compounds:

   a
   hydrogen chloride

   b
   ammonia

4. Explain the difference between a *mixture* of iron and sulfur and the *compound* iron sulfide.
5. List the signs of chemical change, when you fry an egg.
6. Magnesium ribbon burns in oxygen with a dazzling white light. Magnesium oxide forms as a white ash. What signs are there that a chemical change has taken place?
7. What do the + and ⟶ in a word equation mean?
8. Write the word equation for the reaction in question **6**.
9. Is it a chemical change or a physical change? Give reasons.
   a a glass bottle breaking
   b dough being baked into bread, in an oven
   c wool being knitted to make a scarf

3.2 Why do atoms form bonds?

Objectives: explain why atoms form bonds; draw diagrams to show how sodium and chlorine atoms obtain full outer shells; define *ion* and give examples

The reaction between sodium and chlorine

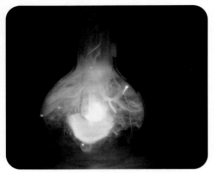

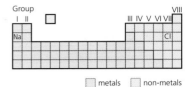

▲ Where sodium and chlorine are, in the Periodic Table.

Sodium and chlorine are both **elements**. When sodium is heated and placed in a jar of chlorine, it burns with a bright flame.

The result is a white solid that has to be scraped from the sides of the jar. It looks completely different from the sodium and chlorine.

So a chemical reaction has taken place. The white solid is **sodium chloride**. Atoms of sodium and chlorine have **bonded** (joined together) to form a **compound**. The word equation for the reaction is:

sodium + chlorine ⟶ sodium chloride

Why do atoms form bonds?

Like sodium and chlorine, the atoms of most elements form bonds.

Why? We get a clue by looking at the elements of Group VIII, **the noble gases**. Their atoms *do not* form bonds.

This is because their atoms have a full outer shell of electrons. A full outer shell makes an atom stable. So the noble gases are **unreactive**.

▲ The noble gas neon is safe to use in light tubes, because it is unreactive.

helium atom:
full outer shell of
2 electrons – *stable*

neon atom:
full outer shell of
8 electrons – *stable*

argon atom:
full outer shell of
8 electrons – *stable*

2   2,8   2,8,8

▲ Welding is often carried out in an atmosphere of the noble gas argon. It will not react with hot metals (unlike oxygen).

And that gives us the answer to our question:

Atoms bond with each other in order to obtain a full outer shell of electrons, like the atoms in Group VIII.

In other words, they bond in order to obtain 8 electrons in their outer shell (or 2, if they have only one shell).

How sodium atoms obtain a full outer shell

Sodium is in Group I of the Periodic Table. This tells you that a sodium atom has just 1 electron in its outer shell. To obtain a full outer shell of 8 electrons, it loses this electron to another atom. It becomes a **sodium ion**:

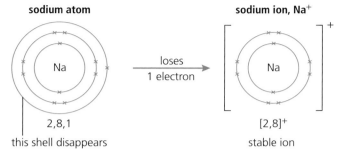

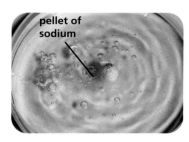

▲ Sodium reacts violently with cold water, thanks to its strong drive to lose electrons and form sodium ions.

The sodium ion has 11 protons but only 10 electrons. So it has a charge of 1+, as you can see from the panel on the right.

The symbol for sodium is Na, so the symbol for the sodium ion is **Na$^+$**. The + means *1 positive charge*. Na$^+$ is a **positive ion**.

The charge on a sodium ion
charge on 11 protons   11+
charge on 10 electrons 10−
total charge   1+

How chlorine atoms obtain a full outer shell

Chlorine is in Group VII of the Periodic Table. This tells you that a chlorine atom has 7 electrons in its outer shell. It can reach 8 electrons by accepting 1 electron from another atom. It becomes a chloride ion:

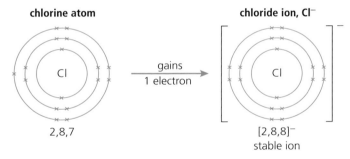

The charge on a chloride ion
charge on 17 protons   17+
charge on 18 electrons 18−
total charge   1−

The chloride ion has a charge of 1− It is a **negative ion**. Its symbol is **Cl$^-$**.

More about ions

Atoms become ions when they lose or gain electrons.
An ion is a charged particle. It is charged because it has an unequal number of protons and electrons.

Positive ions are also called **cations**.   (Remember: *positive cat*.)
Negative ions are also called **anions**.   (Remember: *negative an*.)

Shorthand for an ion
You can describe an ion in the same way as an atom. Just add its charge! Look:

$^{23}_{11}$Na$^+$ $^{35}_{17}$Cl$^-$

1 Atoms of Group VIII elements do not form bonds. Why not?
2 Explain why all other atoms form bonds.
3 Draw a diagram showing how this obtains a full outer shell:
   a a sodium atom   b a chlorine atom
4 Define the term *ion*.
5 Explain why the charge on a chloride ion is 1−.
6 Atoms of Group VIII elements do *not* form ions. Why not?
7 Give another term for: a a positive ion b a negative ion
8 From this list, write down: a the anions b the cations
   Br$^-$ C   K$^+$ O^{2-} Mg^{2+} Ar
9 The ion on the right is from Group VII. Suggest a name for it. (There's a Periodic Table on page 15.) $^{79}_{35}$Br$^-$

3.3 The ionic bond

Objectives: describe how metals react with non-metals to form ionic compounds, with examples; define *ionic bond*; describe the lattice structure of sodium chloride

How sodium and chlorine atoms bond together

As you saw on page 29, a sodium atom must lose one electron, and a chlorine atom must gain one, to obtain a full outer shell of 8 electrons.

So when a sodium atom and a chlorine atom react together, the sodium atom loses its electron *to the chlorine atom*, and two ions are formed.

Here, sodium electrons are shown as • and chlorine electrons as ×:

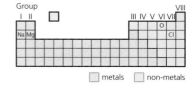

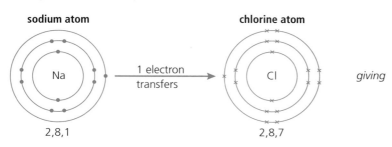

The two ions have opposite charges, so they attract each other. The force of attraction between them is strong. It is called an **ionic bond**.

The ionic bond is a strong electrostatic attraction between ions of opposite charge.

How solid sodium chloride is formed

When sodium reacts with chlorine, trillions of sodium and chloride ions form. But they do not stay in pairs. They form a **lattice** – a regular arrangement of alternating positive and negative ions. Look:

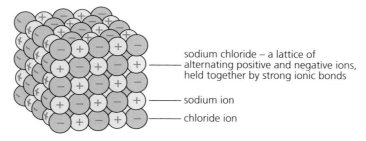

sodium chloride – a lattice of alternating positive and negative ions, held together by strong ionic bonds

sodium ion

chloride ion

The lattice grows to form a giant 3-D structure. It is called 'giant' because it contains a very large number of ions. This giant structure is the compound **sodium chloride**, or **common salt**.

Since it is made of ions, sodium chloride is called an **ionic compound**. It contains one Na^+ ion for each Cl^- ion, so its formula is **NaCl**.

The compound has no overall charge, because the charges on its ions cancel each other out:

the charge on each sodium ion is   1+
the charge on each chloride ion is   1−
              total charge   0

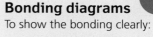

Bonding diagrams
To show the bonding clearly:
- use dots and crosses (o, •, ×) for electrons from atoms of different elements
- write the symbol for the element in the centre of each atom.

▲ These balloons were given opposite charges. So they attract each other, like the ions in sodium chloride. The attraction is called **electrostatic** because the charges do not move. They do not flow like a current.

ATOMS COMBINING

Other ionic compounds
Sodium is a metal. Chlorine is a non-metal. They react together to form an ionic compound. Other metals and non-metals follow the same pattern.

An ionic compound forms when metal atoms lose electrons to non-metal atoms. This gives positive metal ions and negative non-metal ions, with full outer electron shells. They form a lattice.

The compound has no overall charge.

Below are two more examples.

▲ The ionic compound sodium chloride. We call it **salt**! The lattice forms **crystals**. Each contains many trillions of ions.

Magnesium oxide
Magnesium is in Group II of the Periodic Table. Oxygen is in Group VI. A magnesium atom has 2 outer shell electrons, and an oxygen atom has 6. On reacting, each magnesium atom loses its 2 outer shell electrons to an oxygen atom. Magnesium ions and oxide ions form, with full outer shells:

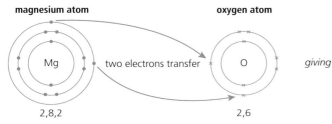

The ions attract each other because of their opposite charges. They form a lattice. The compound is called **magnesium oxide**. It has one magnesium ion for each oxide ion, so its formula is **MgO**. It has no overall charge.

The charge on magnesium oxide
charge on each magnesium ion 2+
charge on each oxide ion 2−
total charge 0

Magnesium chloride
When magnesium reacts with chlorine, each magnesium atom transfers an electron to *two* chlorine atoms, to form **magnesium chloride**.

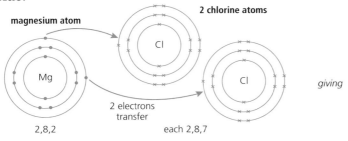

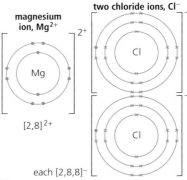

The ions form a lattice with two chloride ions for each magnesium ion. So the formula of the compound is **MgCl$_2$**. It has no overall charge.

1 Draw a diagram to show what happens to the electrons, when a sodium atom reacts with a chlorine atom.
2 What is an *ionic bond*?
3 Describe the structure of solid sodium chloride, and explain why its formula is NaCl.
4 a Explain why a magnesium ion has a charge of 2+.
 b Why do the ions in magnesium oxide stay together?
 c Explain why the formula of magnesium chloride is MgCl$_2$.
5 The metal aluminium is in Group III of the Periodic Table. Suggest a formula for the compound it forms with chlorine.

3.4 More about ions

Objectives: work out the names and formulae for ionic compounds; recall that most transition elements form more than one ion; define *compound ion*

Ions of the first 20 elements
Not every element forms ions during reactions. Of the first 20 elements in the Periodic Table, only 12 easily form ions. Here they are:

Group I	II		III	IV	V	VI	VII	VIII
		H^+ hydrogen						none
Li^+ lithium	Be^{2+} beryllium					O^{2-} oxide	F^- fluoride	none
Na^+ sodium	Mg^{2+} magnesium		Al^{3+} aluminium			S^{2-} sulfide	Cl^- chloride	none
K^+ potassium	Ca^{2+} calcium	transition elements						

Note that:
- The metals, and the non-metal hydrogen, lose electrons and form **positive ions**. These ions have the same names as the atoms.
- The other non-metals form **negative ions**, with names ending in **-ide**.
- The elements in Groups IV and V do not usually form ions. The atoms would have to gain or lose several electrons. It takes too much energy!
- The Group VIII elements do not form ions. Their atoms already have full outer shells.

Naming an ionic compound
To name an ionic compound, you just put the names of the ions together, with the positive one (cation) first:

Ions in compound	Name of compound
K^+ and F^-	potassium fluoride
Ca^{2+} and S^{2-}	calcium sulfide

Writing the formula for an ionic compound
Follow these four steps to work out the formula:
1 Write down the name of the ionic compound.
2 Write down the symbols for its ions.
3 The compound must have no overall charge, so balance the ions until the positive and negative charges add up to zero.
4 Write down the formula without the charges.

Example 1
1 Lithium fluoride.
2 The ions are Li^+ and F^-.
3 One Li^+ is needed for every F^-, to make the total charge zero.
4 The formula is LiF.

Example 2
1 Sodium sulfide.
2 The ions are Na^+ and S^{2-}.
3 Two Na^+ ions are needed for every S^{2-} ion, to make the total charge zero: Na^+ Na^+ S^{2-}.
4 The formula is Na_2S. (What does the ₂ show?)

▲ Bath time. Bath salts contain ionic compounds such as magnesium sulfate (Epsom salts) and sodium hydrogen carbonate (baking soda). Plus scent!

> **The numbers in a formula**
> To show there is more than one of a given atom in a formula …
> - write the number after the symbol for the atom …
> - and a bit lower, like this: $MgCl_2$.

ATOMS COMBINING

Some metals form more than one ion
Turn to the Periodic Table on page **15**. Find the **transition elements**. This block includes many common metals, such as iron and copper.

Some transition elements form only one ion:
- silver forms only Ag^+ ions
- zinc forms only Zn^{2+} ions.

But most transition elements can form more than one ion. For example, copper and iron can each form two:

Ion	Name	Example of compound
Cu^+	copper(I) ion	copper(I) oxide, Cu_2O
Cu^{2+}	copper(II) ion	copper(II) oxide, CuO
Fe^{2+}	iron(II) ion	iron(II) chloride, $FeCl_2$
Fe^{3+}	iron(III) ion	iron(III) chloride, $FeCl_3$

The Roman numeral (II) in a name tells you that the ion has a charge of 2+. What do (I) and (III) tell you?

▲ The two oxides of copper.

Compound ions
All the ions you met so far have been formed from single atoms. But ions can also be formed from a group of bonded atoms.
These are called **compound ions**.

Some common ones are shown on the right. Remember, each is just a single ion, even though it contains more than one atom.

The formulae for their compounds can be worked out as before. Some examples are shown below.

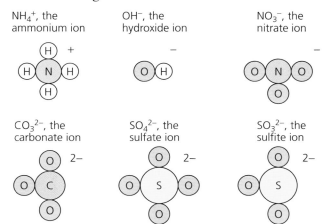

Example 3
1. Sodium carbonate.
2. The ions are Na^+ and CO_3^{2-}.
3. Two Na^+ are needed to balance the charge on one CO_3^{2-}.
4. The formula is Na_2CO_3.

Example 4
1. Calcium nitrate.
2. The ions are Ca^{2+} and NO_3^-.
3. Two NO_3^- are needed to balance the charge on one Ca^{2+}.
4. The formula is $Ca(NO_3)_2$. Note that brackets are put round the NO_3, before the $_2$ is put in.

You will learn how to test for all these compound ions, in the lab tests for anions and cations in Chapter 19.

1. Which do the metals form: anions or cations?
2. Hydrogen is a non-metal. What is unusual about its ion?
3. Carbon (Group IV) does not usually form ions. Why not?
4. Write down the symbols for the ions in:
   a potassium chloride   b calcium sulfide
   c lithium sulfide   d magnesium fluoride
5. Now work out the formula for each compound in **4**.
6. Most ____ elements form ____ ____ one ion. Complete!
7. Work out the formula for:
   a copper(II) chloride   b iron(III) oxide
8. Name each compound:
   $CuCl$ FeS $Mg(NO_3)_2$ NH_4NO_3 $CaSO_3$
9. Work out the formula for: a sodium sulfate
   b potassium hydroxide   c silver nitrate

3.5 The covalent bond

Objectives: explain what a covalent bond is; define *molecule*; describe the covalent bonding in the elements hydrogen, chlorine, oxygen, and nitrogen

Why atoms bond: a reminder
As you saw in Unit 3.3, atoms bond in order to gain a full outer shell of electrons, like the noble gas atoms. So when sodium and chlorine react together, each sodium atom gives up an electron to a chlorine atom.

But that is not the only way. Atoms can also gain full outer shells by *sharing* electrons with each other.

Sharing electrons
When two non-metal atoms react together, *both* need to gain electrons to obtain full outer shells. They manage this by sharing electrons.

We will look at **non-metal elements** in this unit, and at **non-metal compounds** in the next unit. Atoms can share only their outer shell electrons, so the diagrams will show only these.

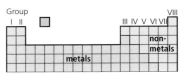

▲ Atoms of non-metals do not give up electrons to obtain a full shell, because they would have to lose so many. It would take too much energy to overcome the pull of the positive nucleus. Instead, they share electrons with each other.

Hydrogen
A hydrogen atom has only one shell, with one electron. The shell can hold two electrons. When two hydrogen atoms get close enough, their shells overlap and then they can share electrons. Like this:

So each has gained a full shell of two electrons, like helium atoms.

The bond between the atoms
Each hydrogen atom has a positive nucleus. Both nuclei attract the shared electrons – and this strong force of attraction holds the two atoms together. This force of attraction is called a **covalent bond**.

A single covalent bond is formed when two atoms share a pair of electrons, leading to full outer shells.

▲ A model of the hydrogen molecule. The molecule can also be shown as H–H. The line represents a single bond.

Molecules
The two bonded hydrogen atoms above form a **molecule**.

A molecule is two or more atoms held together by covalent bonds.

Hydrogen is called a **molecular** element. Its formula is H_2. The $_2$ tells you there are 2 hydrogen atoms in each molecule.

Many other non-metals are also molecular. For example:

iodine, I_2 oxygen, O_2 nitrogen, N_2
chlorine, Cl_2 sulfur, S_8 phosphorus, P_4

Elements with molecules containing two atoms are called **diatomic**. So iodine and oxygen are diatomic. Can you give two other examples?

▲ Chlorine: a gas at room temperature. How many atoms are in each molecule?

Chlorine

A chlorine atom needs a share in one more electron, to obtain a full outer shell of eight electrons. So two chlorine atoms bond covalently like this:

▲ A model of the chlorine molecule.

Since only one pair of electrons is shared, the bond between the atoms is called a **single covalent bond**, or just a **single bond**. You can show it in a short way by a single line, like this: Cl–Cl.

Oxygen

An oxygen atom has six outer shell electrons, so needs a share in *two* more. So two oxygen atoms share two electrons each, giving molecules with the formula O_2. Each atom has a full outer shell of eight electrons:

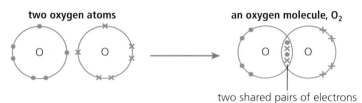

▲ A model of the oxygen molecule.

Since the oxygen atoms share two pairs of electrons, the bond between them is called a **double bond**. You can show it like this: O=O.

Nitrogen

A nitrogen atom has five outer shell electrons, so needs a share in *three* more. So two nitrogen atoms share three electrons each, giving molecules with the formula N_2. Each atom has a full outer shell of eight electrons:

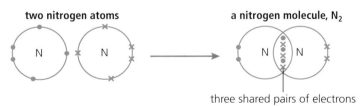

▲ A model of the nitrogen molecule.

Since the nitrogen atoms share three pairs of electrons, the bond between them is called a **triple bond**. You can show it like this: N≡N.

1. **a** Many non-metal atoms share electrons with each other. What is the purpose of this?
   b Name the bond between two atoms that share one pair of electrons.
2. **a** What is a *molecule*?
   b What term is used for molecules with two atoms?
3. Name five molecular elements, and give their formulae.
4. Draw a diagram to show the bonding in:
   a hydrogen   **b** chlorine
5. Nitrogen is in Group V. Explain why it forms a *triple* bond.
6. Atoms of non-metal elements do not form *ionic bonds* with each other. Why not?

3.6 Covalent compounds

Objectives: understand that non-metals react with each other to form molecular compounds; describe the bonding in the compounds named in the tables

Covalent compounds

In the last unit you saw that many non-metal elements exist as molecules. A huge number of non-metal *compounds* also exist as molecules.

In a molecular compound, atoms of *different* elements share electrons. The compounds are called **covalent compounds**. Here are four examples.

> **Most are molecular ...** !
> Most non-metal elements and their compounds exist as molecules.

Covalent compound	Description	Model of the molecule
hydrogen chloride, HCl 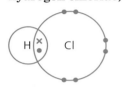 a molecule of hydrogen chloride	The chlorine atom shares *one* electron with the hydrogen atom. Both atoms now have a full outer shell of electrons: 2 for hydrogen (like the helium atom) and 8 for chlorine (like the other noble gas atoms).	
water, H_2O 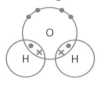 a molecule of water	The oxygen atom shares *two* electrons, one with each hydrogen atom. All three atoms now have a full outer shell of electrons: 2 for hydrogen and 8 for oxygen.	
ammonia, NH_3 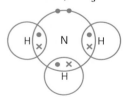 a molecule of ammonia	The nitrogen atom shares *three* electrons, one with each hydrogen atom. All four atoms now have a full outer shell of electrons: 2 for hydrogen and 8 for nitrogen.	
methane, CH_4 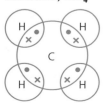 a molecule of methane	The carbon atom shares *four* electrons, one with each hydrogen atom. All five atoms now have a full outer shell of electrons: 2 for hydrogen and 8 for carbon.	

ATOMS COMBINING

More examples of covalent compounds
This table shows three more examples of covalent compounds.

Covalent compound	Description	Model of the molecule
methanol, CH_3OH 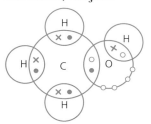 a molecule of methanol	The carbon atom shares four electrons, three with hydrogen atoms and one with an oxygen atom. All the atoms gain full outer shells of electrons. Look at the shape of the molecule: a little like methane, but changed by the presence of the oxygen atom.	
carbon dioxide, CO_2 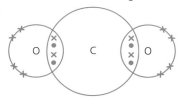 a molecule of carbon dioxide	The carbon atom shares four electrons, two with each oxygen atom. All three atoms gain full outer shells. The two sets of bonding electrons repel each other, giving a **linear** molecule. All the bonds are double bonds, so we can show the molecule like this: $O=C=O$.	
ethene, C_2H_4 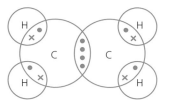 a molecule of ethene	This molecule has *two* carbon atoms. Each shares four electrons. It shares two with hydrogen atoms, and two with the other carbon atom, giving a carbon-carbon double bond. So the molecule is usually drawn as here:	

The shapes of molecules
The pairs of electrons around an atom repel each other, and move as far apart as they can. This dictates the **shapes** of molecules.

In methane, the carbon atom shares electrons with four hydrogen atoms. The four shared electron pairs move as far apart as possible. This gives the molecule a tetrahedral shape. Look at the model on the right.

In a water molecule, the four electron pairs around oxygen also form a tetrahedral arrangement, giving the molecule a bent shape.

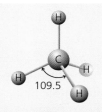

▲ A model of the methane molecule. The rods represent bonds. All angles are equal.

1. What is a *covalent compound*?
2. Draw a diagram to show the bonding in a molecule of:
   a methane   b water   c ammonia
3. How many electrons can chlorine share with other atoms?
4. Draw a diagram to show the bonding in carbon dioxide.
5. Name the two covalent compounds in this unit, where all the outer electrons of the atoms are used in forming bonds.
6. How many electrons are needed to form a $C=C$ bond?

3.7 Comparing ionic and covalent compounds

Objectives: understand that the bonding in a compound dictates many of its properties; compare the general properties of ionic and covalent compounds

Remember
Metals and non-metals react together to form **ionic compounds**.
Non-metals react with each other to form **covalent compounds**.
The covalent compounds you have met so far exist as **molecules**.

Comparing the structures of the solids
In Chapter 1, you met the idea that solids are a **regular lattice** of particles. In ionic compounds, these particles are **ions**. In the covalent compounds you have met so far, they are **molecules**. Let's compare their lattices.

A solid ionic compound   Sodium chloride is a typical ionic compound:

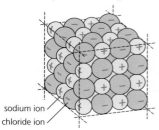

In sodium chloride, the ions are held in a regular lattice like this. They are held by strong ionic bonds.

The lattice grows in all directions, giving a crystal of sodium chloride. This one is greatly magnified.

The crystals look white and shiny. We add them to food, as **salt**, to bring out its taste.

A solid molecular covalent compound   Water is a molecular covalent compound. When you cool it below 0°C it becomes a solid: ice.

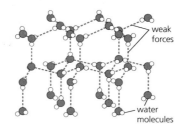

In ice, the water molecules are held in a regular lattice like this. But the forces between them are weak.

The lattice grows in all directions, giving a crystal of ice. An ice cube may be a single crystal!

We use ice to keep drinks cool, and food fresh. (The reactions that cause decay are slower in the cold.)

So both types of compounds have a regular lattice structure in their solid state, and form crystals. But they differ in two key ways:
- In ionic solids the forces between the ions are strong.
- In molecular covalent solids the molecules are not charged, and the forces between them are weak.

These differences lead to very different properties, as you will see next.

> **About crystals**
> - A regular arrangement of particles in a lattice always leads to crystals.
> - The particles can be atoms, ions, or molecules.

ATOMS COMBINING

The properties of ionic compounds
1 **Ionic compounds have high melting and boiling points.**
   For example:

Compound	Melting point/°C	Boiling point/°C
sodium chloride, NaCl	801	1413
magnesium oxide, MgO	2852	3600

This is because the ionic bonds are very strong. It takes a lot of heat energy to break up the lattice. So ionic compounds are solid at room temperature.

Look again at the values in the table. Magnesium oxide has a far higher melting and boiling point than sodium chloride does. This is because its ions have double the charge (Mg^{2+} and O^{2-} compared with Na^+ and Cl^-), so its ionic bonds are stronger.

▲ Magnesium oxide is used to line furnaces, because of its high melting point, 2852 °C.

2 **Ionic compounds are usually soluble in water.**
   The water molecules are able to separate the ions from each other. The ions then move apart, surrounded by water molecules.

3 **Ionic compounds conduct electricity, when melted or dissolved in water.**
   A solid ionic compound will not conduct electricity, because the ions are not free to move. But when it melts, or dissolves in water, the ions can move. Since they are charged, they can then conduct electricity.

> **! Conducting electricity**
> For a substance to conduct electricity, it must have charged particles – electrons or ions – that are free to move, carrying the current.

The properties of covalent compounds
1 **Molecular covalent compounds have low melting and boiling points.**
   For example:

Compound	Melting point/°C	Boiling point/°C
carbon monoxide, CO	−199	−191
hexane, C_6H_{14}	−95	69

This is because the forces of attraction between the molecules – the **intermolecular forces** – are weak. So it does not take much energy to break up the lattice and separate the molecules. That explains why many molecular compounds are liquids or gases at room temperature – and why many of the liquids are **volatile** (evaporate easily).

2 **Covalent compounds tend to be insoluble in water.**
   But they do dissolve in some solvents, for example, tetrachloromethane.

3 **Covalent compounds do not conduct electricity.**
   They have no charged particles to carry a current.

▲ The covalent compound carbon monoxide is formed when petrol burns in the limited supply of air in a car engine. And it is poisonous.

1 The particles in solids usually form a *regular lattice*. Explain what that means, in your own words.
2 Which type of particles make up the lattice, in:
   a ionic compounds?   b molecular compounds?
3 Solid sodium chloride will not conduct electricity, but a solution of sodium chloride will conduct. Explain this.
4 a A compound melts at 20°C. What is its structure?
   b Will the compound conduct electricity at 25°C?
   c Name a compound with a completely different structure.
   d Explain the reasoning behind your answers in **a** and **b**.
5 Describe the arrangement of the molecules in ice. How will the arrangement change as the ice warms up?

3.8 Giant covalent structures

Objectives: describe the giant covalent structures of diamond, graphite, and silicon(IV) oxide; explain how their structures dictate their properties and uses

Not all covalent solids are molecular
All these substances have covalent bonds. Compare their melting points:

Substance	Melting point / °C
ice	0
phosphorus	44
sulfur	115
silicon(IV) oxide (silicon dioxide or silica)	1710
carbon (as diamond)	3550

The first three substances are molecular solids. Their molecules are held in a lattice by weak forces – so the solids melt easily, at low temperatures.

But diamond and silicon(IV) oxide are different. Their melting points show that *they* are not molecular solids with weak lattices. In fact they exist as **giant covalent structures.**

▲ Diamond: so hard that it is used to edge wheels for cutting stone.

Diamond – a giant covalent structure
Diamond is made of carbon atoms, held in a strong lattice:

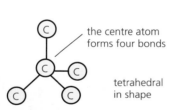

A carbon atom forms covalent bonds to *four* others, as shown above. Each outer atom then bonds to three more, and so on.

Eventually many trillions of carbon atoms are bonded together, in a giant covalent structure. This shows part of it.

The result is a single crystal of diamond. This one has been cut, shaped, and polished, to make it sparkle.

Diamond has these properties:

1 It is very hard, because each atom is held in place by four strong covalent bonds. In fact it is the hardest known natural substance.
2 For the same reason it has a very high melting point, 3550°C.
3 It does not conduct electricity because there are no ions or free electrons to carry the charge.

Silicon(IV) oxide: similar to diamond
Silicon(IV) oxide, SiO_2, occurs naturally as the mineral **quartz**, in **sand**. Like diamond, it forms a giant covalent structure, as shown on the right.

Each silicon atom bonds to four oxygen atoms. Each oxygen atom bonds to two silicon atoms. The result is a solid with properties similar to diamond: hard, a high melting point, and does not conduct electricity.

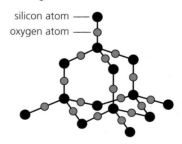

▲ Silicon (IV) oxide is made up of oxygen atoms ● and silicon atoms ○. Trillions of them bond together like this, to give a giant structure. It is the main compound in sand.

ATOMS COMBINING

Graphite – a very different giant structure

Like diamond, graphite is made only of carbon atoms. So diamond and graphite are **allotropes** of carbon – two forms of the same element.

Diamond is a very hard substance. But graphite is one of the softest! This difference is a result of their very different structures:

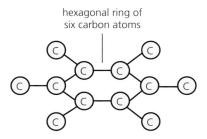

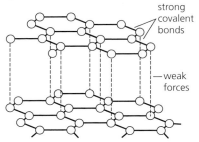

> **Harder than diamond?**
> Scientists continue to search for natural materials that are harder than diamond. We can be sure they will have strong bonds!

In graphite, each carbon atom forms covalent bonds to *three* others. This gives rings of *six* atoms.

The rings form flat sheets. Trillions of these lie on top of each other, held together by weak forces.

Under a microscope, you can see the layered structure of graphite quite clearly.

Graphite has these properties:
1 Unlike diamond, it is soft and slippery. That is because the sheets can slide over each other easily.
2 Unlike diamond, it is a good conductor of electricity. That is because each carbon atom has four outer shell electrons, but forms only three bonds. The fourth electron is free to move through the graphite, forming an electric current.

Making use of these giant structures

Different properties lead to different uses, as this table shows.

Substance	Properties	Uses
diamond	hardest known natural substance; can cut into other substances	in tools for drilling and cutting
graphite	soft and slippery	as a lubricant for engines and locks
	conducts electricity	for electrodes (which are used to carry a current into liquids, for electrolysis)
silicon(IV) oxide	hard, will scratch other substances	in sandpaper
	high melting point	in bricks for lining furnaces

▲ Pencil lead is a mixture of graphite and clay.

1 The covalent compound ethanol melts at −114 °C. Is it a molecular compound, or a giant structure? Explain.
2 Diamond and graphite are allotropes of carbon. Define *allotrope*. (Glossary?)
3 a Why is diamond called a *giant covalent structure*?
  b Give one use of diamond, that results from its strong bonds.
4 Graphite is also a giant covalent structure. Explain why it is:
  a soft   b able to conduct electricity
5 Suggest *two* reason why graphite is used in pencil lead.
6 State two other uses of graphite that are linked to the properties given in **4**.
7 Silicon(IV) oxide and diamond share these properties:
  • hard
  • high melting point
  • does not conduct electricity
  Using what you know about their structures, explain why they share each property.

3.9 The bonding in metals

Objectives: understand that metals form giant structures; describe the metallic bond; explain the properties of metals in terms of their structures

Clues from melting points

Compare these melting points:

Structure	Examples	Melting point / °C
molecular	carbon dioxide water	−56 0
giant ionic	sodium chloride magnesium oxide	801 2852
giant covalent	diamond silicon(IV) oxide	3550 1610
metal	iron copper	1535 1083

The table shows clearly that:

- **molecular substances have low melting points.** That is because the forces between the molecules in the lattice are weak.
- **giant ionic and covalent structures have much higher melting points.** That is because the bonds within them are very strong.

Now look at the metals. They too have high melting points. This suggests that they too are giant structures. And so they are, as you'll see below.

The structure of metals

In metals, the atoms are packed tightly together in a regular lattice. The tight packing allows outer electrons to separate from their atoms. The result is a lattice of positive ions in a 'sea' of delocalised electrons that move freely. Look at copper, for example:

▲ Most of the elements are metals. They have many different uses. For example, copper is used for water pipes and electrical wiring in homes.

> **Delocalised electrons** !
> The electrons that move freely in the metal lattice are not tied to any one ion. That is why they are called **delocalised**.

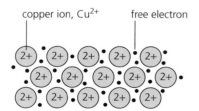
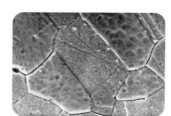
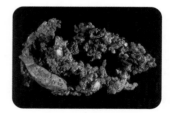

The copper ions are held together by their electrostatic attraction to the free electrons. These strong forces of attraction are called **metallic bonds**.

The regular arrangement of ions results in **crystals** of copper. This shows the crystals in a piece of copper, magnified many times. (They are all at different angles.)

The copper crystals are called **grains**. A lump of copper like this one consists of trillions of grains joined together. You would need a microscope to see them.

Metals form a giant lattice structure, with strong metallic bonds. Metallic bonds are the electrostatic attraction between the positive metal ions and a 'sea' of delocalised electrons.

All metals share the same kind of structure and bonding as copper.

Some key properties of metals

These are key properties of metals, explained in terms of their structure.

1. **Metals usually have high melting points.**
   That is because it takes a lot of heat energy to overcome the strong electrostatic attraction between the metal ions and the sea of delocalised electrons in the lattice. Copper melts at 1083 °C, and nickel at 1455 °C.

   (But there are exceptions. Sodium melts at only 98 °C, for example. And mercury melts at −39 °C, so is a liquid at room temperature.)

2. **Metals are malleable and ductile.**
   Malleable means they can be bent and pressed into shape. *Ductile* means they can be drawn out into wires. That is because the layers of metal ions can slide over each other. Look:

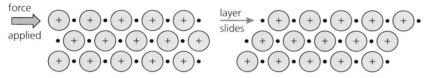

▲ Metals: malleable, ductile, and sometimes very glamorous – like this silver bracelet.

   New metallic bonds then form between metal ions and electrons in the new position. That is why metals do not break easily.

3. **Metals are good conductors of heat.** That is because the free electrons take in heat energy, which makes them move faster. They quickly transfer the heat through the metal structure:

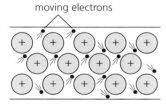

4. **Metals are good conductors of electricity.**
   That is because the free electrons move through the lattice as a current of electricity, when a voltage is applied across the metal.

   Silver is the best conductor of all the metals. Copper is next – but it is used much more than silver because it is cheaper.

▲ What uses of metals can you see in this scene?

---

1. Describe the structure of a solid metal. Include the term *delocalised* in your answer.
2. Define *metallic bond*.
3. What does *malleable* mean?
4. Explain why metals can be drawn out into wires without breaking.
5. a Explain why metals can conduct electricity.
   b Would you expect molten metals to conduct? Explain.
6. Metals are malleable, and conduct heat. So we use some metals for making saucepans. Give two other examples of the uses of metals, that depend on:
   a their malleability   b their ductility
   c their ability to conduct electricity
7. Mercury forms ions with a charge of 2+. It solidifies (freezes) at −39 °C. Draw a diagram to show the structure of solid mercury.

Checkup on Chapter 3

Revision checklist

Core syllabus content
Make sure you can …
- [ ] explain the difference between:
  - an *element* and a *compound*
  - a *compound* and a *mixture*
- [ ] state the signs of a chemical change
- [ ] explain why:
  - atoms of Group VIII elements do not form bonds
  - atoms of other elements do form bonds
- [ ] explain the difference between an *ionic bond* and a *covalent bond*
- [ ] draw a diagram to show how an ionic bond forms between atoms of sodium and chlorine
- [ ] work out the name for an ionic compound
- [ ] explain why the names of the ionic compounds of most transition elements include Roman numerals
- [ ] give the properties of ionic compounds
- [ ] state that non-metal atoms form covalent bonds with each other (except for the noble gas atoms)
- [ ] explain what a *molecule* is
- [ ] draw diagrams to show the covalent bonding in:
  hydrogen   chlorine   water
  methane   hydrogen chloride   ammonia
- [ ] give the properties of molecular compounds
- [ ] describe the giant covalent structures of graphite and diamond, and sketch them
- [ ] explain why diamond and graphite have such different properties
- [ ] give uses for diamond and graphite

Extended syllabus content
Make sure you can also …
- [ ] show how ionic bonds form between atoms of other metals and non-metals (besides sodium and chlorine)
- [ ] describe the lattice structure of ionic compounds
- [ ] work out the formulae of ionic compounds, from the charges on the ions
- [ ] draw diagrams to show the covalent bonding in:
  nitrogen   oxygen   methanol   carbon dioxide
  ethene
- [ ] describe the structure of silicon(IV) oxide
- [ ] explain why silicon(IV) oxide and diamond have similar properties
- [ ] describe metallic bonding, and draw a sketch for it
- [ ] explain how their structure and bonding enables metals to be *malleable*, *ductile*, and *good conductors of heat and electricity*

Questions

Core syllabus content

1 This question is about the ionic bond formed between the metal lithium (proton number 3) and the non-metal fluorine (proton number 9).
  a How many electrons does a lithium atom have? Draw a diagram to show its electron structure.
  b How does a metal atom obtain a stable outer shell of electrons?
  c Draw the structure of a lithium ion, and write the symbol for it, showing its charge.
  d How many electrons does a fluorine atom have? Draw a diagram to show its electron structure.
  e How does a non-metal atom become an ion?
  f Draw the structure of a fluoride ion, and write a symbol for it, showing its charge.
  g Draw a diagram to show what happens when a lithium atom reacts with a fluorine atom.
  h Write a word equation for the reaction between lithium and fluorine.

2 This diagram represents a molecule of a certain gas.
  a Name the gas, and give its formula.
  b What do the symbols • and × represent?
  c Which type of bonding holds the atoms together?
  d Name another compound with this type of bonding.

3 Hydrogen bromide is a compound of the two elements hydrogen and bromine. It melts at −87 °C and boils at −67 °C. It has the same type of bonding as hydrogen chloride.
  a Is hydrogen bromide a solid, a liquid, or a gas at room temperature (20 °C)?
  b Is hydrogen bromide molecular, or does it have a giant structure? What is your evidence?
  c i Which type of bond is formed between the hydrogen and bromine atoms, in hydrogen bromide?
    ii Draw a diagram of the bonding between the atoms, showing only the outer electrons.
  d Write a formula for hydrogen bromide.
  e i Name two other compounds with bonding similar to that in hydrogen bromide.
    ii Write formulae for these two compounds.

ATOMS COMBINING

4

Substance	Melting point /°C	Electrical conductivity of the solid	Electrical conductivity of the liquid	Solubility in water
A	−112	poor	poor	insoluble
B	680	poor	good	soluble
C	−70	poor	poor	insoluble
D	1495	good	good	insoluble
E	610	poor	good	soluble
F	1610	poor	poor	insoluble
G	660	good	good	insoluble

a Which of the seven substances are metals? Give reasons for your choice.
b i Which of the substances are ionic compounds? Give reasons for your choice.
ii What type of forces hold the ions together in an ionic compound?
c Two of the substances have very low melting points, compared with the rest. Explain why these could *not* be ionic compounds.
d Two of the substances are molecular. Which two are they?
e Which substance is a giant covalent structure?
f Name the type of bonding found in:
i B ii C iii E iv F

Extended syllabus content

5 Aluminium and nitrogen react to form an ionic compound called aluminium nitride. These show the electron arrangement for the two elements:

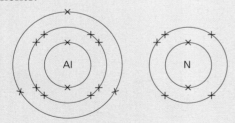

a Answer these questions for an aluminium atom.
i Does it gain or lose electrons, to form an ion?
ii How many electrons are transferred?
iii Is the ion formed positive, or negative?
iv What charge does the ion have?
b Now repeat **a**, but for a nitrogen atom.
c i Give the electronic structure for the ions formed by the two atoms. (2,…)
ii What do you notice about these electronic structures? Explain it.
d Name another non-metal that will form an ionic compound with aluminium, in the same way that nitrogen does.

6 Silicon lies directly below carbon in Group IV of the Periodic Table. Here is some data for silicon, carbon (in the form of diamond), and their oxides.

Substance	Symbol or formula	Melting point/°C	Boiling point/°C
carbon	C	3730	4530
silicon	Si	1410	2400
carbon dioxide	CO_2	(sublimes at −78°C)	
silicon(IV) oxide	SiO_2	1610	2230

a In which state are the two *elements* at room temperature (20°C)?
b Which type of structure does carbon (diamond) have: giant covalent, or molecular?
c Which type of structure would you expect to find in silicon? Give reasons.
d Carbon dioxide has a molecular structure. Are the *intermolecular forces* in it strong, or weak?
e Does silicon(IV) oxide have the same structure as carbon dioxide? What is your evidence?

7 The compound zinc sulfide has a structure like this:

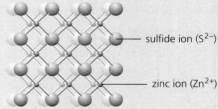

a Which does the diagram represent: a giant structure, or a molecular structure?
b Which type of bonding does zinc sulfide have?
c i In the diagram above, each zinc ion is joined to four sulfide ions. How many zinc ions is each sulfide ion joined to? (Look in the centre.)
ii Deduce the ratio of zinc ions to sulfide ions.
d i From **c**, deduce the formula of zinc sulfide.
ii Is this formula consistent with the charges on the two ions? Explain your answer.
e Name another metal and non-metal that will form a compound with a similar formula.

8 The properties of metals can be explained by the structure and bonding within the metal lattice.
a Describe the bonding in metals.
b Explain why metals:
i are good conductors of electricity
ii are ductile.

4.1 The names and formulae of compounds

Objectives: recall the rules for naming compounds; deduce a formula from a structure; know how to work out formulae by balancing valencies, and charges

So far ...

You have already met the names and formulae for many compounds. In Unit 3.4 we looked at ionic compounds in particular. Now we review all the ideas you met earlier, and add new ones.

The names of compounds

Many compounds contain just two elements. If you know which elements they are, you can usually name the compound. Just follow these rules:

1 **When the compound contains a metal and a non-metal** (or in other words, when it is an ionic compound):
   - give the name of the metal first
   - and then the name of the non-metal, but ending with *-ide*.
   Examples: sodium chloride, magnesium oxide, iron sulfide.

2 **When the compound is made of two non-metals**
   - if one is hydrogen, give its name first
   - otherwise name the one with the lower group number first
   - then give the name of the second non-metal, but ending with *-ide*.
   Examples: hydrogen chloride, carbon dioxide (carbon is in Group IV and oxygen in Group VI).

▲ Rain: that very common compound, water. Which elements does it contain? What is its correct chemical name?

> **Some common names**
> - Some compounds have everyday names that give no clue about the elements in them.
> - Examples are water, ammonia, and methane.
> - You just have to remember their formulae!

The formulae of compounds

The formula of a compound contains the symbols for the elements in it. There are three ways to work out a formula: from the structure of the compound, using valency, and by balancing charges.

1 Finding the formula from the structure of a compound

A molecular compound

The molecules on the right are drawn in different styles.

A molecule of water has one oxygen atom and two hydrogen atoms. So the formula for water is **H_2O**.

A molecule of ammonia has one nitrogen atom and three hydrogen atoms. So the formula for ammonia is **NH_3**.

The formula for a molecular compound tells you the type and *number* of each atom in one molecule.

water molecules

ammonia molecules

An ionic compound

The structure of the ionic compound potassium iodide is shown on the right, in two different styles.

Each lattice has one K^+ ion for every I^- ion. So the formula for potassium iodide is **KI**.

The formula for an ionic compound tells you the type and *ratio* of the ions in the compound.

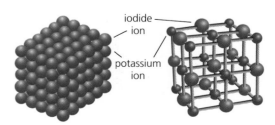

the structure of potassium iodide

REACTING MASSES, AND CHEMICAL EQUATIONS

2 Finding a formula using valency

The valency of an element is the number of electrons its atoms lose, gain or share, in forming compounds.

It is linked to the group numbers in the Periodic Table (page 15). Look:

Elements	In forming a compound, the atoms ...	So the valency of the element is ...	Examples of compounds formed (those in blue are covalent, with shared electrons)
Group I	lose 1 electron	1	sodium chloride, NaCl
Group II	lose 2 electrons	2	magnesium chloride, $MgCl_2$
Group III	lose 3 electrons	3	aluminium chloride, $AlCl_3$
Group IV	share 4 electrons	4	methane, CH_4
Group V	gain or share 3 electrons	3	ammonia, NH_3
Group VI	gain or share 2 electrons	2	magnesium oxide, MgO; water, H_2O
Group VII	gain or share 1 electron	1	sodium chloride, NaCl; hydrogen chloride, HCl
Group VIII	(do not form compounds)	–	none
hydrogen	lose or share 1 electron	1	hydrogen bromide, HBr

How to write a formula using valency
1 Find the valency of each element in the compound, from the table.
2 Write their symbols in the same order as the elements in the name.
3 If the elements have different valencies, write numbers *after* the symbols to balance the valencies.

Example 1 What is the formula for hydrogen sulfide?
1 Valencies: hydrogen, 1; sulfur 2 (Group VI); they are different
2 HS
3 The formula is **H_2S** (the valencies are balanced: 2 × 1 and 2)

Example 2 What is the formula for aluminium oxide?
1 Valencies: aluminium 3 (Group III); oxygen 2 (Group VI); different
2 AlO
3 The formula is **Al_2O_3** (the valencies are balanced: 2 × 3 and 3 × 2)

3 Finding a formula by balancing charges

In an ionic compound, the total charge is zero. So you can also work out the formula of an ionic compound by balancing the charges on its ions.

For example, the magnesium ion is Mg^{2+} and the chloride ion is Cl^-. So a magnesium ion needs two chloride ions to balance its charge. So the formula for magnesium chloride is **$MgCl_2$**. The charges add up to zero.

For more detail on balancing charges, please turn to Unit 3.4.

▲ Hydrogen sulfide is a very poisonous colourless gas. It smells of rotten eggs.

Q

The Periodic Table on page 15 will help you with these.
1 Name the compounds containing only these elements:
  a sodium and fluorine **b** fluorine and hydrogen
  c sulfur and hydrogen **d** bromine and beryllium
2 Deduce the formula: **a**   **b** (F)(F)(C)(F)(F)
3 Is the formula correct? If not, write it correctly.
  a HBr_2 **b** NaF_2 **c** MgO_2 **d** $CaCl_2$ **e** Be_2O
4 Give the formula for: **a** barium iodide **b** hydrogen oxide
5 Suggest a name and formula for a compound that forms when phosphorus reacts with chlorine.
6 The formula for silicon(IV) oxide is SiO_2. Why? (Page 40.)

4.2 Equations for chemical reactions

Objectives: understand that reactions can be described by both word and symbol equations; recall the state symbols; know how to write balanced symbol equations

Equations for two sample reactions

1. **The reaction between carbon and oxygen** Carbon and oxygen react to form carbon dioxide. Carbon and oxygen are the **reactants**. Carbon dioxide is the **product**. You could show the reaction like this:

 + ⟶

1 atom of carbon   1 molecule of oxygen   1 molecule of carbon dioxide

or by a **word equation**, like this:

*carbon + oxygen ⟶ carbon dioxide

or by a **symbol equation**, which gives symbols and formulae:

$C + O_2 \longrightarrow CO_2$

2. **The reaction between hydrogen and oxygen** Hydrogen reacts with oxygen to give water. Compare these ways to show that:

  + ⟶

2 molecules of hydrogen   1 molecule of oxygen   2 molecules of water

hydrogen + oxygen ⟶ water

$2H_2 + O_2 \longrightarrow 2H_2O$

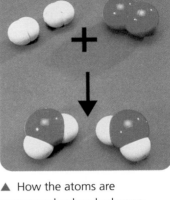

▲ How the atoms are rearranged, when hydrogen burns in oxygen. The *number* of atoms does not change. Which colour is oxygen?

Symbol equations must be balanced

Now look at the atoms on each side of this equation:

$2H_2 + O_2 \longrightarrow 2H_2O$

On the left:		On the right:
4 hydrogen atoms	⟶	4 hydrogen atoms
2 oxygen atoms		2 oxygen atoms

The number of each type of atoms is the same on both sides of the arrow. This is because atoms do not *disappear* during a reaction – they are just *rearranged*, as shown in the photo above.

When the number of each type of atom is the same on both sides, the symbol equation is **balanced**. If it is not balanced, it is not correct.

Adding state symbols

Reactants and products may be solids, liquids, gases, or in solution. You can show their states by adding **state symbols** to the equations:

(*s*) for solid   (*l*) for liquid
(*g*) for gas   (*aq*) for aqueous solution (a solution in water)

For the two reactions above, the equations with state symbols are:

$C\ (s) + O_2\ (g) \longrightarrow CO_2\ (g)$
$2H_2\ (g) + O_2\ (g) \longrightarrow 2H_2O\ (l)$

▲ The reaction between hydrogen and oxygen gives out so much energy that it is used to power rockets. The reactants are carried as liquids in the fuel tanks.

REACTING MASSES, AND CHEMICAL EQUATIONS

How to write the equation for a reaction
These are the steps to follow, when writing an equation:
1. Write the equation in words.
2. Now write it using symbols. Make sure all the formulae are correct.
3. Check that the equation is balanced, for each type of atom in turn. *Make sure you do not change any formulae.*
4. Add the state symbols.

Example 1 Calcium, a metal, burns in chlorine gas to form calcium chloride, a solid. Write an equation for the reaction, using the steps above.
1. calcium + chlorine $\longrightarrow$ calcium chloride
2. $Ca + Cl_2 \longrightarrow CaCl_2$
3. Ca: 1 atom on the left and 1 atom on the right.
   Cl: 2 atoms on the left and 2 atoms on the right.
   The equation is balanced.
4. $Ca\,(s) + Cl_2\,(g) \longrightarrow CaCl_2\,(s)$

▲ Calcium chloride absorbs water, so is used to dry gases. It is packed into this glass cylinder. Gas is pumped through.

Example 2 Hydrogen chloride is formed by burning hydrogen in chlorine. Write an equation for the reaction.
1. hydrogen + chlorine $\longrightarrow$ hydrogen chloride
2. $H_2 + Cl_2 \longrightarrow HCl$
3. H: 2 atoms on the left and 1 atom on the right.
   Cl: 2 atoms on the left and 1 atom on the right.
   The equation is *not* balanced. It needs another molecule of hydrogen chloride on the right. So a 2 is put *in front of* the HCl.
   $H_2 + Cl_2 \longrightarrow 2HCl$
   The equation is now balanced. Do you agree?
4. $H_2\,(g) + Cl_2\,(g) \longrightarrow 2HCl\,(g)$

▲ Hydrogen chloride (a gas) dissolves in water to give hydrochloric acid.

Example 3 Magnesium burns in oxygen to form magnesium oxide, a white solid. Write an equation for the reaction.
1. magnesium + oxygen $\longrightarrow$ magnesium oxide
2. $Mg + O_2 \longrightarrow MgO$
3. Mg: 1 atom on the left and 1 atom on the right.
   O: 2 atoms on the left and 1 atom on the right.
   The equation is *not* balanced. Try this:
   $Mg + O_2 \longrightarrow 2MgO$ (The 2 goes *in front of* the MgO.)
   Another magnesium atom is now needed on the left:
   $2Mg + O_2 \longrightarrow 2MgO$
   The equation is balanced.
4. $2Mg\,(s) + O_2\,(g) \longrightarrow 2MgO\,(s)$

▲ The reaction of grey magnesium metal with oxygen gives this!

1. What do + and $\longrightarrow$ mean, in an equation?
2. Balance these equations:
   a $Na\,(s) + Cl_2\,(g) \longrightarrow NaCl\,(s)$
   b $H_2\,(g) + I_2\,(g) \longrightarrow HI\,(g)$
   c $Na\,(s) + H_2O\,(l) \longrightarrow NaOH\,(aq) + H_2\,(g)$
   d $NH_3\,(g) \longrightarrow N_2\,(g) + H_2\,(g)$
   e $Al\,(s) + O_2\,(g) \longrightarrow Al_2O_3\,(s)$
3. Now write a word equation for each of the symbol equations you balanced in **2**.
4. Which reaction or reactions described in **2** involved:
   a only gases?
   b a solid, a liquid, a gas, and a solution?
5. Aluminium burns in chlorine to form aluminium chloride, $AlCl_3$, a solid. Write a balanced equation for the reaction.

49

4.3 The masses of atoms, molecules, and ions

Objectives: understand that an atom's mass is set by comparing the atom with the ^{12}C atom; explain A_r and M_r; calculate M_r for molecular and ionic substances

So far ...
We looked at the masses of atoms in Unit 2.4. Let's summarise the ideas you met so far. Make sure you understand them!

^{12}C – the standard atom
- An atom is so tiny that you cannot see it, or weigh it in grams. So its mass is given in special units: **atomic mass units**.
- The carbon-12 atom, or ^{12}C, is taken as the standard, and is assigned an atomic mass of 12 units. It is shown on the right.
- The ^{12}C atom has 12 nucleons (6 protons + 6 neutrons), so the mass of a nucleon is taken as 1. That is, 1/12 of the mass of the atom.
- The masses of all other atoms are found by comparing them to 1/12 the mass of a ^{12}C atom. So they are *relative* masses. (*Relative to* means *compared to*.)

6 protons
6 electrons
6 neutrons

▲ An atom of ^{12}C. It is the main isotope of carbon. (See page 20.)

Isotopes
- Most elements have isotopes that occur naturally.
- **Isotopes are atoms of the same element, with the same number of protons, but different numbers of neutrons.**
- So each isotope has a different mass. The two isotopes of chlorine are shown on the right. You can work out their relative atomic masses.

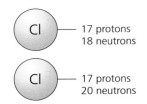

Cl — 17 protons, 18 neutrons

Cl — 17 protons, 20 neutrons

▲ The two natural isotopes of chlorine. They are ^{35}Cl and ^{37}Cl. Which is which?

Relative atomic mass
- To find the average mass of the atoms in an element, you must take its isotopes into account. This gives the relative atomic mass, A_r.
- **The relative atomic mass of an element, A_r, is the average mass of its isotopes, compared to 1/12 the mass of the ^{12}C atom.**

A_r values for some common elements
Hydrogen has three natural isotopes, and its A_r is 1.008. Magnesium also has three, and its A_r is 24.305. But A_r values are usually rounded off to the nearest whole number – except for chlorine, which has a value halfway between two whole numbers.

Compare these A_r values:

Element	Symbol	A_r	Element	Symbol	A_r
hydrogen	H	1	magnesium	Mg	24
carbon	C	12	sulfur	S	32
nitrogen	N	14	chlorine	Cl	35.5
oxygen	O	16	zinc	Zn	65
sodium	Na	23	iodine	I	127

Now turn to the Periodic Table on page 15. The number below each symbol is the A_r for that element. A_r values are also given on page 313.

▲ A mass spectrometer. Atoms are turned into ions and shot through a magnetic field. Their path depends on their mass, with ^{12}C as standard.

REACTING MASSES, AND CHEMICAL EQUATIONS

The masses of molecular and ionic substances

So now you know how the average relative mass of the atoms in an element is found.

But most substances do not exist as single atoms. They exist as molecules, or as ionic compounds.

Using A_r values, you can easily work out relative masses for these too. Here are the rules:

- For a molecule, add up the A_r values of the atoms in the molecule. This gives the **relative molecular mass**, or M_r for short.
- An ion has the same mass as its atom. So for an ionic compound, add up the A_r values of the atoms in its formula. This gives the **relative formula mass**, which is also shortened to M_r.

> **The mass of an ion**
> - To form an ion, an atom gains or loses electrons.
> - But an electron has no mass.
> - So the mass of an ion is the same as the mass of its atom.

Look at these examples:

A molecule of hydrogen contains two hydrogen atoms. So the relative molecular mass of hydrogen is **2**.
($2 \times 1 = 2$)

A molecule of water contains 2 hydrogen atoms and 1 oxygen atom. So the relative molecular mass of water is **18**.
($2 \times 1 + 16 = 18$)

mass of ion = mass of its atom

Sodium chloride has 1 sodium ion for every chloride ion. Its formula is NaCl. So its relative formula mass is **58.5**.
($23 + 35.5 = 58.5$)

More examples
Look at these further examples. Check that you understand the working.

Substance	Formula	Atoms in formula	A_r of atoms	M_r
oxygen (molecular element)	O_2	2O	O = 16	$2 \times 16 = 32$
ammonia (molecular compound)	NH_3	1N 3H	N = 14 H = 1	$1 \times 14 = 14$ $3 \times 1 = \underline{3}$ Total = $\underline{17}$
magnesium nitrate (ionic compound)	$Mg(NO_3)_2$	1Mg 2N 6O	Mg = 24 N = 14 O = 16	$1 \times 24 = 24$ $2 \times 14 = 28$ $6 \times 16 = \underline{96}$ Total = $\underline{148}$

▲ Hydrogen: the element with the lowest mass of all. Here it is being made by reacting magnesium with dilute hydrochloric acid. It is scarce on Earth, but is the most abundant gas in the Universe. The gas planet Jupiter is mainly hydrogen.

1 a What does the term *relative atomic mass* mean?
   b Why does it contain the word *relative*?
   c Write down the abbreviation used for it.
2 The carbon-12 atom has a special role in chemistry. Explain.
3 Where in the Periodic Table are A_r values shown?
4 Explain why an ion has the same mass as its atom.
5 Calculate the relative molecular mass (M_r) for:
   a chlorine, Cl_2 b sulfur dioxide, SO_2 c butane, C_4H_{10}
6 Calculate the relative formula mass (M_r) for:
   a aluminium oxide, Al_2O_3
   b ammonium sulfate, $(NH_4)_2SO_4$

4.4 Calculations about mass and %

Objectives: state the two laws of chemistry that underlie calculations; be able to calculate quantities that react, % composition, and % purity

Two laws of chemistry

If you know the *actual* amounts of two substances that react, you can:
- predict other amounts that will react
- say how much product will form.

You just need to remember these two laws of chemistry:

1 **Elements always react in the same ratio, to form a given compound.**
   For example, when carbon burns in oxygen to form carbon dioxide:
   6 g of carbon combines with **16 g** of oxygen, so
   12 g of carbon will combine with **32 g** of oxygen, and so on.
2 **The total mass does not change, during a chemical reaction.**
   So **total mass of reactants = total mass of products**.
   So **6 g** of carbon and **16 g** of oxygen give **22 g** of carbon dioxide.
   12 g of carbon and **32 g** of oxygen give **44 g** of carbon dioxide.

▲ A model of the carbon dioxide molecule. The amounts of carbon and oxygen that react to give this compound are always in the same ratio.

Calculating quantities

Calculating quantities is quite easy, using the laws above.

Example 64 g of copper reacts with 16 g of oxygen to give the black compound copper(II) oxide.

a What mass of copper will react with 32 g of oxygen?
   64 g of copper reacts with 16 g of oxygen, so
   2 × 64 g or **128 g** of copper will react with 32 g of oxygen.
b What mass of oxygen will react with 32 g of copper?
   16 g of oxygen reacts with 64 g of copper, so
   $\frac{16}{2}$ or **8 g** of oxygen will react with 32 g of copper.
c What mass of copper(II) oxide will be formed, in **b**?
   40 g of copper(II) oxide will be formed. (32 + 8 = 40)
d How much copper and oxygen will give 8 g of copper(II) oxide?
   64 g of copper and 16 g of oxygen give 80 g of copper(II) oxide,
   so $\frac{64}{10}$ of copper and $\frac{16}{10}$ g of oxygen will give 8 g of copper(II) oxide,
   so **6.4 g of copper** and **1.6 g of oxygen** are needed.

Percentages: a reminder

Calculations in chemistry often involve percentages. Remember:
- **The full amount of anything is 100%.**
- **To change a fraction to a %, just multiply it by 100.**

Example 1 Change the fractions $\frac{1}{2}$ and $\frac{18}{25}$ to percentages.
$\frac{1}{2} \times 100 = \mathbf{50\%}$ $\frac{18}{25} \times 100 = \mathbf{72\%}$

Example 2 Give 19 % as a fraction. $19\% = \frac{19}{100}$

▲ The French scientist Antoine Lavoisier (1743–1794) was the first to state that the total mass does not change, during a reaction. He was executed during the French Revolution, when he was 51.

52

REACTING MASSES, AND CHEMICAL EQUATIONS

Calculating % composition of a compound

The **percentage composition** of a compound tells you how much of each element it contains, *as a percentage of the total mass*. This is how to work it out:

1. Write down the formula of the compound.
2. Using A_r values, work out its molecular or formula mass (M_r).
3. Write the mass of the element as a fraction of the M_r.
4. Multiply the fraction by 100, to give a percentage.

Example Calculate the percentage of oxygen in sulfur dioxide.
1. The formula of sulfur dioxide is SO_2.
2. A_r values: S = 32, O = 16. So its M_r is 64. (32 + 2 × 16)
3. Mass of oxygen as a fraction of the total = $\frac{32}{64}$
4. Mass of oxygen as a percentage of the total = $\frac{32}{64} \times 100 = 50\%$
   So the compound is **50% oxygen**.
   This means it is also 50% sulfur (100% − 50% = 50%).

▲ The idea of % composition is used for mixtures too. This elephant is made of **brass**, a mixture of about 70% copper and 30% zinc. But the % composition of brass can be varied. The % composition of a compound *never* varies.

Calculating % purity

A **pure** substance has nothing else mixed with it.
But substances often contain unwanted substances, or **impurities**.
Purity is usually given as a percentage. This is how to work it out:

% purity of a substance = $\dfrac{\text{mass of pure substance in it}}{\text{total mass}} \times 100\%$

Example Impure copper is refined (purified), to obtain pure copper for use in computers. 20 tonnes of copper gave 18 tonnes of pure copper, on refining.

a What was the % purity of the copper before refining?

% purity of the copper = $\dfrac{18 \text{ tonnes}}{20 \text{ tonnes}} \times 100\% = \mathbf{90\%}$

So the copper was **90% pure**.

b How much pure copper will 50 tonnes of the impure copper give?
The impure copper is 90% pure.
90% is $\dfrac{90}{100}$.
So 50 tonnes of it will give $\dfrac{90}{100} \times 50$ tonnes or **45 tonnes** of pure copper.

▲ For some purposes, a high % purity is essential. The silicon wafers used to make chips for computers and mobiles must be 99.9999% pure. Silicon is obtained from the silicon(IV) oxide in sand.

1. In an experiment, 24 g of magnesium reacted with 71 g of chlorine, to give magnesium chloride, $MgCl_2$.
   a How much magnesium chloride was obtained?
   b How much chlorine will react with 12 g of magnesium?
   c How much magnesium chloride will form, in **b**?
2. 8 g of hydrogen burned in oxygen, giving 72 g of water. How many grams of oxygen:
   a combined with the hydrogen?
   b would be needed to make 36 g of water?
3. Find the % of: a carbon b hydrogen in methane, CH_4.
4. Calculate the % of each element present in the compound calcium carbonate, $CaCO_3$.
5. A sample of lead(II) bromide weighed 15 grams. But it was impure. It contained only 13.5 g of lead(II) bromide.
   a Calculate the % purity of the sample.
   b What mass of impurity did the sample contain?
6. A sample of compound X was prepared. Its mass was 30 g. But it was found to be only 80% pure.
   a How many grams of X were present in the sample?
   b What mass of the sample would contain 12 g of X?

Checkup on Chapter 4

Revision checklist

Core syllabus content
Make sure you can …
- [ ] name a simple compound, when you are given the names of the two elements that form it
- [ ] work out the formulae of molecular substances from drawings of their molecules
- [ ] write the equation for a reaction:
  - as a word equation
  - as a symbol equation
- [ ] balance a symbol equation
- [ ] say what the state symbols mean: (s), (l), (g), (aq)
- [ ] explain that the ^{12}C atom is taken as the standard, for working out masses of atoms
- [ ] state that ^{12}C can also be written as carbon-12
- [ ] define *isotope* (first met in Unit 2.4)
- [ ] define *relative atomic mass* (first met in Unit 2.4)
- [ ] state what these two symbols represent: A_r M_r
- [ ] explain the difference between *relative formula mass* and *relative molecular mass* (both known as M_r for short)
- [ ] work out the M_r for a molecular element, and for any compound, given the formula and A_r values
- [ ] recall that elements always react in the same ratio, to form a given compound
- [ ] recall that the total mass does not change, during a chemical reaction
- [ ] calculate quantities of reactants and products in a reaction, from information you are given

Extended syllabus content
Make sure you can also …
- [ ] work out the formula for an ionic compound from a drawing of its lattice structure
- [ ] work out the formula of a simple compound by balancing the valencies of the elements in it
- [ ] work out the formula of an ionic compound by balancing the charges of the ions
- [ ] calculate the % composition of a compound, using the formula and A_r values
- [ ] calculate the % purity of a substance from the information you are given

Questions

Core syllabus content
If you are not sure about symbols for the elements, you can check the Periodic Table on page 312.

1. Write the formulae for these compounds:
   a. water
   b. carbon monoxide
   c. carbon dioxide
   d. sulfur dioxide
   e. sulfur trioxide
   f. sodium chloride
   g. magnesium chloride
   h. hydrogen chloride
   i. methane
   j. ammonia

2. You can work out the formula of a compound from the ratio of the different atoms in it. Sodium carbonate has the formula Na_2CO_3 because it contains 2 atoms of sodium for every 1 atom of carbon and 3 atoms of oxygen. Deduce the formula for each compound **a** to **h**:

	Compound	Ratio in which the atoms are combined in it
a	lead oxide	1 of lead, 2 of oxygen
b	lead oxide	3 of lead, 4 of oxygen
c	potassium nitrate	1 of potassium, 1 of nitrogen, 3 of oxygen
d	nitrogen oxide	2 of nitrogen, 1 of oxygen
e	nitrogen oxide	2 of nitrogen, 4 of oxygen
f	sodium hydrogen carbonate	1 of sodium, 1 of hydrogen, 1 of carbon, 3 of oxygen
g	sodium sulfate	2 of sodium, 1 of sulfur, 4 of oxygen
h	sodium thiosulfate	2 of sodium, 2 of sulfur, 3 of oxygen

3. For each compound, write down the ratio of atoms present:
   a. copper(II) oxide, CuO
   b. copper(I) oxide, Cu_2O
   c. aluminium chloride, $AlCl_3$
   d. nitric acid, HNO_3
   e. calcium hydroxide, $Ca(OH)_2$
   f. ethanoic acid, CH_3COOH
   g. ammonium nitrate, NH_4NO_3
   h. ammonium sulfate, $(NH_4)_2SO_4$
   i. sodium phosphate, $Na_3(PO_4)_2$
   j. hydrated iron(II) sulfate, $FeSO_4.7H_2O$
   k. hydrated cobalt(II) chloride, $CoCl_2.6H_2O$

4 Write the chemical formulae for the compounds with the structures shown below:
a H—Br
b Cl—P—Cl
     |
    Cl
c H—O—O—H
d H—C≡C—H
e H₂N=NH₂ (H-N=N-H with H's)
f XeF₆ structure
g H—O—S(=O)(=O)—O—H
h H—O—S(=O)—O—H

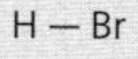

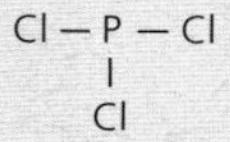

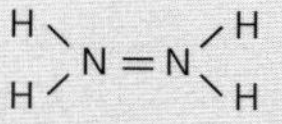

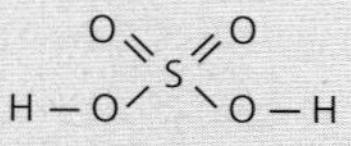

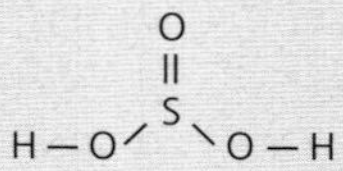

5 This shows the structure of a molecular compound.
a Name the compound. 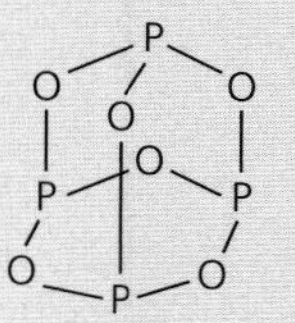
b Write its molecular formula.

6 Write these as word equations:
a $Zn + 2HCl \longrightarrow ZnCl_2 + H_2$
b $Na_2CO_3 + H_2SO_4 \longrightarrow Na_2SO_4 + CO_2 + H_2O$
c $2Mg + CO_2 \longrightarrow 2MgO + C$
d $ZnO + C \longrightarrow Zn + CO$
e $Cl_2 + 2NaBr \longrightarrow 2NaCl + Br_2$
f $CaO + 2HNO_3 \longrightarrow Ca(NO_3)_2 + H_2O$

7 Balance these equations:
a $N_2 + \ldots O_2 \longrightarrow \ldots NO_2$
b $K_2CO_3 + \ldots HCl \longrightarrow \ldots KCl + CO_2 + H_2O$
c $C_3H_8 + \ldots O_2 \longrightarrow \ldots CO_2 + 4H_2O$
d $Fe_2O_3 + \ldots CO \longrightarrow \ldots Fe + \ldots CO_2$
e $Ca(OH)_2 + \ldots HCl \longrightarrow CaCl_2 + \ldots H_2O$
f $2Al + \ldots HCl \longrightarrow 2AlCl_3 + \ldots H_2$

8 Zinc oxide reacts with carbon as follows:
$ZnO + C \longrightarrow Zn + CO_2$
a For these reactants and products, give:
  i the A_r values for the two elements
  ii the M_r values for the two compounds
  (A_r values are given on page 313.)
b Starting with 6 g of carbon, calculate:
  i the mass of zinc that could be obtained
  ii the mass of zinc oxide that would be used up

9 Calculate M_r for these molecular compounds. (A_r values are given on page 313.)
a ethanol, CH_3CH_2OH
b sulfuric acid, H_2SO_4
c hydrogen chloride, HCl
d phosphorus(V) oxide, P_2O_5

10 Calculate the relative formula mass for these ionic compounds. (A_r values are given on page 313.)
a magnesium oxide, MgO
b calcium fluoride, CaF_2
c ammonium sulfate, $(NH_4)_2SO_4$
d potassium carbonate, K_2CO_3
e hydrated iron(II) sulfate, $FeSO_4.7H_2O$

Extended syllabus content

11 This shows the structure of an ionic compound. 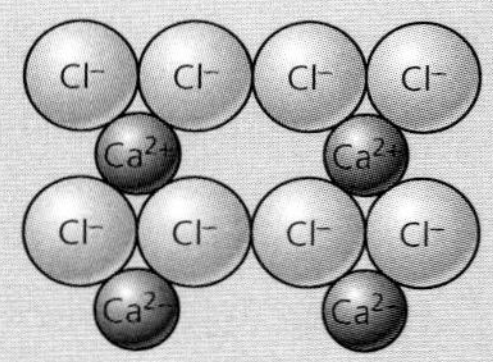
a Name the compound.
b What is its formula?

12 Iron reacts with excess sulfuric acid to give iron(II) sulfate. The equation for the reaction is:
$Fe + H_2SO_4 \longrightarrow FeSO_4 + H_2$
5 g of iron gives 13.6 g of iron(II) sulfate.
a Using excess acid, how much iron(II) sulfate can be obtained from:
  i 10 g of iron?   ii 1 g of iron?
b How much iron will be needed to make 136 g of iron(II) sulfate?
c A 10 g sample of impure iron(II) sulfate contains 8 g of iron(II) sulfate. Calculate the percentage purity of the iron(II) sulfate.

13 Aluminium is extracted from bauxite, which is impure aluminium oxide. 1 tonne (1000 kg) of bauxite was found to have this composition:
aluminium oxide 825 kg
iron(III) oxide 100 kg
sand 75 kg
a What percentage of this ore is impurities?
b 1 tonne of the ore gives 437 kg of aluminium.
  i How much aluminium will be obtained from 5 tonnes of the ore?
  ii What mass of sand is in this 5 tonnes?
c What will the percentage of aluminium oxide in the ore be, if all the iron(III) oxide is removed, leaving only the aluminium oxide and sand?

14 a Using the values obtained in question 10, calculate the percentage composition by mass:
  i of magnesium in magnesium oxide
  ii of nitrogen in ammonium sulfate
  iii of iron in hydrated iron (II) sulfate
b If the water is removed from hydrated iron(II) sulfate, how does the % of iron change?
  i increases
  ii decreases
  iii stays the same

5.1 The mole

Objectives: explain what *a mole* of an element or compound is; state *the Avogadro constant*; be able to convert moles to grams, and grams to moles

Meet the mole!
Look at these examples:

Carbon is made of carbon **atoms**. Its symbol is C. Its A_r is **12**.	Water is made of water **molecules**. Its formula is H_2O. Its M_r is **18**.	Sodium chloride is made of **ions** with 1 chloride ion for each sodium ion. Its formula is NaCl. Its M_r is **58.5**.
 The photo above shows **12 grams** of carbon. This amount contains **1 mole** of carbon atoms.	 The beaker contains **18 grams** of water, or **1 mole** of water molecules.	 This is **58.5 grams** of sodium chloride, or **1 mole** of sodium chloride 'units' ($Na^+ + Cl^-$).

So you can see that:
One mole of particles is the number of particles obtained by weighing out the A_r or M_r of the substance, in grams.

How big is a mole?
A mole is an enormous number: 602 000 000 000 000 000 000 000. This number is called **the Avogadro constant**.
It is written in a short way as 6.02×10^{23}. (The 10^{23} tells you to move the decimal point 23 places to the right, to get the full number.)
One mole of a substance is 6.02×10^{23} particles of that substance. The symbol for the mole is *mol*.

> **Remember, it's a number** !
> - A mole is just a number – but a very big one: 6.02×10^{23}.
> - Its symbol is *mol* – just like the symbol for gram is *g*.

What particles?
The particles can be atoms, or molecules, or ions. Look at this table:

Substance	It is …	It is made of these particles …	1 mol of it is …
helium, He	an element	helium atoms	6.02×10^{23} helium atoms
chlorine, Cl_2	an element	chlorine molecules	6.02×10^{23} chlorine molecules
water, H_2O	a compound	water molecules	6.02×10^{23} water molecules
sodium chloride, NaCl	a compound	sodium and chloride ions	6.02×10^{23} 'units' of sodium chloride ($Na^+ + Cl^-$)
magnesium choride, $MgCl_2$	a compound	magnesium and chloride ions	6.02×10^{23} 'units' of magnesium chloride ($Mg^{2+} + 2\,Cl^-$)

Now look at the last two rows. Here a particle is the 'unit' of ions that represents their ratio in the compound – just as the formula does.

USING MOLES

Finding the molar mass

The molar mass of a substance is the mass of 1 mole of its particles.

You can find the molar mass of any substance by following these steps:
1 Write down the symbol or formula of the substance.
2 Find its A_r or M_r.
3 Express that mass in grams (g).

Look at these examples:

Substance	Symbol or formula	A_r	M_r	Molar mass
helium	He	He = 4	exists as single atoms	4 grams
oxygen	O_2	O = 16	$2 \times 16 = 32$	32 grams
ethanol	C_2H_5OH	C = 12 H = 1 O = 16	$2 \times 12 = 24$ $6 \times 1 = 6$ $1 \times 16 = \underline{16}$ 46	46 grams

Some calculations on the mole

The two equations below will help you. Try the calculation triangle on the right too.

The calculation triangle

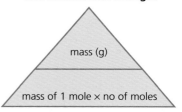

> **mass of a given number of moles = mass of 1 mole × number of moles**
>
> **number of moles in a given mass** = $\dfrac{\text{mass}}{\text{mass of 1 mole}}$

▲ Cover the thing you want to find – and you will see how to calculate it.

Example 1 Calculate the mass of 0.5 mol of bromine molecules.
Bromine exists as molecules. Its formula is Br_2.
The A_r of bromine is 80, so its M_r is 160.
So 1 mol of bromine molecules has a mass of 160 g
So 0.5 mol of bromine molecules has a mass of 0.5 × 160 g, or **80 g**.

$A_r = 80$
$M_r = 160$

Example 2 Calculate the mass of 0.5 mol of bromine *atoms*.
The A_r of bromine is 80, so 1 mol of bromine atoms has a mass of 80 g.
So 0.5 mol of bromine atoms has a mass of 0.5 × 80 g, or **40 g**.

Example 3 How many moles of oxygen molecules are in 64 g of oxygen?
The M_r of oxygen is 32, so 32 g of oxygen is 1 mol of oxygen molecules.
Therefore 64 g is $\dfrac{64}{32}$ mol, or **2 mol** of oxygen molecules.

$A_r = 16$
$M_r = 32$

Note: the A_r values you need are on page 313.
1 How many atoms are in 1 mole of atoms?
2 How many molecules are in 2 moles of molecules?
3 What name is given to the number 6.02×10^{23}?
4 Find the molar mass of:
  a hydrogen atoms b iodine atoms
  c chlorine atoms d chlorine molecules
5 Find the mass of 2 mol of:
  a oxygen atoms b oxygen molecules
6 Find the mass of 4 mol of ammonia, NH_3.
7 Find the mass of 3 mol of ethanol, C_2H_5OH.
8 How many moles of molecules are there in:
  a 18 grams of hydrogen, H_2? b 54 grams of water?
9 3 moles of sodium hydroxide has a mass of 120 g. Find the M_r of sodium hydroxide.
10 The M_r of magnesium chloride ($MgCl_2$) is 95. How many of these are in 95 grams of magnesium chloride?
  a 'units' of magnesium chloride b Mg^{2+} ions
  c Cl^- ions

5.2 Calculations from equations

Objectives: from the equation, state the number of moles taking part in a reaction; convert moles to masses; deduce equations from reacting masses

What an equation tells you
When carbon burns in oxygen, the reaction can be shown like this:

 + ⟶ 1 molecule of carbon dioxide

1 atom of carbon   1 molecule of oxygen   1 molecule of carbon dioxide

or in a short way, using the symbol equation:

$$C\ (s) + O_2\ (g) \longrightarrow CO_2\ (g)$$

This equation tells you that:

| 1 carbon atom | reacts with | 1 molecule of oxygen | to give | 1 molecule of carbon dioxide |

Now suppose there is 1 *mole* of carbon atoms. Then we can say that:

| 1 mole of carbon atoms | reacts with | 1 mole of oxygen molecules | to give | 1 mole of carbon dioxide molecules |

So from the equation, we can tell how many moles react.
But moles can be changed to grams, using A_r and M_r.
The A_r values are: C = 12, O = 16.
So the M_r values are: O_2 = 32, CO_2 = (12 + 32) = 44, and we can write:

| 12 g of carbon | reacts with | 32 g of oxygen | to give | 44 g of carbon dioxide |

Since substances always react in the same ratio, this also means that:

| 6 g of carbon | reacts with | 16 g of oxygen | to give | 22 g of carbon dioxide |

and so on. So, with the help of the mole, you can learn a great deal from a chemical equation.

From the equation for a reaction you can tell:
- how many moles of each substance take part
- how many grams of each substance take part.

Reminder: the total mass does not change
Look what happens to the total mass, during the reaction above:

    mass of carbon and oxygen at the start: 12 g + 32 g = **44 g**
    mass of carbon dioxide at the end: **44 g**

The total mass has not changed, during the reaction. This is because no atoms have disappeared. They have just been rearranged.
That is one of the two laws of chemistry that you met on page 52:
The total mass does not change, during a chemical reaction.

▲ Iron and sulfur reacting: the total mass is the same before and after.

58

Calculating masses from equations

An equation tells you the number of *moles* taking part in a reaction. To calculate *masses*, follow these steps:

1. Write the balanced equation for the reaction.
2. Write down the A_r or M_r for each substance that takes part.
3. Using A_r or M_r, change the moles in the equation to grams.
4. This gives the *theoretical* masses for the reaction. From these, you can then find any *actual* mass.

Example Hydrogen burns in oxygen to form water. What mass of oxygen is needed for 1 g of hydrogen, and what mass of water is obtained?

1. The equation for the reaction is: $2H_2(g) + O_2(g) \longrightarrow 2H_2O(l)$
2. A_r: H = 1, O = 16. M_r: H_2 = 2, O_2 = 32, H_2O = 18.
3. So, for the equation, the amounts in grams are:

   $2H_2(g)\ +\ O_2(g)\ \longrightarrow\ 2H_2O(l)$
   2 × 2 g   32 g   2 × 18 g   or
   4 g   32 g   36 g

4. But you start with only 1 g of hydrogen, so the *actual* masses are:
   1 g   32/4 g   36/4 g   or
   1 g   8 g   9 g

So **1 g** of hydrogen needs **8 g** of oxygen to burn, and gives **9 g** of water.

Working out equations, from masses

You can also do the reverse! If you know the masses that react, you can work out the equation for the reaction. Just change masses to moles.

Example Iron reacts with a solution of copper(II) sulfate ($CuSO_4$) to give copper and a solution of iron sulfate. 1.4 g of iron gave 1.6 g of copper. The formula of the iron sulfate is either $FeSO_4$ or $Fe_2(SO_4)_3$. Write the correct equation for the reaction.

1. A_r: Fe = 56, Cu = 64.
2. Change the masses to moles of atoms:
   $\frac{1.4}{56}$ moles of iron atoms gave $\frac{1.6}{64}$ moles of copper atoms, or

   0.025 moles of iron atoms gave 0.025 moles of copper atoms, so 1 mole of iron atoms gave 1 mole of copper atoms.
3. So the balanced equation for the reaction must be:
   $Fe\ +\ CuSO_4\ \longrightarrow\ Cu\ +\ FeSO_4$
4. Add the state symbols to complete it:
   $Fe(s)\ +\ CuSO_4(aq)\ \longrightarrow\ Cu(s)\ +\ FeSO_4(aq)$

▲ Hydrogen burning in oxygen. The result: water.

The limiting reactant

The equation in step 3 on the left below shows that **1 mole** of Fe reacts with **1 mole** of $CuSO_4$. So **0.1 moles** of Fe will react with **0.1 moles** of $CuSO_4$.

Now look at the reactants above:
0.08 moles of Fe (4.48 ÷ 56) and **0.1 moles** of $CuSO_4$ [(200 ÷ 1000) × 0.5].

So the copper(II) sulfate is in excess. When they react, 0.02 moles of copper(II) sulfate will be left over.

Iron is called the **limiting reactant** for the quantities used here, because it limits how much copper will form. The reaction stops when it is used up.

1. The reaction between magnesium and oxygen is:
   $2Mg(s) + O_2(g) \longrightarrow 2MgO(s)$
   a. How many moles of magnesium atoms react with 1 mole of oxygen molecules?
   b. The A_r values are: Mg = 24, O = 16. How many grams of oxygen react with:
      i 48 g of magnesium?   ii 12 g of magnesium?

2. Iron reacts with sulfur like this, on heating:
   $Fe(s) + S(s) \longrightarrow FeS(s)$
   a. i 5.6 g of iron and 4.0 g of sulfur are heated together. Which is the limiting reactant? (A_r: Fe = 56, S = 32)
      ii What mass of iron(II) sulfide will be formed?
   b. Calculate the mass of sulfur that will react with 11.2 g of iron.

5.3 Reactions involving gases

Objectives: define *molar gas volume*; state the molar gas volume of any gas at rtp; calculate the volume of a gas from grams, moles, and equations

A closer look at some gases

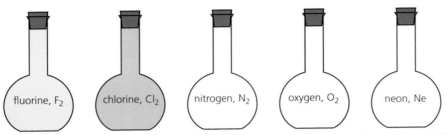

Imagine five huge flasks, each with a volume of 24 dm³. Each is filled with a different gas. Each gas is at room temperature and pressure, or **rtp**.

(We take **room temperature and pressure** as the standard conditions for comparing gases; rtp is 20 °C and 1 atmosphere.)

If you weighed the gas in each flask, and converted its mass to moles, you would find something amazing. There is exactly 1 mole of each gas!

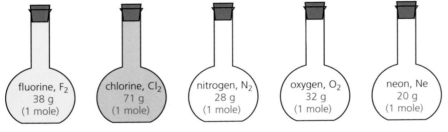

So we can conclude that:

1 mole of every gas has the same volume, at a given temperature and pressure. At room temperature and pressure, this volume is 24 dm³.

This was discovered by Avogadro, in 1811. So it is often called **Avogadro's Law**. It does not matter whether a gas exists as atoms or molecules, or whether its atoms are large or small. The law still holds.

The molar gas volume
The volume occupied by 1 mole of a gas is called its **molar gas volume**. **The molar gas volume at rtp is 24 dm³.**

Another way to look at it
Look at the two gas jars on the right.
A contains nitrogen dioxide, NO_2. **B** contains oxygen, O_2.

The two gas jars have identical volumes, and the gases are at the same temperature and pressure.

You cannot see the gas molecules – let alone count them. But, from Avogadro's Law, you can say that the two jars contain the same number of molecules. Brilliant!

> **Remember**
> - 1 dm³ = 1000 cm³
> - To convert dm³ to cm³, multiply by 1000.

▲ Think of a ball of diameter 36 cm. Its volume is about 24 dm³.

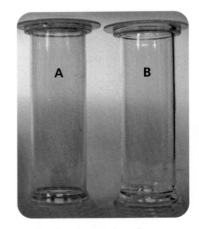

▲ Gas jar **A** contains nitrogen dioxide, NO_2; **B** contains oxygen, O_2.

Calculating gas volumes from moles and grams

Avogadro's Law makes it easy to work out the volumes of gases.

Example 1 What volume does 0.25 mol of a gas occupy at rtp?
1 mol occupies 24 dm³ (the molar gas volume) so
0.25 mol occupies 0.25 × 24 dm³ = 6 dm³
so 0.25 mol of any gas occupies **6 dm³** (or **6000 cm³**) at rtp.

Example 2 What volume does 22 g of carbon dioxide occupy at rtp?
M_r of carbon dioxide = 44, so
44 g = 1 mol, so
22 g = 0.5 mol
so the volume occupied = 0.5 × 24 dm³ = **12 dm³**.

Now check that you get the same answers using the calculation triangle.

The calculation triangle

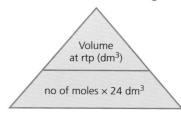

▲ Cover the thing you want to find – and you will see how to calculate it.

Calculating gas volumes from equations

From the equation for a reaction, you can tell how many *moles* of a gas take part. Then you can use Avogadro's Law to work out its *volume*. In these examples, all volumes are measured at rtp.

Example 1 What volume of hydrogen will react with 24 dm³ of oxygen to form water?

1. The equation for the reaction is: $2H_2\ (g) + O_2\ (g) \longrightarrow 2H_2O\ (l)$
2. So 2 volumes of hydrogen react with 1 of oxygen, or
   2 × 24 dm³   react with 24 dm³, so
   48 dm³ of hydrogen will react.

Example 2 When sulfur burns in air it forms sulfur dioxide. What volume of this gas is produced when 1 g of sulfur burns? (A_r: S = 32.)

1. The equation for the reaction is: $S\ (s) + O_2\ (g) \longrightarrow SO_2\ (g)$
2. 32 g of sulfur atoms = 1 mol of sulfur atoms, so
   1 g = $\frac{1}{32}$ mol or 0.03125 mol of sulfur atoms.
3. 1 mol of sulfur atoms gives 1 mol of sulfur dioxide molecules so
   0.03125 mol   gives 0.03125 mol.
4. 1 mol of sulfur dioxide molecules has a volume of 24 dm³ at rtp so
   0.03125 mol has a volume of 0.03125 × 24 dm³ at rtp, or 0.75 dm³.
   So **0.75 dm³** (or **750 cm³**) of sulfur dioxide are produced.

▲ Sulfur dioxide is one of the gases given out in volcanic eruptions. These scientists are collecting gas samples on the slopes of an active volcano.

(A_r: O = 16, N = 14, H = 1, C = 12.)

1. What does *rtp* mean? What values does it have?
2. What does *molar gas volume* mean?
3. What is the molar gas volume of neon gas at rtp?
4. For any gas, calculate the volume in dm³ at rtp of:
   a 7 mol   b 0.5 mol   c 0.001 mol
5. Calculate the volume in dm³ at rtp of:
   a 16 g of oxygen (O_2)   b 1.7 g of ammonia (NH_3)
6. Convert to cm³: a 3 dm³   b 0.1 dm³ (Panel on page 60!)
7. Convert to dm³: a 2000 cm³   b 500 cm³
8. You burn 6 grams of carbon in plenty of air:
   $C\ (s) + O_2\ (g) \longrightarrow CO_2\ (g)$
   a What volume of gas will form (in dm³ at rtp)?
   b What volume of oxygen will be used up?
9. If you burn 6 grams of carbon in *limited* air, the reaction is different: $2C\ (s) + O_2\ (g) \longrightarrow 2CO\ (g)$
   a What volume of gas will form this time?
   b What volume of oxygen will be used up?

5.4 The concentration of a solution

Objectives: explain the terms *solute*, *solvent*, and *solution*; define *concentration of a solution*, and give two units for it; be able to do calculations on concentration

What does *concentration* mean?

A

B

C
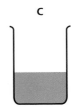

Solution **A** contains 2.5 grams of copper(II) sulfate in 1 dm³ of water. So its concentration is **2.5 g/dm³**.

Solution **B** contains 25 grams of copper(II) sulfate in 1 dm³ of water. So its concentration is **25 g/dm³**.

Solution **C** contains 125 grams of copper(II) sulfate in 0.5 dm³ of water. So its concentration is **250 g/dm³**.

The concentration of a solution is the amount of solute that is dissolved in 1 dm³ of solution. You can express it in grams, or moles.

> **!** Solute, solvent, solution
> - The *solute* is the substance you dissolve in a liquid, to make a solution.
> - The *solvent* is the liquid you use. (Usually water!)

Finding the concentration in moles

This is how to find the concentration of a solution in moles per dm³.

1 First, find the M_r of the solute.
2 Then use this equation:

$$\text{concentration (mol/dm}^3\text{)} = \frac{\text{amount of solute (mol)}}{\text{volume of solution (dm}^3\text{)}}$$

Example Find the concentrations of A, B, and C above in moles per dm³.

1 The M_r of copper(II) sulfate is 250, as shown on the right.
   So 1 mol of the compound has a mass of 250 g.
2 Using the equation above:
   Solution A has 2.5 g of the compound in 1 dm³ of solution.
   2.5 g = 0.01 mol, so its concentration is $\frac{0.01}{1}$ or **0.01 mol/dm³**.
   Solution B has 25 g of the compound in 1 dm³ of solution.
   25 g = 0.1 mol, so its concentration is $\frac{0.01}{1}$ or **0.1 mol/dm³**.
   Solution C has 125 g of the compound in 0.5 dm³, which means 250 g in 1 dm³ of solution.
   250 g = 1 mol, so its concentration is $\frac{1}{1}$ or **1 mol/dm³**.

> **!** M_r for copper(II) sulfate
> Its formula is $CuSO_4.5H_2O$.
> This has 1 Cu, 1 S, 9 O, and 10 H.
> So the formula mass is:
> 1 Cu = 1 × 64 = 64
> 1 S = 1 × 32 = 32
> 9 O = 9 × 16 = 144
> 10 H = 10 × 1 = 10
> Total = 250

Use the equation above to check that the last column in this table is correct:

Amount of solute (mol)	Volume of solution (dm³)	Concentration of solution (mol/dm³)
1.0	1.0	1.0
0.2	0.1	2.0
0.5	0.2	2.5
1.5	0.3	5.0

Finding the amount of solute in a solution

If you know the concentration of a solution, and its volume:
- you can work out how much solute it contains, in moles. Just rearrange the equation from the last page:
  amount of solute (mol) = concentration (mol/dm³) × volume (dm³)
- you can then convert moles to grams, by multiplying the number of moles by M_r.

The calculation triangle is useful too. Try it out!

The calculation triangle

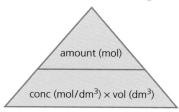

▲ Cover the thing you want to find – and you will see how to calculate it. To draw this triangle, remember that **a**lligators **c**hew **v**isitors!

Sample calculations

The table shows four solutions, with different volumes and concentrations. The mass of solute is calculated for each.

Solution	sodium hydroxide NaOH	sodium thiosulfate $Na_2S_2O_3$	lead nitrate $Pb(NO_3)_2$	silver nitrate $AgNO_3$
	2 dm³	250 cm³	100 cm³	25 cm³
Concentration (mol/dm³)	1	2	0.1	0.05
Amount of solute present (moles)	1 × 2 = 2	2 × $\frac{250}{1000}$ = 0.5	0.1 × $\frac{100}{1000}$ = 0.01	0.05 × $\frac{25}{1000}$ = 0.00125
M_r	40	158	331	170
Mass of solute present (g)	80	79	3.31	0.2125

Check that you understand the working that leads to the answers in the last row.

1. For the solutions at the top of page 62, name:
   a. the solute
   b. the solvent
2. Define *concentration of a solution*.
3. Convert these volumes to cm³:
   a. 0.5 dm³   b. 2 dm³
4. Calculate the number of moles of solute in:
   a. 500 cm³ of a solution of concentration 2 mol/dm³
   b. 2 dm³ of a solution of concentration 0.5 mol/dm³
5. What is the concentration of a solution containing:
   a. 4 moles in 2 dm³ of solution?
   b. 0.3 moles in 200 cm³ of solution?
6. A solution of X has a concentration of 4 mol/dm³. What volume of the solution contains 2 mol of X?
7. A solution of Y has a concentration of 6 mol/dm³. What volume of the solution contains 0.03 mol of Y?
8. The M_r of sodium hydroxide is 40. How many grams of sodium hydroxide are there in:
   a. 500 cm³ of a solution of concentration 1 mol/dm³
   b. 25 cm³ of a solution of concentration of 0.5 mol/dm³?
9. Using A_r values from page 313, calculate the concentration in mol/dm³ of an aqueous solution that contains:
   a. 53 g of sodium carbonate (Na_2CO_3) in 1000 cm³
   b. 62.5 g of copper(II) sulfate ($CuSO_4.5H_2O$) in 1000 cm³

5.5 Finding the empirical formula

Objectives: define *empirical formula*; calculate it from the masses of elements that combine; describe an experiment to find an empirical formula

Reminder: what a formula tells you

The formula of carbon dioxide is **CO_2**. This formula tells you that:

1 carbon atom	combines with	2 oxygen atoms
1 mole of carbon atoms	combines with	2 moles of oxygen atoms

so

A_r: C=12, O=16

Moles can be changed to grams, using A_r and M_r. So we can write:

12 g of carbon	combines with	32 g of oxygen

In the same way:
6 g of carbon combines with 16 g of oxygen
24 kg of carbon combines with 64 kg of oxygen, and so on.
The masses of substances that combine are *always in the same ratio*.

Therefore, from the formula of a compound, you can tell:

- how many moles of the different atoms combine
- how many grams of the different elements combine.

Finding the empirical formula

You can also do things the other way round.
If you know the masses that combine, you can work out the formula.

These are the steps:

Find the masses that combine (in **grams**) by experiment.	→	Change grams to **moles of atoms**.	→	This tells you the **ratio** in which atoms combine.	→	So you can write a **formula**.

A formula found in this way is called the **empirical formula**.
The empirical formula shows the simplest ratio in which atoms combine to form a compound.

Example 1 32 grams of sulfur combine with 32 grams of oxygen to form an oxide of sulfur. What is its empirical formula?

Draw up a table like this:

elements that combine	sulfur	oxygen
masses that combine	32 g	32 g
relative atomic masses (A_r)	32	16
moles of atoms that combine	$\frac{32}{32} = 1$	$\frac{32}{16} = 2$
ratio in which atoms combine	1 : 2	
empirical formula	SO_2	

So the empirical formula of the oxide that forms is **SO_2**.

▲ Sulfur combines with oxygen as it burns, to form an oxide.

64

USING MOLES

Example 2 An experiment shows that compound Y is 80% carbon and 20% hydrogen. What is its empirical formula?

Y is 80% carbon and 20% hydrogen. So 100 g of Y contains 80 g of carbon and 20 g of hydrogen. Draw up a table:

elements that combine	carbon	hydrogen
masses that combine	80 g	20 g
relative atomic masses (A_r)	12	1
moles of atoms that combine	$\frac{80}{12} = 6.67$ (answer rounded off)	$\frac{20}{1} = 20$
ratio in which atoms combine	6.67 : 20 or 1 : 3 (allowing for the rounding off)	
empirical formula	CH_3	

So the empirical formula of Y is **CH_3**.

▲ Empirical formulae are found by experiment – and it usually involves weighing.

But we can tell right away that the *molecular* formula for Y must be different. (A carbon atom does not bond to only 3 hydrogen atoms.) You will learn how to find the molecular formula from the empirical formula in the next unit.

An experiment to find the empirical formula

To work out the empirical formula, you need to know the masses of elements that combine. *The only way to do this is by experiment*.

For example, magnesium combines with oxygen to form magnesium oxide. The masses that combine can be found like this:

1 Weigh a crucible and lid, empty. Then add a coil of magnesium ribbon and weigh it again, to find the mass of the magnesium.
2 Heat the crucible. Raise the lid carefully at intervals to let oxygen in. The magnesium burns brightly.
3 When burning is complete, let the crucible cool (still with its lid on). Then weigh it again. The increase in mass is due to oxygen.

The results showed that 2.4 g of magnesium combined with 1.6 g of oxygen. Draw up a table again:

> **Rounding off**
> - Often, answers to calculations will not be whole numbers.
> - They may have a string of digits after the decimal point. For example, 80/12 = 6.6666666.....
> - Use your common sense about rounding such numbers off!

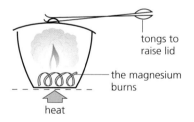

elements that combine	magnesium	oxygen
masses that combine	2.4 g	1.6 g
relative atomic masses (A_r)	24	16
moles of atoms that combine	$\frac{2.4}{24} = 0.1$	$\frac{1.6}{16} = 0.1$
ratio in which atoms combine	1 : 1	
empirical formula	MgO	

So the empirical formula for the oxide is **MgO**.

---

1 a How many atoms of hydrogen combine with one carbon atom to form methane, CH_4?
  b How many grams of hydrogen combine with 12 grams of carbon to form methane?
2 What does the word *empirical* mean? (Check the glossary?)
3 56 g of iron combine with 32 g of sulfur to form iron sulfide. Find the empirical formula for iron sulfide. (A_r: Fe = 56, S = 32.)
4 An oxide of sulfur is 40% sulfur and 60% oxygen. What is its empirical formula?

5.6 From empirical to final formula

Objectives: know that the formula and empirical formula are the same, for an ionic compound; find a molecular formula, given the empirical formula and other data

The formula of an ionic compound

The empirical formula shows the *simplest ratio* in which atoms combine.

The diagram on the right shows the structure of sodium chloride. The sodium and chlorine atoms are in the ratio 1 : 1 in this compound. So its empirical formula is NaCl.

The formula of an ionic compound is the same as its empirical formula.

On page 65, the empirical formula for magnesium oxide was found to be MgO. So the formula for magnesium oxide is also **MgO**.

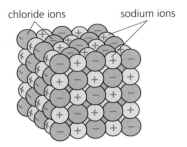

▲ The structure of sodium chloride.

The formula of a molecular compound

The gas ethane is from the alkane family of compounds. Alkanes contain only hydrogen and carbon. An ethane molecule is drawn on the right.

From the drawing you can see that the ratio of carbon to hydrogen atoms in ethane is 2 : 6. The simplest ratio is therefore 1 : 3. So the *empirical* formula of ethane is CH_3. (It is compound Y on page 65.) But its *molecular* formula is C_2H_6.

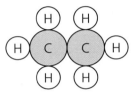

▲ An ethane molecule.

The molecular formula shows the *actual* numbers of atoms that combine to form a molecule.

The molecular formula is more useful than the empirical formula, because it gives you more information.

For some molecular compounds, both formulae are the same. For others they are different. Compare them for the alkanes in the table on the right. What do you notice?

alkane	empirical formula	molecular formula
methane	CH_4	CH_4
ethane	CH_3	C_2H_6
propane	C_3H_8	C_3H_8
butane	C_2H_5	C_4H_{10}
pentane	C_5H_{12}	C_5H_{12}
hexane	C_3H_7	C_6H_{14}

How to find the molecular formula

To find the molecular formula for an unknown compound, you need to know three things:

- the **relative molecular mass** of the compound (M_r). This can be found using a mass spectrometer.
- its **empirical formula**. This is found by experiment, as on page 65.
- its **empirical mass**. This is the mass calculated using the empirical formula and A_r values.

Once you know all that, you can work out the molecular formula by following these steps:

> **To find the molecular formula:**
>
> i Calculate $\dfrac{M_r}{\text{empirical mass}}$ for the compound. This gives a number, n.
>
> ii Multiply the numbers of atoms in the empirical formula by n. (The new number is placed after each symbol.)

Next we will look at two examples.

▲ Butane: empirical formula C_2H_5, molecular formula C_4H_{10}. It is burned as a fuel.

USING MOLES

Calculating the molecular formula

Example 1 A molecular compound has the empirical formula HO. Its relative molecular mass is 34. What is its molecular formula? (A_r: H = 1, O = 16.)

For the empirical formula HO, the empirical mass = 17. But M_r = 34.

So $\dfrac{M_r}{\text{empirical mass}} = \dfrac{34}{17} = 2$

So the molecular formula is 2 × HO, or **H_2O_2**.
So the compound is hydrogen peroxide.
Note how the numbers are placed after the symbols.

▲ Using hydrogen peroxide solution to clean a hospital floor. Hydrogen peroxide acts as a bleach, and kills germs.

Example 2 Like methane, octane is an alkane. It contains only carbon and hydrogen. It is 84.2% carbon and 15.8% hydrogen by mass. Its M_r is 114. What is its molecular formula?

1 **First find the empirical formula for the compound.**
   From the %, we can say that in 100 g of octane, 84.2 g is carbon and 15.8 g is hydrogen.
   So 84.2 g of carbon combines with 15.8 g of hydrogen.

   Changing masses to moles:

   $\dfrac{84.2}{12}$ mol of carbon atoms combine with $\dfrac{15.8}{1}$ mol of hydrogen atoms,

   so

   7.02 mol of carbon atoms combine with 15.8 mol of hydrogen atoms,

   or

   1 mol of carbon atoms combines with $\dfrac{15.8}{7.02}$ or **2.25** mol of hydrogen atoms.

   So the atoms combine in the ratio of 1 : 2.25 or **4 : 9**. (The ratio is given as whole numbers, since whole atoms combine.) To find the simplest whole number ratio, you multiply by a suitable factor, in this case 4.)
   The empirical formula of octane is therefore **C_4H_9**.

2 **Then use M_r to find the molecular formula.**
   For the empirical formula (C_4H_9), the empirical mass = 57.
   But M_r = 114.

   So $\dfrac{M_r}{\text{empirical mass}} = \dfrac{114}{57} = 2$

   So the molecular formula of octane is 2 × C_4H_9 or **C_8H_{18}**.

▲ Octane is one of the main ingredients in gasoline (petrol). When it burns in the engine, it gives out lots of energy to move that car.

1 In the ionic compound magnesium chloride, magnesium and chlorine atoms combine in the ratio 1 : 2. What is the formula of magnesium chloride?
2 In the ionic compound aluminium fluoride, aluminium and fluorine atoms combine in the ratio 1 : 3. What is the formula of aluminium fluoride?
3 What is the difference between an *empirical formula* and a *molecular formula*? Can they ever be the same?
4 What is the empirical formula of benzene, C_6H_6?
5 A compound has the empirical formula CH_2. Its M_r is 28. What is its molecular formula?
6 A hydrocarbon is 84% carbon, by mass. Its relative molecular mass is 100. Find:
   a its empirical formula   b its molecular formula
7 An oxide of phosphorus has an M_r of 220. It is 56.4% phosphorus. Find its molecular formula. (A_r of phosphorus, 31.)

67

5.7 Finding % yield and % purity

Objectives: explain the meanings of the terms *% yield* and *% purity*; recall the equations for working out % yield and % purity, and use them in calculations

Yield and purity

The **yield** is the amount of product you obtain from a reaction. Suppose you own a factory that makes paint or fertilisers. You will want the highest yield possible, for the lowest cost!

Now imagine your factory makes medical drugs, or flavouring for foods. The yield will still be important – but the **purity** of the product may be even more important. Impurities could harm people.

In this unit, you'll learn how to calculate the % yield from a reaction. We will also look at % purity again. You met it already in Unit 4.4 – but this time, we will use the mole.

▲ Everything is carefully controlled in a chemical factory, to give a high yield – and as quickly as possible.

Finding the % yield

You can work out % yield like this:

$$\% \text{ yield} = \frac{\text{actual mass of product obtained}}{\text{theoretical mass of product}} \times 100\%$$

Example The medical drug aspirin is made from salicyclic acid. 1 mole of salicylic acid gives 1 mole of aspirin:

$$C_7H_6O_3 \xrightarrow{\text{chemicals}} C_9H_8O_4$$

salicylic acid → aspirin

In a trial, 100.0 grams of salicylic acid gave 121.2 grams of aspirin. What was the % yield?

1. A_r: C = 12, H = 1, O = 16.
   So M_r: salicyclic acid = 138, aspirin = 180.
2. 138 g of salicylic acid = 1 mol
   so 100 g = $\frac{100}{138}$ mole = 0.725 mol.
3. 1 mol of salicylic acid gives 1 mol of aspirin (from the equation above)
   so 0.725 mol give 0.725 mol of aspirin
   or 0.725 × 180 g = 130.5 g
   So the **theoretical mass** for the reaction is **130.5 g**.
4. But the **actual mass** obtained in the trial was 121.2 g.
   So % yield = $\frac{121.2}{130.5}$ g × 100 = **92.9%**

This is a high yield – so it is worth continuing with those trials.

> **The theoretical mass**
> - The theoretical mass of a product in a reaction is the mass obtained by calculation, from the equation.
> - It assumes a yield of 100%.

Finding the % purity

When you make something in a chemical reaction, it will have impurities mixed with it – for example, small amounts of unreacted substances.

You can work out the % purity of the product you obtained like this:

$$\% \text{ purity of a product} = \frac{\text{mass of the pure product}}{\text{mass of the impure product obtained}} \times 100\%$$

▲ For some products, a very high level of purity is essential – for example, when you are creating new medical drugs.

USING MOLES

Below are examples of how to work out the % purity.

Example 1 Aspirin is an acid. (Its full name is acetylsalicylic acid.) It is neutralised by sodium hydroxide in this reaction:

$$C_9H_8O_4\ (aq) + NaOH\ (aq) \longrightarrow C_9H_7O_4Na\ (aq) + H_2O\ (l)$$

Some aspirin was prepared in the lab. Through titration, it was found that 4.00 g of the aspirin were neutralised by 17.5 cm³ of 1M sodium hydroxide solution. How pure was the aspirin sample?

1. M_r of $C_9H_8O_4$ = 180 (A_r: C = 12, H = 1, O = 16)
2. 17.5 cm³ of 1M sodium hydroxide contain $\frac{17.5}{1\,000}$ moles or 0.0175 moles of NaOH
3. 1 mole of NaOH reacts with 1 mole of $C_9H_8O_4$ so 0.0175 moles react with 0.0175 moles.
4. 0.0175 moles of $C_9H_8O_4$ = 0.0175 × 180 g or 3.15 g of aspirin.
5. But the mass of the aspirin sample was 4 g.
   So % purity of the aspirin = $\frac{3.15}{4}$ × 100% or **78.75%**.

This is far from acceptable for medical use. The aspirin could be purified by crystallisation. (See page 242.) Repeated crystallisation might be needed.

Example 2 Chalk is almost pure calcium carbonate. 10 g of chalk was reacted with an excess of dilute hydrochloric acid. 2280 cm³ of carbon dioxide gas was collected at room temperature and pressure (rtp). What was the purity of the chalk?

(The hydrochloric acid is *in excess,* so chalk is the *limiting reactant.*)

You can work out the purity of the chalk from the volume of carbon dioxide given off. The equation for the reaction is:

$$CaCO_3\ (s) + 2HCl\ (aq) \longrightarrow CaCl_2\ (aq) + H_2O\ (l) + CO_2\ (g)$$

1. M_r of $CaCO_3$ = 100 (A_r: Ca = 40, C = 12, O = 16.)
2. 1 mole of $CaCO_3$ gives 1 mole of CO_2 and
   1 mole of gas has a volume of 24 000 cm³ at rtp.
3. So 24 000 cm³ of gas is produced by 100 g of calcium carbonate and
   2280 cm³ is produced by $\frac{2280}{24\,000}$ × 100 g or **9.5 g**.
   So there is 9.5 g of calcium carbonate in the 10 g of chalk.
   So the % purity of the chalk = $\frac{9.5}{10}$ g × 100 = **95%**.

Purity check!
You can check whether a sample is pure by measuring its melting and boiling points, and comparing them with the values for the pure product.

- Impurities lower the melting point and raise the boiling point.
- The more impurity present, the greater the change.

▲ White chalk cliffs on the Danish island of Mon. Chalk forms in the ocean floor, over many millions of years, from the hard parts of tiny marine organisms.

1. Define the term: **a** % yield **b** % purity
2. Suppose 100 g of salicylic acid gave only 100 g of aspirin, in the reaction on page 68. What was the % yield this time?
3. 17 kg of aluminium was produced from 51 kg of aluminium oxide (Al_2O_3) by electrolysis. What was the percentage yield? (A_r: Al = 27, O = 16.)
4. Some seawater is evaporated. The sea salt obtained is found to be 86% sodium chloride. How much sodium chloride could be obtained from 200 g of this salt?
5. A 5.0 g sample of dry ice (solid carbon dioxide) turned into 2400 cm³ of carbon dioxide gas at rtp. What was the % purity of the dry ice? (M_r of CO_2 = 44.)

Checkup on Chapter 5

Revision checklist

Extended syllabus content
Make sure you can …
- [ ] explain what a *mole* of atoms or molecules or ions is, and give examples
- [ ] state what the *Avogadro constant* is
- [ ] recognise *mol* as the symbol for mole
- [ ] define *molar mass*
- [ ] do these calculations, using A_r and M_r:
  - find the molar mass of a substance
  - change moles to masses
  - change masses to moles
- [ ] use the idea of the mole to:
  - calculate the masses of reactants or products from the equation for a reaction
  - work out the equation for a reaction, given the masses of the reactants and products
- [ ] define *molar gas volume* and *rtp*
- [ ] calculate the volume that a gas will occupy at rtp, from its mass, or number of moles
- [ ] calculate the volume of gas produced in a reaction, given the equation and the mass of one substance
- [ ] explain what *concentration of a solution* means and give examples, using grams and moles
- [ ] state the units used for concentration
- [ ] work out:
  - the concentration of a solution, when you know the amount of solute dissolved in it
  - the amount of solute dissolved in a solution, when you know its concentration
- [ ] explain what the *empirical formula* of a substance is
- [ ] work out the empirical formula, from the masses that react
- [ ] explain why the empirical formula and the formula are the same, for an ionic compound
- [ ] work out a molecular formula, using the empirical formula, empirical mass, and M_r
- [ ] define *% yield*
- [ ] calculate the % yield for a reaction, from the equation and the actual mass of product obtained
- [ ] define *% purity*
- [ ] calculate the % purity of a product, given the mass of the impure product, and the mass of pure product it contains

Questions

Extended syllabus content

1 Iron is obtained by reducing iron(III) oxide using the gas carbon monoxide. The reaction is:
$Fe_2O_3\ (s) + 3CO\ (g) \longrightarrow 2Fe\ (s) + 3CO_2\ (g)$
  a Write a word equation for the reaction.
  b What is the formula mass of iron(III) oxide? (A_r: Fe = 56, O = 16.)
  c How many moles of Fe_2O_3 are there in 320 kg of iron(III) oxide? (1 kg = 1000 g.)
  d How many moles of Fe are obtained from 1 mole of Fe_2O_3?
  e From **c** and **d**, find how many moles of iron atoms are obtained from 320 kg of iron(III) oxide.
  f How much iron (in kg) is obtained from 320 kg of iron(III) oxide?

2 With strong heating, calcium carbonate undergoes thermal decomposition:
$CaCO_3\ (s) \longrightarrow CaO\ (s) + CO_2\ (g)$
  a Write a word equation for the change.
  b How many moles of $CaCO_3$ are in 50 g of calcium carbonate? (A_r: Ca = 40, C = 12, O = 16.)
  c **i** What mass of calcium oxide is obtained from the thermal decomposition of 50 g of calcium carbonate, assuming a 40% yield?
  ii What mass of carbon dioxide will be given off at the same time?
  iii What volume will this gas occupy at rtp?

3 Nitroglycerine is used as an explosive. The equation for the explosion reaction is:
$4C_3H_5(NO_3)_3\ (l) \longrightarrow$
$12CO_2\ (g) + 10H_2O\ (l) + 6N_2\ (g) + O_2\ (g)$
  a How many moles does the equation show for:
  i nitroglycerine?
  ii *gas* molecules produced?
  b How many moles of gas molecules are obtained from 1 mole of nitroglycerine?
  c What is the total volume of gas (at rtp) obtained from 1 mole of nitroglycerine?
  d What is the mass of 1 mole of nitroglycerine? (A_r: H = 1, C = 12, N = 14, O = 16.)
  e What will be the total volume of gas (at rtp) from exploding 1 kg of nitroglycerine?
  f Using your answers above, try to explain *why* nitroglycerine is used as an explosive.

4 Nitrogen monoxide reacts with oxygen like this:
$2NO\ (g) + O_2\ (g) \longrightarrow 2NO_2\ (g)$
  a How many moles of oxygen molecules react with 1 mole of nitrogen monoxide molecules?

b What volume of oxygen will react with 50 cm³ of nitrogen monoxide?
c If the volumes in **b** are used, what is:
  i the total volume of the two reactants?
  ii the volume of nitrogen dioxide formed?

5 2 g of iron was added to 50 cm³ of 0.5 mol/dm³ dilute sulfuric acid. When the reaction was over, the mixture was filtered. 0.6 g of iron was found unreacted. (A_r: Fe = 56.)
  a Name the limiting reactant in this reaction.
  b How many moles of iron atoms took part?
  c How many moles of sulfuric acid reacted?
  d Write the equation for the reaction, and deduce the charge on the iron ion that formed.
  e What volume of hydrogen (calculated at rtp) bubbled off during the reaction?

6 27 g of aluminium burns in chlorine to form 133.5 g of aluminium chloride. (A_r: Al = 27, Cl = 35.5.)
  a What mass of chlorine is present in 133.5 g of aluminium chloride?
  b How many moles of chlorine atoms is this?
  c How many moles of aluminium atoms are present in 27 g of aluminium?
  d Use your answers for parts **b** and **c** to find the simplest formula of aluminium chloride.
  e 1 dm³ of an aqueous solution is made using 13.35 g of aluminium chloride. What is its concentration in mol/dm³?

7 You need solutions of concentration 2 mol/dm³, with 10 g of solute in each. What volume of solution will you prepare, for each solute below? (A_r: H = 1, Li = 7, N = 14, O = 16, Mg = 24, S = 32.)
  a lithium sulfate, Li_2SO_4
  b magnesium sulfate, $MgSO_4$
  c ammonium nitrate, NH_4NO_3

8 Phosphorus forms two oxides, with empirical formulae P_2O_3 and P_2O_5. (A_r: P = 31, O = 16.)
  a Which oxide contains a higher percentage of phosphorus?
  b What mass of phosphorus will combine with 1 mole of oxygen molecules (O_2) to form P_2O_3?
  c What is the molecular formula of the oxide that has a formula mass of 284?
  d Suggest a molecular formula for the other oxide.

9 Zinc and phosphorus react to give zinc phosphide. 9.75 g of zinc combines with 3.1 g of phosphorus.
  a Find the empirical formula for the compound. (A_r: Zn = 65, P = 31.)
  b Calculate the percentage of phosphorus in it.

10 110 g of manganese was extracted from 174 g of manganese oxide. (A_r: Mn = 55, O = 16.)
  a What mass of oxygen is there in 174 g of manganese oxide?
  b How many moles of oxygen atoms is this?
  c How many moles of manganese atoms are there in 110 g of manganese?
  d Give the empirical formula of manganese oxide.
  e What mass of manganese can obtained from 1000 g of manganese oxide?

11 Find the molecular formulae for these compounds. (A_r: H = 1, C = 12, N = 14, O = 16.)

Compound	M_r	Empirical formula	Molecular formula
a hydrazine	32	NH_2	
b cyanogen	52	CN	
c nitrogen oxide	92	NO_2	
d glucose	180	CH_2O	

12 Hydrocarbons A and B both contain 85.7% carbon. The mass of one mole is 42 g for A and 84 g for B.
  a Which elements does a hydrocarbon contain?
  b Calculate the empirical formulae of A and B.
  c Calculate the molecular formulae of A and B.

13 Mercury(II) oxide breaks down on heating:
$2HgO\ (s) \longrightarrow 2Hg\ (l) + O_2\ (g)$
  a Calculate the mass of 1 mole of mercury(II) oxide. (A_r: O = 16, Hg = 201)
  b How much mercury and oxygen *could* be obtained from 21.7 g of mercury(II) oxide?
  c Only 19.0 g of mercury was collected. Calculate the % yield of mercury for this experiment.

14 A 5-g sample of impure magnesium carbonate is reacted with an excess of hydrochloric acid:
$MgCO_3\ (s) + 2HCl\ (aq) \longrightarrow$
$MgCl_2\ (aq) + H_2O\ (l) + CO_2\ (g)$
1200 cm³ of carbon dioxide is collected at rtp.
  a How many moles of CO_2 are produced?
  b What mass of pure magnesium carbonate would give this volume of carbon dioxide? (A_r: C = 12, O = 16, Mg = 24.)
  c Calculate the % purity of the 5-g sample.

6.1 Oxidation and reduction

Objectives: define *oxidation* and *reduction* in terms of oxygen gain and loss; identify what is oxidised, and what is reduced, in a reaction; explain the term *redox*

Different groups of reactions

Thousands of different reactions go on in labs, and factories, and homes. We can divide them into different groups. For example, **neutralisation reactions** and **precipitation reactions**.

One big group is the **redox reactions**, in which **oxidation** and **reduction** occur. We focus on those in this chapter.

Oxidation: oxygen is gained

Magnesium burns in air with a dazzling white flame. A white ash is formed. The reaction is:

magnesium + oxygen ⟶ magnesium oxide
$2Mg\,(s) + O_2\,(g) \longrightarrow 2MgO\,(s)$

The magnesium has gained oxygen. We say it has been **oxidised**.

A gain of oxygen is called *oxidation*. The substance has been oxidised.

Reduction: oxygen is lost

When hydrogen is passed over heated copper(II) oxide in the apparatus below, the black compound turns pink:

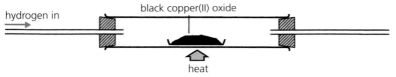

This reaction is taking place:

copper(II) oxide + hydrogen ⟶ copper + water
$CuO\,(s) + H_2\,(g) \longrightarrow Cu\,(s) + H_2O\,(l)$

So the copper(II) oxide is *losing* oxygen. It is being **reduced**.

A loss of oxygen is called *reduction*. The substance is reduced.

▲ Magnesium burning in oxygen. The magnesium is being oxidised.

▲ Iron occurs naturally on Earth as iron(III) oxide, Fe_2O_3. Here the oxide is being **reduced** to iron in the blast furnace.

▲ And here, iron is being **oxidised** to iron(III) oxide again! We call this **rusting**. The formula for rust is $Fe_2O_3.2H_2O$.

REDOX REACTIONS

Oxidation and reduction take place together

Look again at the reaction between copper(II) oxide and hydrogen.

Copper(II) oxide loses oxygen, and hydrogen gains oxygen:

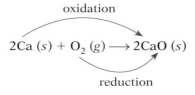

$$CuO\ (s) + H_2\ (g) \longrightarrow Cu\ (s) + H_2O\ (l)$$

(oxidation from CuO to H₂O; reduction from CuO to Cu)

So the copper(II) oxide is reduced, and the hydrogen is oxidised.
**Oxidation and reduction *always* take place together.
So the reaction is called a *redox reaction*.**

▲ A redox reaction that cooks our food. The gas (methane) reacts with the oxygen in air, giving out heat.

Two more examples of redox reactions

The reaction between calcium and oxygen Calcium burns in air with a red flame, to form the white compound calcium oxide. It is easy to see that calcium has been oxidised. But oxidation and reduction *always* take place together, which means oxygen has been reduced:

$$2Ca\ (s) + O_2\ (g) \longrightarrow 2CaO\ (s)$$

(oxidation from Ca to CaO; reduction from O₂ to CaO)

The reaction between hydrogen and oxygen Hydrogen reacts explosively with oxygen, to form water. Hydrogen is oxidised, and oxygen is reduced:

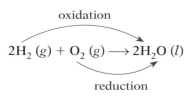

$$2H_2\ (g) + O_2\ (g) \longrightarrow 2H_2O\ (l)$$

▲ Rushing to school on redox. The burning of petrol is a redox reaction. So is the 'burning' of glucose in our body cells. It reacts with oxygen to give us energy, in a process called **respiration**.

> **Those burning reactions**
> - Another name for burning is **combustion**.
> - Combustion is a redox reaction.
> - For example, when an element burns in oxygen, it is oxidised to its oxide.

1 Copy and complete the statements:
   a Oxidation means …
   b Reduction means …
   c Oxidation and reduction always …
2 Magnesium reacts with sulfur dioxide like this:
   $$2Mg\ (s) + SO_2\ (g) \longrightarrow 2MgO\ (s) + S\ (s)$$
   Copy the equation, and use labelled arrows to show which substance is oxidised, and which is reduced.
3 Explain where the term *redox* comes from.
4 Many people cook with natural gas, which is mainly methane, CH_4. The equation for its combustion is:
   $$CH_4\ (g) + 2O_2\ (g) \longrightarrow CO_2\ (g) + 2H_2O\ (l)$$
   Show that this is a redox reaction.
5 Write down the equation for the reaction between magnesium and oxygen. Use labelled arrows to show which element is oxidised, and which is reduced.

6.2 Redox and electron transfer

Objectives: define *oxidation* and *reduction* in terms of electron transfer; identify what is oxidised, and what is reduced; write half-equations and ionic equations

Another definition for oxidation and reduction

When magnesium burns in oxygen, magnesium oxide is formed:

$$2Mg\,(s) + O_2\,(g) \longrightarrow 2MgO\,(s)$$

The magnesium has clearly been oxidised. Oxidation and reduction *always* take place together, so the oxygen must have been reduced. But how?

Let's see what is happening to the electrons:

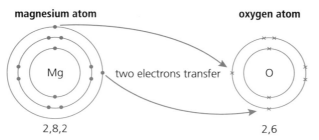

During the reaction, each magnesium atom loses two electrons and each oxygen atom gains two. This leads us to a new definition:

If a substance loses electrons during a reaction, it has been oxidised.

If it gains electrons, it has been reduced.

The reaction is a redox reaction.

Writing half-equations to show the electron transfer

You can use **half-equations** to show the electron transfer in a reaction. One half-equation shows electron loss, and the other shows electron gain.

This is how to write the half-equations for the reaction above:

1. **Write down each reactant, with the electrons it gains or loses.**
   magnesium: $\quad Mg \longrightarrow Mg^{2+} + 2e^-$
   oxygen: $\quad O + 2e^- \longrightarrow O^{2-}$

2. **Check that each substance is in its correct form (ion, atom or molecule) on each side of the arrow. If it is not, correct it.**
   Oxygen is not in its correct form on the left above. It exists as molecules, so you must change O to O_2. That means you must also double the number of electrons and oxide ions:
   oxygen: $\quad O_2 + 4e^- \longrightarrow 2O^{2-}$

3. **The number of electrons must be the same in both equations. If it is not, multiply one (or both) equations by a number, to balance them.**
   So we must multiply the magnesium half-equation by 2.
   magnesium: $\quad 2Mg \longrightarrow 2Mg^{2+} + 4e^-$
   oxygen: $\quad O_2 + 4e^- \longrightarrow 2O^{2-}$
   The equations are now balanced, each with 4 electrons.

> **Remember OILRIG!**
> **O**xidation **I**s **L**oss of electrons.
> **R**eduction **I**s **G**ain of electrons.

> **Showing oxidation**
> You can show oxidation (the loss of electrons) in two ways:
> $Mg \longrightarrow Mg^{2+} + 2e^-$
> or
> $Mg - 2e^- \longrightarrow Mg^{2+}$
> Both are correct!

▲ Magnesium oxide is used as a filler inside the rings on electric cookers. It is a non-conductor when solid, has a high melting point, and transfers heat well.

REDOX REACTIONS

Redox without oxygen
Our definition of redox reactions is now much broader:
Any reaction in which electron transfer takes place is a redox reaction.
So the reaction does not have to include oxygen! Look at these examples:

1 **The reaction between sodium and chlorine**
   The equation is:
   $2Na\ (s) + Cl_2\ (g) \longrightarrow 2NaCl\ (s)$

   The sodium atoms give electrons to the chlorine atoms, forming ions as shown on the right. So sodium is oxidised, and chlorine is reduced.

   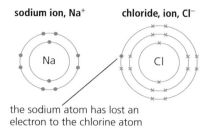
   the sodium atom has lost an electron to the chlorine atom

   So the reaction is a redox reaction. Look at the half-equations:
   sodium: $2Na \longrightarrow 2Na^+ + 2e^-$ (oxidation)
   chorine: $Cl_2 + 2e^- \longrightarrow 2Cl^-$ (reduction)

2 **The reaction between chlorine and potassium bromide**
   When chlorine gas is bubbled through a colourless solution of potassium bromide, the solution goes orange due to this reaction:

   $Cl_2\ (g) + 2KBr\ (aq) \longrightarrow 2KCl\ (aq) + Br_2\ (aq)$
        colourless   orange

   Bromine has been pushed out or **displaced** from the compound. Chlorine has taken its place.

   The half-equations for the reaction are:
   chlorine: $Cl_2 + 2e^- \longrightarrow 2Cl^-$ (reduction)
   bromide ion: $2Br^- \longrightarrow Br_2 + 2e^-$ (oxidation)

From half-equations to the ionic equation
Adding the balanced half-equations gives the **ionic equation** for the reaction.
An ionic equation shows the ions that take part in the reaction.

For example, for the reaction between chlorine and potassium bromide:

$$Cl_2 + 2e^- \longrightarrow 2Cl^-$$
$$2Br^- \longrightarrow Br_2 + 2e^-$$
$$\overline{Cl_2 + \cancel{2e^-} + 2Br^- \longrightarrow 2Cl^- + Br_2 + \cancel{2e^-}}$$

The electrons cancel, giving the ionic equation for the reaction:
$$Cl_2 + 2Br^- \longrightarrow 2Cl^- + Br_2$$

Redox: a summary
Oxidation is gain of oxygen, or loss of electrons.
Reduction is loss of oxygen, or gain of electrons.
Oxidation and reduction always take place together, in a **redox reaction**.

▲ Bromine being displaced by chlorine from a colourless solution of potassium bromide. It is a redox reaction.
The solution goes orange.

1 Give a *full* definition for: a oxidation b reduction
2 What does a *half-equation* show?
3 Like magnesium, calcium is in Group II. It burns in oxygen to give calcium oxide.
   a Write the balanced equation for the reaction.
   b Use the idea of electron transfer to explain why this is a redox reaction.
   c Write the balanced half-equations for the reaction.
4 Potassium and chlorine react to form potassium chloride.
   a Write the balanced equation for this redox reaction.
   b Now write the balanced half-equations for it.
5 Bromine displaces iodine from a solution of potassium iodide.
   a Write the balanced half-equations for this reaction.
   b Add the half-equations, to give the ionic equation for the reaction.

6.3 Redox and oxidation numbers

Objectives: define *oxidation number*; give the oxidation number for each element in a compound; use changes in oxidation number to identify redox reactions

What does *oxidation number* mean?
Oxidation number tells you how many electrons each atom of an element has lost, gained, or shared, in forming a compound.

The rules for oxidation number
1 Each atom in a formula has an oxidation number.
2 The oxidation number is usually given as a Roman numeral. Look:
   number   0   1   2   3   4   5   6   7
   Roman numeral 0   I   II III IV V   VI VII

▲ The element sodium. As for all uncombined elements, its oxidation number is 0 (zero).

3 The oxidation number of an uncombined element is 0 (**zero**).
4 The oxidation number of a simple (monatomic) ion such as Na^+ or Cl^- is the same as its charge.
5 And, related to **4**, many elements have the same oxidation number in most or all of their compounds. Note the link with group numbers:

Element	Usual oxidation number in compounds
hydrogen	+I
sodium and the other Group I metals	+I
calcium and the other Group II metals	+II
aluminium, Group III	+III
oxygen (except in peroxides) Group VI	−II
chlorine and the other Group VII non-metals, in compounds without oxygen	−I

In covalent compounds ...
- Electrons are *shared* in covalent compounds, not lost or gained.
- But the atoms of one element in the compound will attract the shared electrons more strongly, so its oxidation number in that compound will have a − sign.
- For example, the oxidation number of oxygen in the covalent compound H_2O is −II.

6 Most of the transition elements can have different oxidation numbers in their compounds. Look at these:

Element	Common oxidation numbers in compounds
iron	+II and +III
copper	+I and +II
manganese	+II, +IV, and +VII
chromium	+III and +VI

For these elements, the oxidation number is included in the compound's name. For example, iron(III) chloride, copper(II) oxide.

7 The oxidation numbers in the formula of a compound add up to **zero**. Look at the formula for magnesium chloride, for example:

$MgCl_2$

+II   2 × −I   Total = zero

So oxidation numbers can be used to check that formulae are correct.

▲ Copper and its oxidation numbers:
A – copper metal, 0
B – copper(I) oxide, +I
C – copper(II) chloride, +II

REDOX REACTIONS

Oxidation numbers change during redox reactions

Look at the equation for the reaction between sodium and chlorine:

$$2Na\,(s) + Cl_2\,(g) \longrightarrow 2NaCl\,(s)$$
$$0 0 +I\,-I$$

The oxidation numbers are also shown, using the rules on page 76. Notice how they have changed during the reaction.

Each sodium atom loses an electron during the reaction, to form an Na^+ ion. So sodium is oxidised, and its oxidation number rises from 0 to +I. Each chlorine atom gains an electron, to form a Cl^- ion. So chlorine is reduced, and its oxidation number falls from 0 to −I.

If oxidation numbers change during a reaction, it is a redox reaction.

▲ A redox reaction: sodium burning in chlorine to form sodium chloride.

```
       A rise in oxidation number means oxidation has occurred. →
−IV   −III   −II   −I    0    +I    +II   +III   +IV
  ← A fall in oxidation number means reduction has occurred.
```

Using oxidation numbers to identify redox reactions

Example 1 Iron reacts with sulfur to form iron(II) sulfide:

$$Fe\,(s) + S\,(s) \longrightarrow FeS\,(s)$$
$$0 0 +II\,-II$$

The oxidation numbers are shown, using the rules on page 76. There is a change in oxidation numbers. So this is a redox reaction.

Example 2 When chlorine is bubbled through a solution of iron(II) chloride, iron(III) chloride forms. The equation and oxidation numbers are:

$$2FeCl_2\,(aq) + Cl_2\,(aq) \longrightarrow 2FeCl_3\,(aq)$$
$$+II\,-I 0 +III\,-I$$

There is a change in oxidation numbers. So this is a redox reaction.

Example 3 When ammonia and hydrogen chloride gases mix, they react to form ammonium chloride. The equation and oxidation numbers are:

$$NH_3\,(g) + HCl\,(g) \longrightarrow NH_4Cl\,(s)$$
$$-III\,+I +I\,-I -III\,+I\,-I$$

There is no change in oxidation numbers. So this is *not* a redox reaction.

▲ A redox reaction: iron filings reacting with sulfur to form iron(II) sulfide. Heat is needed only to start the reaction off – and then it gives out heat.

Q

1. Define *oxidation number*.
2. Give the oxidation number for each element in:
   a Fe   b HCl   c $CaCl_2$ d NaOH
   e $CaCO_3$
3. a Write the word equation for this reaction:
   $2H_2\,(g) + O_2\,(g) \longrightarrow 2H_2O\,(l)$
   b Now copy out the chemical equation from **a**. Below each symbol write the oxidation number of the atoms.
   c Use oxidation numbers to show that it is a redox reaction.
   d State which substance is oxidised, and which is reduced.
4. Repeat the steps in question **3** for each of these equations:
   i $2KBr\,(s) \longrightarrow 2K\,(s) + Br_2\,(l)$
   ii $2KI\,(aq) + Cl_2\,(g) \longrightarrow 2KCl\,(aq) + I_2\,(aq)$
5. a Carbon burns in oxygen to form carbon dioxide. Write the chemical equation for the reaction.
   b Use oxidation numbers to show that the reaction in **a** is a redox reaction.
6. *Every reaction between two elements is a redox reaction.*
   Do you agree with this statement? Explain.

6.4 Oxidising and reducing agents

Objectives: define *oxidising agent* and *reducing agent*; explain why these react as they do; describe tests for them; give everyday uses of oxidising agents

What are oxidising and reducing agents?

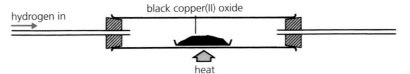

Redox reactions: a summary

A + B ⟶ products
oxidising   reducing
agent   agent
(is reduced) (is oxidised)

In the experiment above, the copper(II) oxide and hydrogen react like this:

copper(II) oxide + hydrogen ⟶ copper + water
$CuO\ (s)\ +\ H_2\ (g)\ \longrightarrow\ Cu\ (s)\ +\ H_2O\ (l)$

The copper(II) oxide is **reduced** to copper by reaction with hydrogen. So hydrogen acts as a **reducing agent**.

The hydrogen is itself **oxidised** to water. So copper(II) oxide acts as an **oxidising agent**.

Every redox reaction has an oxidising agent and a reducing agent. The oxidising agent oxidises the reducing agent, and is itself reduced. The reducing agent reduces the oxidising agent, and is itself oxidised.

Look at the summary in the box on the right above.

Oxidising and reducing agents in the lab

Some substances are strong oxidising agents, because they have a strong drive to gain electrons. They readily oxidise other substances, by taking electrons from them. Oxygen and chlorine are examples.

Some substances are strong reducing agents, readily giving up electrons. For example, hydrogen, and reactive metals like sodium.

Some oxidising and reducing agents show a colour change when they react. This makes them useful in lab tests. Let's look at two examples.

1 Potassium manganate(VII): an oxidising agent

Manganese, a transition element, can have different oxidation numbers.

Potassium manganate(VII) is a purple compound. Its formula is $KMnO_4$. Here, manganese has the oxidation number +VII. But it is much more stable with the oxidation number +II. So it has a strong drive to reduce its oxidation number to +II, by taking electrons from other substances.

So potassium manganate(VII) acts as a powerful oxidising agent, in the presence of a little acid. It is itself reduced, with a colour change:

$MnO_4^-\ (aq)$ $\xrightarrow{\text{reduction}}$ $Mn^{2+}\ (aq)$
manganate(VII) ion   manganese(II) ion
(purple)   (colourless)

Suppose you have an unknown liquid. You want to test it, to see if it contains a reducing agent. Add some acidifed potassium manganate(VII) solution. If a reducing agent is present, the purple colour will fade.

Remember OILRIG!
Oxidation **I**s **L**oss of electrons.
Reduction **I**s **G**ain of electrons.

▲ If a reducing agent is present, the strong purple colour of potassium manganate(VII) will fade, as seen in the test-tube on the right.

REDOX REACTIONS

2 Potassium iodide: a reducing agent

When potassium iodide solution is added to hydrogen peroxide, in the presence of sulfuric acid, this redox reaction takes place:

H_2O_2 (aq) + 2KI (aq) + H_2SO_4 (aq) ⟶ I_2 (aq) + K_2SO_4 (aq) + $2H_2O$ (l)
hydrogen   potassium   iodine   potassium
peroxide   iodide   sulfate

You can see that the hydrogen peroxide loses oxygen: it is reduced. The potassium iodide acts as a reducing agent. At the same time the potassium iodide is oxidised to iodine. This causes a colour change:

$2I^-$ (aq) —oxidation→ I_2 (aq)
colourless   red-brown

So potassium iodide is used to test for the presence of an oxidising agent. (Hydrogen peroxide is an oxidising agent.)

▲ The test-tube shows the red-brown colour you get when potassium iodide is oxidised by an oxidising agent.

Oxidising agents outside the lab

Strong oxidising agents have many uses outside the lab. For example:
- they kill bacteria and moulds, so are widely used in household cleaning products.
- they are also used in water treatment plants, to kill bacteria; **chlorine** is the main one used for this.
- they turn coloured compounds into colourless compounds by oxidising them, so they are used in bleaches for textiles, recycled paper, and hair.

▲ Yes. My hair met hydrogen peroxide. (An oxidising agent, formula H_2O_2.)

▲ Somewhere near you there is a shop that sells products to clean and disinfect homes. And at least some of them will contain oxidising agents.

▲ These tablets release chlorine when added to water. They are for use where people do not have a supply of clean water. Chlorine kills bacteria and viruses, and makes the water safe to drink.

1. What is:
   a an oxidising agent?   b a reducing agent?
2. Identify the oxidising and reducing agents in these reactions, by looking at the gain and loss of oxygen:
   a 2Mg (s) + O_2 (g) ⟶ 2MgO (s)
   b Fe_2O_3 (s) + 3CO (g) ⟶ 2Fe (l) + $3CO_2$ (g)
3. Now identify the oxidising and reducing agents in these:
   a 2Fe + $3Cl_2$ ⟶ $2FeCl_3$
   b Fe + $CuSO_4$ ⟶ $FeSO_4$ + Cu
4. Explain why:
   a potassium manganate(VII) is a powerful oxidising agent
   b potassium iodide is used to test for oxidising agents

Checkup on Chapter 6

Revision checklist

Core syllabus content
Make sure you can …
- [ ] define *oxidation* as a gain of oxygen and *reduction* as a loss of oxygen
- [ ] explain that oxidation and reduction always occur together, and give an example
- [ ] explain what the term *redox reaction* means
- [ ] say what is being oxidised, and what is being reduced, in reactions involving oxygen
- [ ] explain what the Roman numeral in the name of a compound tells you

Extended syllabus content
Make sure you can also …
- [ ] define oxidation and reduction in terms of electron transfer
- [ ] explain these terms:
  half-equation   *ionic equation*
- [ ] write balanced half-equations for a redox reaction, to show the electron transfer
- [ ] add the balanced half-equations to give the ionic equation for a reaction
- [ ] explain the term *oxidation number*
- [ ] state the rules for assigning oxidation numbers to the atoms in elements and their compounds
- [ ] deduce the oxidation number of an element from the compound's name, for elements that can have different oxidation numbers
- [ ] give the oxidation number for each element present, in the equation for a reaction
- [ ] identify a redox reaction from changes in oxidation numbers, shown in the equation for the reaction
- [ ] define these terms:
  oxidising agent   *reducing agent*
- [ ] explain why some substances are:
  strong oxidising agents   *strong reducing agents*
  and give examples
- [ ] explain why potassium manganate(VII) is used in the lab to test for the presence of reducing agents
- [ ] explain why potassium iodide is used in the lab to test for the presence of oxidising agents

Questions

Core syllabus content

1 If a substance gains oxygen in a reaction, it has been oxidised. If it loses oxygen, it has been reduced. Oxidation and reduction always take place together, so if one substance is oxidised, another is reduced.
  a First, write a word equation for each redox reaction **A** to **F** below.
  b Then, using the ideas above, say which substance is being oxidised, and which is being reduced, in each reaction.
  A $Ca\,(s) + O_2\,(g) \longrightarrow 2\,CaO\,(s)$
  B $2CO\,(g) + O_2\,(g) \longrightarrow 2CO_2\,(g)$
  C $CH_4\,(g) + 2O_2\,(g) \longrightarrow CO_2\,(g) + 2H_2O\,(l)$
  D $2CuO\,(s) + C\,(s) \longrightarrow 2Cu\,(s) + CO_2\,(g)$
  E $4Fe\,(s) + 3O_2\,(g) \longrightarrow 2Fe_2O_3\,(s)$
  F $Fe_2O_3\,(s) + 3CO\,(g) \longrightarrow 2Fe\,(s) + 3CO_2\,(g)$

2 a Is this a redox reaction? Give your evidence.
  A $2Mg\,(s) + CO_2\,(g) \longrightarrow 2MgO\,(s) + C\,(s)$
  B $SiO_2\,(s) + C\,(s) \longrightarrow Si\,(s) + CO_2\,(g)$
  C $NaOH\,(aq) + HCl\,(aq) \longrightarrow NaCl\,(aq) + H_2O\,(l)$
  D $Fe\,(s) + CuO\,(s) \longrightarrow FeO\,(s) + Cu\,(s)$
  E $C\,(s) + PbO\,(s) \longrightarrow CO\,(g) + Pb\,(s)$
  b For each redox reaction you identify, state:
  i what is being oxidised
  ii what is being reduced.

Extended syllabus content

3 All reactions in which electron transfer take place are redox reactions. This diagram shows the electron transfer during one redox reaction.

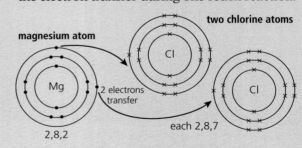

  a What is the product of this reaction?
  b Write a balanced equation for the full reaction.
  c i Which element is being oxidised?
  ii Write a half-equation for the oxidation.
  d i Which element is being reduced?
  ii Write a half-equation for the reduction of this element.

4 Fluorine, from Group VII, reacts with lithium, from Group I, to form a poisonous white compound, in a redox reaction.

a Name the compound.
b Write a balanced equation for the reaction.
c Draw a diagram to show the electron transfer that takes place during the reaction.
d i Which element is oxidised in the reaction?
  ii Write a half-equation for this oxidation.
e Write a half-equation for the reduction of the other element.

5 Chlorine gas is bubbled into a solution containing sodium bromide. The equation for the reaction is:
$Cl_2\ (g) + 2NaBr\ (aq) \rightarrow Br_2\ (aq) + 2NaCl\ (aq)$
a Chlorine takes the place of bromine, in the metal compound. What is this type of reaction called?
b The compounds of Group I metals are white, and give colourless solutions. What would you see as the above reaction proceeds?
c i Write a half-equation for the reaction of the chlorine.
  ii Is the chlorine oxidised, or reduced, in this reaction? Explain.
d Write a half-equation for the reaction of the bromide ion.
e Reactive elements have a strong tendency to exist as ions. Which is more reactive, chlorine or bromine? Explain why you think so.
f i Which halide ion could be used to convert bromine back to the bromide ion?
  ii Write the ionic equation for this reaction.

6 Iodine is extracted from seaweed using acidified hydrogen peroxide, in a redox reaction. The ionic equation for the reaction is:
$2I^-\ (aq) + H_2O_2\ (aq) + 2H^+\ (aq) \rightarrow I_2\ (aq) + 2H_2O\ (l)$
a Give the oxidation number for iodine in seaweed.
b There is a colour change in this reaction. Why?
c i Is the iodide ion oxidised, or reduced?
  ii Write the half-equation for this change.
d In hydrogen peroxide, the oxidation number for hydrogen is +I.
  i What is the oxidation number of the oxygen in hydrogen peroxide?
  ii How does the oxidation number for oxygen change during the reaction?
  iii Copy and complete this half-equation for hydrogen peroxide:
  $H_2O_2\ (aq) + 2H^+\ (aq) + \ldots\ldots \rightarrow 2H_2O\ (l)$

7 Oxidation numbers in a formula add up to zero.
a Give the oxidation state of the underlined atom in each formula below:
  i aluminium oxide, $\underline{Al}_2O_3$
  ii ammonia, $\underline{N}H_3$
  iii $H_2\underline{C}O_3\ (aq)$, carbonic acid
  iv phosphorus trichloride, $\underline{P}Cl_3$
  v copper(I) chloride, $\underline{Cu}Cl$
  vi copper(II) chloride, $\underline{Cu}Cl_2$
b Now comment on the compounds in v and vi.

8 The oxidising agent potassium manganate(VII) can be used to analyse the % of iron(II) present in iron tablets. Below is an **ionic equation**, showing the ions that take part in the reaction:
$MnO_4^-\ (aq) + 8H^+\ (aq) + 5Fe^{2+}\ (aq) \rightarrow Mn^{2+}\ (aq) + 5Fe^{3+}\ (aq) + 4H_2O\ (l)$
a What does the H^+ in the equation tell you about this reaction? (Hint: check page 130.)
b Describe the colour change.
c Which is the reducing reagent in this reaction?
d How could you tell when all the iron(II) had reacted?
e Write the half-equation for the iron(II) ions.

9 Potassium chromate(VI) is yellow. In acid it forms orange potassium dichromate(VI). These are the ions that give those colours:

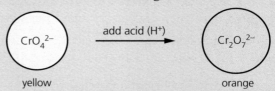

yellow   orange

a What is the oxidation number of chromium in:
  i the yellow compound, K_2CrO_4?
  ii the orange compound, $K_2Cr_2O_7$?
b This reaction of chromium ions is not a redox reaction. Explain why.

10 When solutions of silver nitrate and potassium chloride are mixed, a white precipitate forms.
The ionic equation for the reaction is:
$Ag^+\ (aq) + Cl^-\ (aq) \rightarrow AgCl\ (s)$
a i What is the name of the white precipitate?
  ii Is it a soluble or insoluble compound?
b Is the precipitation of silver chloride a redox reaction or not? Explain your answer.
c When left in light, silver chloride decomposes to form silver and chlorine gas.
Write an equation for the reaction and show clearly that this is a redox reaction.

7.1 Conductors and non-conductors

Objectives: define *a conductor*; understand an electric current as a flow of charged particles; recall which groups of substances conduct, which do not, and why

Batteries and electric current

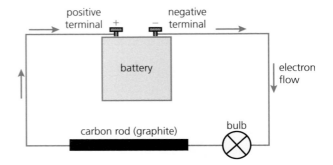

The photograph above shows a battery, a bulb and a carbon rod joined or **connected** to each other by copper wires. This arrangement is called an **electric circuit**.

The bulb is lit. So electricity must be flowing in the circuit. **Electricity is a stream of electrons.**

Now look at the diagram. The battery acts like an electron pump. Electrons leave it through the **negative terminal**. They travel around the circuit, and enter the battery again through the **positive terminal**.

When the electrons stream through the fine wire in the bulb, they cause it to heat up. It gets white-hot and gives out light.

Conductors and non-conductors

In the circuit above, the rod of carbon (graphite) and the copper wire allow electricity to pass through. So they are called **conductors**.

But if the rod is made of plastic or ceramic, the bulb will not light. Plastic and ceramic do not let electricity pass through them. They are **non-conductors** or **insulators**.

▲ The current is carried into the styling wand through copper wire. Then it flows through wire made of nichrome (a nickel-chromium alloy) which heats up. Meanwhile, the plastic protects you.

Some everyday uses of conductors and non-conductors

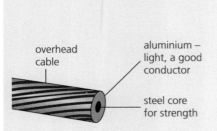
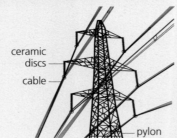
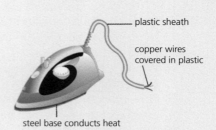

Cables that carry electricity around the country are made of aluminium and steel. Both are conductors. (Aluminium is a better conductor than steel.)

At pylons, ceramic discs support the bare cables. Ceramic is a non-conductor, and prevents the current from running down the pylon. (That would be deadly!)

Copper is used for wiring, at home. It is a very good conductor. The wire sheaths and plug cases are made of plastic, a non-conductor, for safety.

82

ELECTRICITY AND CHEMICAL CHANGE

Testing substances to see if they conduct

You can test any substance to see if it conducts, by connecting it into a circuit like the one on page 82. For example:

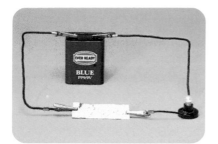  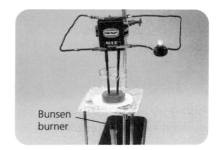

Tin. A strip of tin is connected into the circuit, in place of the carbon rod. The bulb lights, so tin must be a conductor.

Ethanol. The liquid is connected into the circuit by placing carbon rods in it. The bulb does not light, so ethanol is a non-conductor.

Lead(II) bromide. It does not conduct when solid. But if you melt it, it conducts, and gives off a choking brown vapour.

The results These are the results from a range of tests:

1 **The only solids that conduct are the metals and carbon (graphite).**
   These conduct because of their free electrons (pages 41 and 42). The electrons get pumped out of one end of the solid by the battery, while more electrons flow in the other end.

   For the same reason, *molten* metals conduct. (It is hard to test molten graphite, because at room pressure graphite evaporates on heating.)

2 **Molecular substances are non-conductors.**
   This is because they contain no free electrons, or other charged particles, that can flow through them.

   Ethanol (above) is molecular. So are petrol, paraffin, sulfur, sugar, and plastic. These never conduct, whether solid or molten.

3 **Ionic compounds don't conduct when solid, but do conduct when melted or dissolved in water.** *They break down at the same time.*
   An ionic compound contains no free electrons. But it does contain **ions**, which have a charge. They are free to move when the substance is melted or dissolved, so they can carry the current.

   Lead(II) bromide above is an ionic compound. It conducts when it is melted. The vapour that forms is bromine. So the molten compound is broken down or **decomposed** by electricity, to lead and bromine.

   This means that electricity can be used on purpose, to bring about the decomposition of an ionic compound.

   Find out more in the next unit.

▲ Metals conduct, thanks to their free electrons, which form a current.

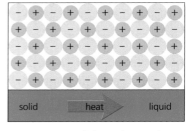

▲ An ionic solid conducts when it melts, because the ions become free to move.

1 What is a *conductor* of electricity?
2 Draw a circuit you could use to see if mercury conducts.
3 Explain why metals are able to conduct electricity.
4 Table salt is sodium chloride. Suggest a way to test whether dry table salt conducts electricity, and predict the outcome.
5 What does *decomposition* mean? (Glossary?)
6 Predict what will happen when electricity is passed through molten sodium chloride.
7 Vegetable oil does not conduct electricity. What can you deduce from this?

7.2 The principles of electrolysis

Objectives: define *electrolysis*; outline the electrolysis of molten lead(II) bromide; predict the products of electrolysis for ionic compounds (molten *and* in solution)

Electrolysis: breaking down by electricity

On page 83 you saw that molten lead(II) bromide breaks down when a current passes through it. So you can make use of electricity as a way to break down compounds. The process is called **electrolysis**.

Electrolysis is the decomposition of an ionic compound, when molten or in aqueous solution, by the passage of an electric current.

The electrolysis of molten lead(II) bromide

The apparatus is shown on the right. It is called an **electrolytic cell**.

- The carbon rods are called **electrodes**.
- The positive electrode (+) is connected to the positive terminal of the battery. It is called the **anode**. The negative electrode (−) is called the **cathode**.
- The liquid that undergoes electrolysis is called the **electrolyte**.
- When the switch is closed, the current flows from the battery. Brown bromine gas bubbles off at the anode. A silvery bead of lead forms below the cathode. The lead(II) bromide is decomposing:

  lead(II) bromide ⟶ lead + bromine
  $PbBr_2 \, (l) \longrightarrow Pb \, (l) + Br_2 \, (g)$

- The carbon electrodes are **inert**: they carry the current, but are otherwise unchanged. (Platinum electrodes would be inert too.)

The electrolysis of other molten ionic compounds

The pattern is the same for all molten ionic compounds of two elements: **Electrolysis decomposes the molten compound down to its elements, giving the metal at the cathode, and the non-metal at the anode.**

Why does decomposition occur?

Lead(II) bromide contains lead ions (Pb^{2+}) and bromide ions (Br^-). Follow the numbers to see what happens when the switch is closed:

5 Electrons flow along the wire from the anode to the positive terminal of the battery.

4 At the anode (+), the Br^- ions give up electrons. Brown bromine vapour bubbles off.

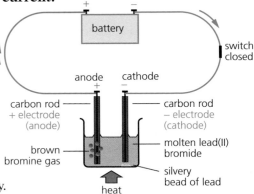

!
Which electrode is positive?
Remember **PA**! (**P**ositive **A**node).

Opposites attract!
During electrolysis:
- positive ions go to the negative electrode (*cations to cathode*)
- negative ions go to the positive electrode (*anions to anode*)

1 **Electrons** flow along the wire from the negative terminal of the battery to the cathode.

2 In the liquid, the **ions** carry the current. They move to the electrode of opposite charge.

3 At the cathode (−), the Pb^{2+} ions accept electrons. A silvery bead of lead appears below the cathode.

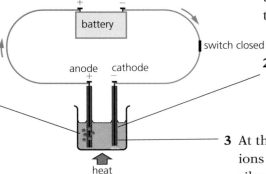

So electrons carry the current through the wires and electrodes. But the ions carry it through the liquid.

ELECTRICITY AND CHEMICAL CHANGE

The electrolysis of aqueous solutions
Electrolysis can also be carried out on a **solution** of an ionic compound in water, because the ions in the solutions are free to move. But the result may be different than for the molten compound. Compare these:

Electrolyte	Observations at the cathode (−) ...	Observations at the anode (+) ...
molten sodium chloride	silvery beads form (sodium)	bubbles of a pale yellow-green gas form (chlorine)
a concentrated solution of sodium chloride	bubbles of a colourless gas form (hydrogen)	bubbles of a pale yellow-green gas form (chlorine)

Why the difference? *Because the water itself produces ions.* Although water is molecular, a tiny % of its molecules are split into ions:

some water molecules ⟶ hydrogen ions + hydroxide ions
$$H_2O\,(l) \longrightarrow H^+\,(aq) + OH^-\,(aq)$$

These ions also take part in the electrolysis, so the products may change.

The rules for electrolysis of a solution
At the cathode (−), either a metal or hydrogen forms.

1. The more reactive an element is, the more it 'likes' to exist as ions. So if a metal is more reactive than hydrogen, its ions stay in solution and hydrogen bubbles off. (Look at the list on the right.)
2. But if the metal is less reactive than hydrogen, the metal forms.

At the anode (+), a non-metal other than hydrogen forms.

1. If it is a *concentrated* solution of a **halide** (a compound containing Cl^-, Br^- or I^- ions), then chlorine, bromine, or iodine form.
2. But if the halide solution is *dilute*, or there is no halide, oxygen forms.

Order of reactivity
potassium
sodium
calcium
magnesium
aluminium
zinc
iron
lead
hydrogen
copper
silver

increasing reactivity

Look at these examples. Do they follow the rules?

Electrolyte	At the cathode (−) you get...	At the anode (+) you get...
dilute sulfuric acid, H_2SO_4	hydrogen	oxygen
a dilute solution of sodium chloride, NaCl	hydrogen	oxygen
a concentrated solution of potassium bromide, KBr	hydrogen	bromine
a dilute solution of potassium bromide, KBr	hydrogen	oxygen
concentrated hydrochloric acid, HCl	hydrogen (H^+ is the only positive ion present)	chlorine

Look again at the three dilute solutions, including dilute sulfuric acid.
In each case the *water* has decomposed!

1. Define: **a** electrolysis **b** electrolyte
2. **a** Which type of compounds can be electrolysed? Why?
   b What form must they be in? Why?
3. **a** What is the purpose of the electrodes, on page 84?
   b What term is used for the negative electrode?
4. The electrodes are *inert*. What does that mean? (Glossary?)
5. What does electrolysis of these molten compounds give?
   a sodium chloride, NaCl **b** lead sulfide, PbS
6. Predict what you will observe, and name the product at each electrode, when this solution undergoes electrolysis:
   a concentrated aqueous sodium chloride
   a dilute sulfuric acid
7. Repeat question **6** for these solutions:
   a a concentrated solution of magnesium chloride, $MgCl_2$
   b dilute aqueous iron(II) bromide, $FeBr_2$
   c a dilute solution of copper(II) sulfate, $CuSO_4$

7.3 The reactions at the electrodes

Objectives: describe how ions move, compete, and react, during electrolysis; explain electrolysis as a redox reaction; write half-equations for electrode reactions

What happens to ions in the molten lead(II) bromide?

In molten lead(II) bromide, the ions are free to move. This shows what happens to them, when the switch in the circuit is closed:

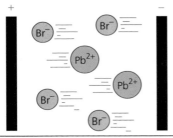 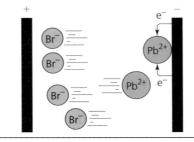 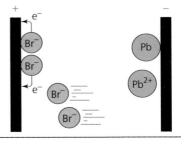

First, the ions move. Opposite charges attract. So the positive lead ions (Pb^{2+}) move to the cathode ($-$). The negative bromide ions (Br^-) move to the anode ($+$). The moving ions carry the current.	**At the cathode ($-$):** the lead ions each receive two electrons and become lead atoms. The **half-equation** is: $Pb^{2+}(l) + 2e^- \longrightarrow Pb(l)$ Lead collects on the electrode and eventually drops off it.	**At the anode ($+$):** the bromide ions each give up an electron, and become atoms. These then pair up to form molecules. The **half-equation** is: $2Br^-(l) \longrightarrow Br_2(g) + 2e^-$ The bromine gas bubbles off.
The free ions move.	Ions gain electrons: reduction.	Ions lose electrons: oxidation.
Remember **OILRIG**: **O**xidation **I**s **L**oss of electrons, **R**eduction **I**s **G**ain of electrons.	Overall, electrolysis is a redox reaction. Reduction takes place at the cathode and oxidation at the anode.	

The reactions for other molten compounds follow the same pattern.

> **Remember RAC!**
> **R**eduction **A**t **C**athode.

For a concentrated solution of sodium chloride

This time, ions from water are also present:

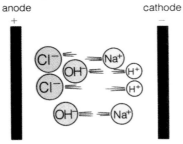 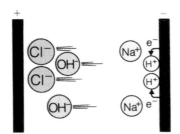 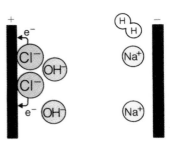

The solution contains Na^+ ions and Cl^- ions from the salt, and H^+ and OH^- ions from water. The positive ions go to the cathode and the negative ions to the anode.

At the cathode, the H^+ ions accept electrons, since hydrogen is less reactive than sodium:
$2H^+(aq) + 2e^- \longrightarrow H_2(g)$
The hydrogen gas bubbles off.

At the anode, the Cl^- ions give up electrons more readily than the OH^- ions do.
$2Cl^-(aq) \longrightarrow Cl_2(aq) + 2e^-$
The chlorine gas bubbles off.

The Na^+ and OH^- ions remain, giving a solution of sodium hydroxide.

For a dilute solution of sodium chloride

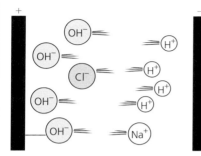
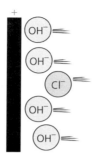
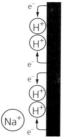
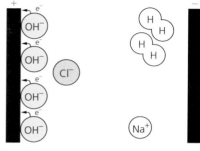

The same ions are present as before. But now the proportion of Na$^+$ and Cl$^-$ ions is lower, since this is a dilute solution. So the result will be different.

At the cathode, hydrogen 'wins' as before, and bubbles off:

$$4H^+ (aq) + 4e^- \longrightarrow 2H_2 (g)$$

(4 electrons are shown, to balance the half-equation at the anode.)

At the anode, OH$^-$ ions give up electrons, since not many Cl$^-$ ions are present. Oxygen bubbles off:

$$4OH^- (aq) \longrightarrow$$
$$O_2 (g) + 2H_2O (l) + 4e^-$$

When the hydrogen and oxygen bubble off, the Na$^+$ and Cl$^-$ ions remain. **So we *still* have a solution of sodium chloride. The overall result is that water has decomposed.**

Half-equations for electrode reactions

A **half-equation** shows the electron transfer at an electrode. This table shows the steps in writing half-equations.

The steps	Example: the electrolysis of molten magnesium chloride
1 First, name the ions present, and the products.	Magnesium ions and chloride ions are present. Magnesium and chlorine form.
2 Write each half-equation correctly. • Give the ion its correct charge. • Remember, positive ions go to the cathode, and negative ions to the anode. • Write the correct symbol for the element that forms. For example, Cl$_2$ for chlorine (not Cl). • The number of electrons in the equation should be the same as the total charge on the ion(s) in it.	Ions: Mg^{2+} and Cl$^-$ **Reduction at the cathode:** Mg^{2+} + 2e$^-$ $\longrightarrow$ Mg **Oxidation at the anode:** 2Cl$^-$ $\longrightarrow$ Cl$_2$ + 2e$^-$ (two Cl$^-$ ions, so a total charge of 2–) Note that it is also correct to write the anode reaction as: 2Cl$^-$ – 2e$^-$ $\longrightarrow$ Cl$_2$
3 You could then add the state symbols.	Mg^{2+} (l) + 2e$^-$ $\longrightarrow$ Mg (s) 2Cl$^-$ (l) $\longrightarrow$ Cl$_2$ (g) + 2e$^-$

1. At which electrode does reduction always take place? Explain why.
2. Give the half-equation for the reaction at the anode during the electrolysis of these molten compounds:
   a potassium chloride   b calcium oxide
3. Hydrogen rather than sodium forms at the cathode, when electrolysis is carried out on any solution of sodium chloride.
   a Explain why. b Where does the hydrogen come from?
4. Compare the solutions left behind, after the electrolysis of concentrated and dilute solutions of sodium chloride.
5. Predict the products for the electrolysis of a concentrated solution of copper(II) chloride, CuCl$_2$.
6. Give the two half-equations for the electrolysis of:
   a a concentrated solution of hydrochloric acid, HCl
   b a dilute solution of sodium hydroxide, NaOH
   c a dilute solution of copper(II) chloride, CuCl$_2$

7.4 Electroplating

Objectives: describe the electrolysis of copper(II) sulfate solution using copper electrodes; describe the process of electroplating and give examples of its uses

When electrodes are not inert
A solution of copper(II) sulfate contains blue Cu^{2+} ions, SO_4^{2-} ions, and H^+ and OH^- ions from water. Electrolysis of the solution will give different results, *depending on the electrodes*. Compare these:

A Using carbon electrodes (inert)

At the cathode Copper ions are **discharged** (become atoms):

$2Cu^{2+} (aq) + 4e^- \longrightarrow 2Cu (s)$

The copper coats the electrode.

At the anode Oxygen bubbles off:

$4OH^- (aq) \longrightarrow 2H_2O (l) + O_2 (g) + 4e^-$

So copper and oxygen are produced. This fits the rules on page 85. The blue colour of the solution fades because the concentration of copper ions in it decreases as the ions are discharged.

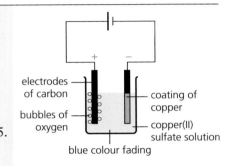

B Using copper electrodes (not inert)

At the cathode Again, copper is formed, and coats the electrode:

$Cu^{2+} (aq) + 2e^- \longrightarrow Cu (s)$

At the anode The anode dissolves, giving copper ions in solution:

$Cu (s) \longrightarrow Cu^{2+} (aq) + 2e^-$

As the anode dissolves, the copper ions move to the cathode, to form copper. **So copper moves from the anode to the cathode.** The solution stays blue because the concentration of copper ions in it does not change.

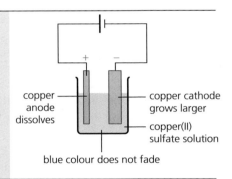

The idea in **B** leads to an important use of electrolysis: **electroplating**.

Metal refining
B leads to another important process too: **metal refining**. *Refining* means removing impurities. Copper needs a very high level of purity for use in mobile phones. This shows the refining process:

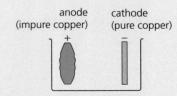

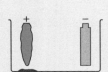

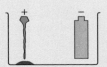

The anode is made of impure copper. The cathode is pure copper. The electrolyte is dilute copper(II) sulfate solution.

The copper in the anode dissolves. But the impurities do not dissolve. They drop to the floor of the cell as a sludge.

A layer of pure copper builds up on the cathode. When the anode is almost gone, the anode and cathode are replaced.

The copper deposited on the cathode is over 99.9% pure. The sludge may contain valuable metals such as platinum, gold, silver, and selenium. These are recovered and sold.

ELECTRICITY AND CHEMICAL CHANGE

▲ A steel tap plated with chromium, to make it look bright and shiny. Chromium does not stick well to steel. So the steel is first electroplated with copper or nickel, and then chromium.

▲ Shiny silvery features on cars are often **plastic**, electroplated with nickel! Plastic is a non-conductor. So first a thin coat of copper is deposited on the plastic, in a chemical bath.

Electroplating

Electroplating means using electricity to coat one metal with another, to make it look better, or to prevent corrosion. Steel taps that are coated with chromium look good, and will not rust. Steel cans are coated with tin to make food cans. Cheap metal jewellery is often coated with silver.

The drawing on the right shows how to electroplate a steel jug with silver. The jug is used as the cathode. The anode is made of silver. The electrolyte is a solution of a soluble silver compound – in this case silver nitrate.

At the anode   The silver dissolves, forming silver ions in solution:

$$Ag\,(s) \longrightarrow Ag^+\,(aq) + e^-$$

At the cathode   The silver ions are attracted to the cathode. There they receive electrons, forming a coat of silver on the jug:

$$Ag^+\,(aq) + e^- \longrightarrow Ag\,(s)$$

When the layer of silver is thick enough, the jug is removed.

▲ Silverplating: electroplating with silver. When the electrodes are connected to a power source, electroplating begins.

Preparing to electroplate

In general, to electroplate an object with metal **X**, the set-up is:
cathode   – object to be electroplated
anode   – metal **X**
electrolyte – a solution of a soluble compound of **X**.

1 Copper(II) ions are blue. When electrolysis is carried out on blue copper(II) sulfate solution, the blue solution:
  a loses its colour when carbon electrodes are used
  b keeps its colour when copper electrodes are used
  Explain each of these observations.
2 *The copper electrodes are not inert, in the electrolysis in B on page 88*. Give evidence to support this statement.
3 What does *electroplating* mean?
4 Give two reasons why electroplating is carried out.
5 The steel used for cans of soup is coated with tin. Why?
6 You plan to electroplate steel cutlery with nickel.
  a What will you use as the anode?
  b What will you use as the cathode?
  c Suggest a suitable electrolyte.

Checkup on Chapter 7

Revision checklist
Core syllabus content
Make sure you can ...
- [ ] define the terms *conductor* and *non-conductor*
- [ ] explain what these terms mean:
  electrolysis   *electrolyte*   *electrode*
  inert electrode   *anode*   *cathode*
- [ ] explain why an ionic compound must be melted, or dissolved in water, for electrolysis
- [ ] outline what happens to an electrolyte, during electrolysis
- [ ] predict what will be observed at each electrode, in the electrolysis of a molten ionic compound
- [ ] explain why the products of electrolysis may be different, when a compound is dissolved in water rather than melted
- [ ] define *halide*
- [ ] give the general rules for the products at the anode and cathode, in the electrolysis of a solution
- [ ] identify the products, and describe what is observed at each electrode, in the electrolysis of:
  - molten lead(II) bromide
  - a concentrated solution of sodium chloride
  - dilute sulfuric acid
- [ ] explain what *electroplating* is, and give two reasons for carrying out this process
- [ ] state which will be the anode, and suggest a suitable electrolyte, for electroplating object X with metal Y

Extended syllabus content
Make sure you can also ...
- [ ] predict the products, for the electrolysis of halides in dilute and concentrated solutions
- [ ] describe the reactions at the electrodes for the electrolyses you met in this chapter, and write half-equations for the electrode reactions
- [ ] describe the differences, when the electrolysis of copper(II) sulfate is carried out:
  - using inert electrodes (carbon or platinum)
  - using copper electrodes

Questions
Core syllabus content

1 Electrolysis of molten lead bromide is carried out:

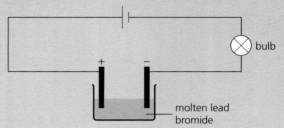

a The bulb will not light until the lead bromide has melted. Why not?
b What will be *seen* at the anode (+)?
c Name the substance that forms at the anode.
d What will form at the cathode (−)?

2 Six substances A to F were dissolved in water, and connected in turn into the circuit below. A represents an ammeter, which is used to measure current. The table shows the results.

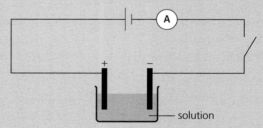

Substance	Current (amperes)	At cathode (−)	At anode (+)
A	0.8	copper	chlorine
B	1.0	hydrogen	chlorine
C	0.0	—	—
D	0.8	hydrogen	chlorine
E	1.2	hydrogen	oxygen
F	0.7	silver	oxygen

a Which solution conducts best?
b Which solution is a non-electrolyte?
c Which solution could be:
  i silver nitrate?   ii copper(II) chloride?
  iii sugar?   iv dilute sulfuric acid?
d i Two of the solutions give the same products at the electrodes. Which two?
  ii Name two chemicals which would give those products, when connected into the circuit as concentrated solutions.

3 The electrolysis below produces two gases, A and B.

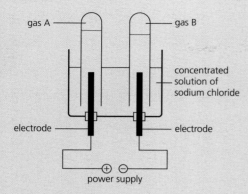

  a Why does the solution conduct electricity?
  b **i** The apparatus uses inert electrodes. Why are they called *inert*?
  ii Name the two elements that can be used to make inert electrodes.
  c Does gas A form at the anode, or the cathode?
  d What would you observe at each electrode, during the electrolysis?
  e Identify each gas A and B, and describe the colour that would be seen in the test-tubes.
  f How would the products differ if molten sodium chloride was electrolysed?

Extended syllabus content

4 a List the ions that are present in concentrated solutions of:
  i sodium chloride   **ii** copper(II) chloride
  b Explain why and how the ions move, when each solution is electrolysed using platinum electrodes.
  c Write the half-equation for the reaction at:
  i the anode   **ii** the cathode
  during the electrolysis of each solution.
  d Explain why the anode reactions for both solutions are the same.
  e **i** The anode reactions will be different if the solutions are made very dilute. Explain why.
  ii Write the half-equations for the anode reactions when very dilute solutions are used.
  f Explain why copper is obtained at the cathode, but sodium is not.
  g Name another solution that will give the same products as the concentrated solution of sodium chloride does, on electrolysis.
  h Which solution in **a** could be the electrolyte in an electroplating experiment?

5 Molten lithium chloride contains lithium ions (Li^+) and chloride ions (Cl^-).
  a Copy the following diagram and use arrows to show which way:
  i the ions move when the switch is closed
  ii the electrons flow in the wires

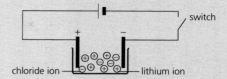

  b **i** Write equations for the reaction at each electrode, and the overall reaction.
  ii Describe each of the reactions using the terms *reduction*, *oxidation* and *redox*.

6 This question is about the electrolysis of a dilute aqueous solution of lithium chloride.
  a Give the names and symbols of the ions present.
  b Say what will be formed, and write a half-equation for the reaction:
  i at the anode   **ii** at the cathode
  c Name another compound that will give the same products at the electrodes.
  d How will the products change, if a concentrated solution of lithium chloride is used?

7 An experiment is needed, to see if an object made of iron can be electroplated with chromium.
  a Suggest a solution to use as the electrolyte.
  b **i** Draw a labelled diagram of the apparatus that could be used for the electroplating.
  ii Show how the *electrons* will travel from one electrode to the other.
  c Write half-equations for the reactions at each electrode.
  d At which electrode does oxidation take place?
  e The concentration of the solution does not change. Why not?

8 Nickel(II) sulfate ($NiSO_4$) is green. A solution of this salt is electrolysed using nickel electrodes.
  a Write a half-equation for the reaction at each electrode.
  b At which electrode does reduction take place? Explain your answer
  c What happens to the size of the anode?
  d The colour of the solution does not change, during the electrolysis. Explain why.
  e Suggest one industrial use for this electrolysis.

8.1 Energy changes in reactions

Objectives: define *exothermic* and *endothermic* reactions; explain *surroundings*; describe how these reactions affect the temperature of their surroundings

Energy changes in reactions
During a chemical reaction, there is *always* an energy change.
Energy is given out or taken in. It is usually in the form of heat, or **thermal energy**. So reactions can be divided into two groups: **exothermic** and **endothermic**.

Exothermic reactions
Exothermic reactions give out thermal energy to their surroundings. So the temperature of the surroundings rises. Look at these examples:

About energy
- Energy takes many forms, including heat, light, sound, chemical energy, electrical energy ... and motion.
- It cannot be created, nor destroyed.
- But it can change from one form to another.

Here iron is reacting with sulfur to give iron(II) sulfide. The test-tube was heated a little, just to start the reaction. Now the mixture is glowing red hot, *all by itself!*

Here a solution of sodium chloride is being added to a solution of silver nitrate. A white precipitate of silver chloride forms – and the water in the beaker gets warmer.

Here calcium oxide is reacting with water to give calcium hydroxide. The temperature of the liquid is being measured. It rises sharply.

The **reactants** in each case are the particles (atoms, molecules, or ions) that react together. The **products** are the particles the reaction produces. The **surroundings** are everything else. For example, the walls of the three containers, the water in **B**, the excess water in **C**, and the air.

In each case, the reaction between the particles gives out thermal energy. So the temperature of the surroundings rises.

Chemical energy
- All chemicals have chemical energy.
- It is stored in their bonds.

Describing exothermic reactions
The reactions above can be described like this:

reactants ⟶ products + thermal energy

The reactants and products have energy. It is stored as chemical energy in their bonds. The *total* energy does not change, during a reaction. It is the same on each side of the arrow. So it follows that the products must have less energy than the reactants do.

In an exothermic reaction, the products have less energy than the reactants do. Energy has been given out.

This is shown on the **energy level diagram** on the right. The arrow represents the energy given out to the surroundings, during the reaction.

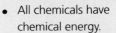

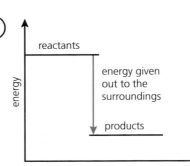

▲ An energy level diagram for an exothermic reaction. The products have less energy than the reactants do.

ENERGY CHANGES IN REACTIONS

Endothermic reactions
Endothermic reactions take in thermal energy from their surroundings.
So the temperature of the surroundings falls. Look at these examples:

The solids barium hydroxide and ammonium chloride react when they are mixed. The reaction takes in so much thermal energy that water *under* the beaker freezes!

The sweet powder called sherbet is a mixture of citric acid and sodium hydrogen carbonate. These react on your tongue, taking in thermal energy – so your tongue cools.

In the crucible, calcium carbonate is breaking down to calcium oxide and carbon dioxide. The reaction needs thermal energy from the flame of the Bunsen burner.

In each case, thermal energy is transferred from the surroundings: in **D** from the air and wet wood, in **E** from the boy's tongue, and in **F** from the burning gas.

So in **D** and **E**, the temperature of the surroundings falls. But the crucible and other surroundings in **F** stay hot because the gas is still burning.

Describing endothermic reactions
The reactions above can be described like this:

reactants + thermal energy ⟶ products

Once again, the *total* energy does not change, during the reaction. It is the same on each side of the arrow. So it follows that the products must have more energy than the reactants do.

In an endothermic reaction, the products have more energy than the reactants do. Energy has been taken in.

The energy level diagram for endothermic reactions is shown on the right. Compare it with diagram ①. What differences can you see?

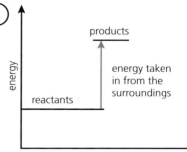

▲ An energy level diagram for an endothermic reaction. The products have more energy than the reactants do.

Some more examples
- *Exothermic*: the burning of coal and other fuels; sodium reacting with water
- *Endothermic*: the reactions that take place in coooking

Remember!
Exo means *out* (think of **Exit**).
Endo means *in*.

1 Give another term for *thermal energy*.
2 Define:
   a an exothermic reaction   b an endothermic reaction
3 In reaction **B**, the water in the beaker gets warmer. Why?
4 In **D**, the water under the beaker freezes. Why?
5 Draw and label an energy level diagram for:
   a an endothermic reaction   b an exothermic reaction
6 All chemicals contain chemical energy. Where is it stored?
7 In *B*, some chemical energy is being changed into thermal energy. Do you think this statement is correct? Explain.

8.2 A closer look at energy changes

Objectives: interpret reaction pathway diagrams; state that bond breaking is endothermic, and bond making is exothermic; define E_a and ΔH for a reaction

Reaction pathway diagrams

In Unit 8.1 you met energy level diagrams. These show that reactants and products are at different energy levels.

A diagram to show how the energy changes *during* a reaction is a bit different. It is called a **reaction pathway diagram**.

Below are two reaction pathway diagrams. ①is for an exothermic reation. ②is for an endothermic reaction.

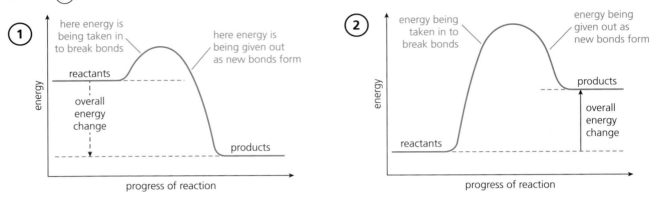

Note how the energy level changes during the reactions. The line first curves upwards – and then down. Let's see why, using the exothermic reaction between hydrogen and oxygen as example:

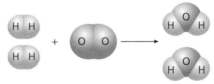

- For reaction to take place, the bonds in the hydrogen and oxygen molecules have to be broken. Energy must be taken in for this – so the energy level rises. The line curves upwards.

- Then new bonds are made, to form the water molecules. This gives out energy. So the energy level falls. The line slopes down.

More on making and breaking bonds

As you saw above, energy must be taken in to break bonds.
Bond breaking is an endothermic process.

Then energy is given out when new bonds are made.
Bond making is an exothermic process.

In ① above, the energy taken in is *less than* the energy given out. So this reaction is exothermic overall.

But in ②, the energy taken in is *more than* the energy given out. So this reaction is endothermic overall.

▲ The reaction between hydrogen and oxygen is so highly exothermic that it is used to fuel space rockets. Superheated steam explodes from the engine, pushing the rocket upwards.

ENERGY CHANGES IN REACTIONS

Activation energy, and enthalpy

Let's introduce two new terms.

Activation energy Before bonds are broken, particles must first collide with each other. That makes sense. But they must also collide *with sufficient energy*. If there is insufficient energy, there will be no reaction!

The activation energy is the minimum energy that colliding particles must have, in order to react. Its symbol is E_a.
Look how it is shown on the reaction pathway diagrams ③ and ④.

Enthalpy change of the reaction This is the name given to the *overall* energy change during a reaction. Its symbol is ΔH. (Say *delta H*.)
It is negative for an exothermic reaction, and positive for an endothermic reaction. Look how it is shown on these diagrams:

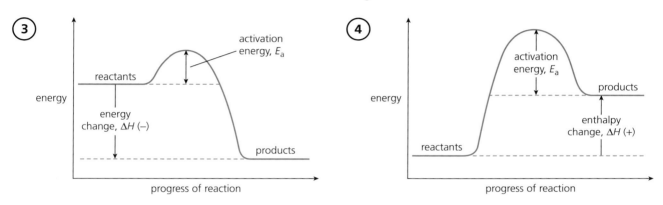

Where does the activation energy come from?

Activation energy is supplied in different ways.

- For some reactions, the energy in the surroundings is sufficient. Just mix the reactants without heating. See reactions **B** and **C** on page 92.
- Some exothermic reactions need heat from a Bunsen burner to start bonds breaking. The energy given out when new bonds form will then break further bonds. For example, reaction **A** on page 92.
- You may not even need a Bunsen. A single spark from a motor could be enough to cause a mixture of hydrogen and oxygen to react explosively.
- But for endothermic reactions like **F** on page 93, which has a high activation energy, you must keep heating until the reaction is complete.

▲ One way to start bonds breaking!

1. What does *a reaction pathway diagram* show?
2. For a reaction to take place, two steps must occur. What are they? (Your answer should include the word *bonds!*)
3. a The line always curves upwards first, in a reaction pathway diagram. Explain why.
   b What else is common to all reaction pathway diagrams?
4. Explain why reaction ④ is endothermic overall, in terms of bond making and bond breaking.
5. Define, and give the symbol for:
   a activation energy   b enthalpy change in a reaction
6. Why is activation energy needed, in a reaction?
7. In a certain reaction, the bonds between atoms within the reactant molecules are weak. Is the activation energy for this reaction likely to be *high*, or *low*?
8. Draw a labelled reaction pathway diagram for either an endothermic OR an exothermic reaction. Include ΔH and E_a.

8.3 Calculating enthalpy changes

Objectives: recall the unit used for energy changes; define *bond energy*; calculate ΔH for a reaction, using the equation for the reaction and bond energy values

The key ideas for calculating enthalpy change
In a chemical reaction:
- the bonds in the reactant particles must be broken; this takes in energy.
- new bonds are made, giving the product particles; this releases energy.
- the overall change in enthalpy of the reaction is found like this:

| enthalpy change for the reaction, ΔH | = | energy in for bond breaking | − | energy out from bond making |

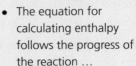

First, energy in ...
- The equation for calculating enthalpy follows the progress of the reaction ...
- So 'energy in' comes first!

The unit of energy
The unit of energy is the **joule**, symbol **J**.
1000 joules = 1 **kilojoule** or 1 **kJ**.
Energy changes in reactions are given as **kilojoules per mole**, or **kJ/mol**.

Bond energy
The energy needed to break bonds *is the same as* the energy given out when these bonds form. It is called the **bond energy**.

The bond energy is the energy needed to break bonds, or released when the same bonds form. It is given in kJ/mol.

It has been worked out by experiment, for all bonds. Look at the list of bond energy values on the right.

242 kJ of energy must be supplied to break the bonds in a mole of chlorine molecules, to give chlorine atoms. If these atoms join again to form chlorine molecules, 242 kJ of energy will be given out.

Bond energy (kJ/mol)

Bond	Energy
H−H	436
Cl−Cl	242
H−Cl	431
C−C	346
C=C	612
C−O	358
C−H	413
O=O	498
O−H	464
N−H	391
N≡N	946

▲ The triple bond in a molecule of nitrogen (N_2) has a high bond energy. (Look at the list above.) So the bond is hard to break. This makes nitrogen unreactive or **inert**. So it is pumped into the packaging for snacks and other foods, to help keep the food fresh.

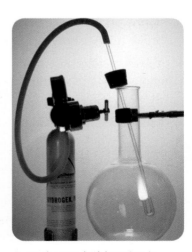

▲ Hydrogen (H_2) burning in chlorine (Cl_2). Which are harder to break, the bonds in hydrogen or in chlorine? (Check the list!)

ENERGY CHANGES IN REACTIONS

Calculating the enthalpy changes in reactions

Let's look at two examples. The bond energies are taken from the list on page 96.

1 The exothermic reaction between hydrogen and chlorine

Hydrogen reacts with chlorine in sunshine, to form hydrogen chloride:
$H_2 + Cl_2 \longrightarrow 2HCl$

Write the equation like this, to show all the bonds:
H—H + Cl—Cl ⟶ 2 H—Cl

Energy in to break each mole of bonds in the reactants:
1 × H—H	436 kJ
1 × Cl—Cl	242 kJ
Total energy in	678 kJ

Energy out from the two moles of bonds that form to give the product:
2 × H—Cl   2 × 431 = 862 kJ

Enthalpy change (ΔH) = energy in − energy out
= 678 kJ − 862 kJ = **−184 kJ**

The − sign shows that it is an exothermic reaction. So 184 kJ of energy is given out when 1 mole of hydrogen and 1 mole of chlorine react

▲ Energy diagram showing the enthalpy change for the exothermic reaction between hydrogen and chlorine.

2 The endothermic decomposition of ammonia

The decomposition of ammonia is carried out by heating:
$2NH_3 \longrightarrow N_2 + 3H_2$

Write the equation like this, to show all the bonds:

2 H—N—H ⟶ N≡N + 3 H—H
   (with H above and below N)

Energy in to break the two moles of bonds:
6 × N—H   6 × 391 = 2346 kJ

Energy out from the four moles of bonds forming:
1 × N≡N		946 kJ
3 × H—H	3 × 436 =	1308 kJ
Total energy out		2254 kJ

Enthalpy change (ΔH) = energy in − energy out
= 2346 kJ − 2254 kJ = **+92 kJ**

The + sign shows that this is an endothermic reaction. So 92 kJ of energy is taken in to decompose 2 moles of ammonia.

▲ Energy diagram showing the enthalpy change for the decomposition of ammonia – an endothermic reaction.

1. Write out a general equation you can use to calculate the enthalpy change in a reaction.
2. The energy needed to break the O=O bonds in oxygen, O_2, is 498 kJ/mol. What does kJ stand for?
3. Hydrogen burns in oxygen, and is used as a fuel. Give the equation for its reaction with oxygen. Then calculate the energy required to break the bonds in the reaction.
4. Ethene reacts with hydrogen like this:

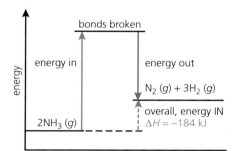

   a List the bond energies for the bonds that break and form.
   b Calculate the energy change for this reaction.
5. Nitrogen (N_2) is unreactive. Use the idea of bond energy to suggest a reason. (Its molecules have the N≡N bond.)

97

8.4 The hydrogen-oxygen fuel cell

Objectives: state that hydrogen and oxygen react in this cell, producing water and a current; give its advantages and disadvantages, compared with the petrol engine

The reaction between hydrogen and oxygen

When hydrogen burns in oxygen, the reaction gives out a great deal of **thermal energy**:

$$2H_2(g) + O_2(g) \rightarrow 2H_2O(l) + \text{thermal energy}$$

As you saw on page 94, this reaction is used to power rockets.

But the reaction can be carried out in a way that produces **electrical energy** instead! This time the equation is:

$$2H_2(g) + O_2(g) \rightarrow 2H_2O(l) + \text{electrical energy} + \text{some thermal energy}$$

The reaction takes place in a **hydrogen-oxygen fuel cell**. The electrical energy produces a current of electricity that can be used to power cars and other vehicles, and to heat and light buildings.

The hydrogen-oxygen fuel cell uses hydrogen and oxygen to produce electricity. The only chemical product is water, which is harmless.

▲ Filling up the tank with hydrogen at a hydrogen filling station in Tokyo, Japan.

Inside the hydrogen-oxygen fuel cell

The diagram below shows a typical hydrogen-oxygen fuel cell used in cars. The hydrogen is supplied from a fuel tank, and the oxygen is taken in from the air. Follow the numbers to see what happens:

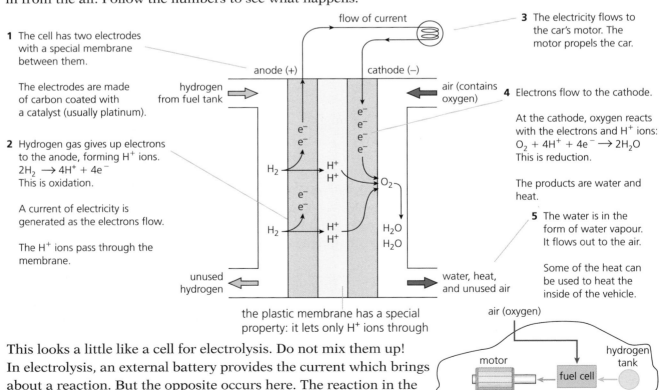

1 The cell has two electrodes with a special membrane between them.

The electrodes are made of carbon coated with a catalyst (usually platinum).

2 Hydrogen gas gives up electrons to the anode, forming H^+ ions.
$2H_2 \rightarrow 4H^+ + 4e^-$
This is oxidation.

A current of electricity is generated as the electrons flow.

The H^+ ions pass through the membrane.

the plastic membrane has a special property: it lets only H^+ ions through

3 The electricity flows to the car's motor. The motor propels the car.

4 Electrons flow to the cathode.

At the cathode, oxygen reacts with the electrons and H^+ ions:
$O_2 + 4H^+ + 4e^- \rightarrow 2H_2O$
This is reduction.

The products are water and heat.

5 The water is in the form of water vapour. It flows out to the air.

Some of the heat can be used to heat the inside of the vehicle.

This looks a little like a cell for electrolysis. Do not mix them up! In electrolysis, an external battery provides the current which brings about a reaction. But the opposite occurs here. The reaction in the cell produces a current. So the cell is acting as a battery.

ENERGY CHANGES IN REACTIONS

A redox reaction
- The reaction in the hydrogen-oxygen fuel cell is a **redox** reaction.
- Hydrogen gives up electrons at the anode. It is **oxidised**: $2H_2 \rightarrow 4H^+ + 4e^-$
- Oxygen accepts electrons at the cathode. It is **reduced**: $O_2 + 4H^+ + 4e^- \rightarrow 2H_2O$

Hydrogen on Earth
- There is only a tiny amount of the element hydrogen in Earth's atmosphere.
- It is so light that most of it has escaped into space.
- But compounds containing hydrogen are common on Earth. For example, water, and methane (natural gas).
- So we have to obtain hydrogen from its compounds.

How does it compare with the petrol engine?
Most cars and other vehicles around the world still have engines that burn petrol (gasoline) in the oxygen in air. The thermal energy that is given out is converted to motion: the car moves.

Let's compare the hydrogen-oxygen fuel cell with the petrol engine.

Staying safe ...
- Hydrogen is as safe to use as petrol.
- The risks are different – but they are well known and understood.

The hydrogen oxygen fuel cell	The petrol engine
Advantages	Advantages
• The only chemical produced is water, which is harmless. • Hydrogen can be made by the electrolysis of water, which is plentiful, on adding a little acid. (See the reactions on page 98.)	• Petrol filling stations are widely available, for drivers. • Petrol is relatively cheap (although this varies, depending on the world supply of petroleum).
Disadvantages	Disadvantages
• Hydrogen is a gas. To reduce its volume and make it practical to use, it must be stored as a liquid under high pressure. • Hydrogen filling stations are not yet widely available. • Most hydrogen is obtained from fossil fuels. And where it is obtained by electrolysis, the electricity is mostly from power stations that burn fossil fuels. These are linked to pollution and climate change. (But see below.)	• The burning of petrol in engines produce harmful substances: – carbon dioxide, which is linked to climate change – oxides of nitrogen – deadly carbon monoxide, if the petrol burns in limited oxygen – unburnt volatile gases from the petrol • Petrol is obtained from oil, a limited and non-renewable resource.

Cars with hydrogen-oxygen fuel cells also cost more than cars with petrol engines. But this, and some other disadvantages, are not permanent!

The cells will get cheaper as the technology develops. Water may become the main source for hydrogen when more cheap clean electricity – from solar, wind, and other renewable sources – is available for electrolysis. More hydrogen filling stations will be provided.

Some experts predict that the hydrogen-oxygen fuel cell will play a big part in the fight against climate change and air pollution.

▲ A hydrogen fuelled bus in Berlin, Germany.

1 In what ways is the reaction that takes place in the hydrogen-oxygen fuel cell:
   a similar to the reaction when hydrogen burns in air?
   b different from the reaction when hydrogen burns in air? Give as many ways as you can.
2 The reaction in the hydrogen-oxygen fuel cell has only one chemical product. Is this product harmful? Explain.
3 Give one use for the hydrogen-oxygen fuel cell.
4 What is the main benefit of the hydrogen-oxygen fuel cell to the environment, compared with the petrol engine?
5 There is very little hydrogen in the atmosphere. It must be obtained from its compounds. Give reasons why water would be a better source than fossil fuels, for hydrogen.
6 Name the process used to obtain hydrogen from water.

Checkup on Chapter 8

Revision checklist

Core syllabus content
Make sure you can …
- [ ] explain these terms:
  exothermic reaction   *endothermic reaction*
- [ ] give examples of exothermic and endothermic reactions
- [ ] describe how
  - an exothermic reaction
  - an endothermic reaction
  affects the temperature of its surroundings
- [ ] state that, during a reaction:
  - bonds in the reactants are broken
  - then new bonds are made, to give the product(s)
- [ ] recall that bond breaking takes in energy, and bond making gives out energy
- [ ] state that the overall energy change in a reaction = energy in − energy out
- [ ] state that the overall energy change in a reaction:
  - has a minus sign for exothermic reactions
  - has a plus sign for endothermic reactions
- [ ] interpret reaction pathway diagrams for exothermic and endothermic reactions
- [ ] state that a hydrogen-oxygen fuel cell uses hydrogen and oxygen to produce electricity, with water as the only chemical product
- [ ] give uses for the hydrogen-oxygen fuel cell

Extended syllabus content
Make sure you can also …
- [ ] state that bond breaking is an endothermic process, and bond making is an exothermic process
- [ ] define these terms, and give their symbols:
  the enthalpy change in a reaction, ΔH
  the activation energy, E_a
- [ ] explain the enthalpy change in a reaction in terms of bond breaking and bond making
- [ ] draw reaction pathway diagrams for exothermic and enothermic reactions, and label them to show reactants, products, ΔH, and E_a
- [ ] define *bond energy* and state the units used for it
- [ ] calculate the enthalpy change in a reaction, given the equation for the reaction, and bond energies
- [ ] outline how the hydrogen-oxygen fuel cell works

- [ ] give advantages and disadvantages of the hydrogen-oxygen fuel cell for cars, compared with the petrol engine

Questions

Core syllabus content

1 Look at this neutralisation reaction:
   NaOH (*aq*) + HCl (*aq*) ⟶ NaCl (*aq*) + H$_2$O (*l*)
   a It is an exothermic reaction. Explain *exothermic*.
   b How will the temperature of the solution change, as the chemicals react?
   c Which have more energy, the reactants or the products?

2 Water at 25 °C was used to dissolve two compounds. The temperature of each solution was measured immediately afterwards.

Compound	Temperature of solution / °C
ammonium nitrate	21
calcium chloride	45

   a List the apparatus needed for this experiment.
   b Calculate the temperature change on dissolving each compound.
   c i   Which compound dissolved exothermically?
      ii How did you decide this?
      iii What can you say about the energy of its ions in solution, compared with in the solid?
   d For each solution, estimate the temperature of the solution if:
      i   the amount of water is halved, but the same mass of compound is used
      ii the mass of the compound is halved, but the volume of water is unchanged
      iii both the mass of the compound, and the volume of water, are halved.

3 Copy this reaction pathway diagram:

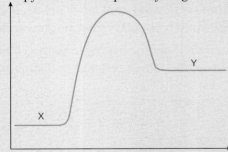

   a On your diagram:
      i   label the vertical axis
      ii give the correct terms for X and Y.
   b Is the reaction *exothermic*, or *endothermic*?
   c Will the temperature of the reaction mixture *rise*, or *fall*, during the reaction?

ENERGY CHANGES IN REACTIONS

Extended syllabus content

4 The fuel *natural gas* is mostly methane. Its combustion in oxygen is exothermic:
$$CH_4 (g) + 2O_2 (g) \longrightarrow CO_2 (g) + 2H_2O (l)$$
  a Explain *why* this reaction is exothermic, in terms of bond breaking and bond making.
  b i Copy and complete this reaction pathway diagram for the reaction, indicating:
   A the overall energy change
   B the energy needed to break bonds
   C the energy given out by new bonds forming.

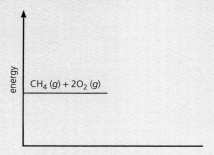

  ii Methane will not burn in air until a spark or flame is applied. Why not?
  c When 1 mole of methane burns in oxygen, the energy change is -890 kJ.
   i What does the $-$ sign tell you?
   ii Which word describes a reaction with this type of energy change?
  d How much energy is given out when 1 gram of methane burns? (A_r: C = 12, H = 1.)

5 A student was asked to draw a reaction pathway diagram for this reaction:
$$C_3H_6 + H_2 \longrightarrow C_3H_8 \quad \Delta H = -124 \text{ kJ/mol}$$
The student's diagram had a number of errors:

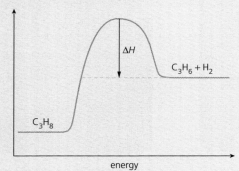

  a Redraw the diagram above, correcting the errors.
  b i What does ΔH stand for?
   ii ΔH has a $-$ sign in this reaction. In terms of the transfer of energy, what does that signify?
  c i Define *activation energy*.
   ii Give the symbol used for activation energy.
   iii Mark the activation energy on your diagram.

6 The gas hydrazine, N_2H_4, burns in oxygen like this:

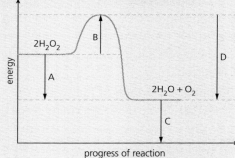

  a Count and list the bonds broken in this reaction.
  b Count and list the new bonds formed.
  c Calculate the total energy:
   i required to break the bonds
   ii released when the new bonds form.
   (The bond energies in kJ/mole are:
   N−H 391; N−N 158; N≡N 945; O−H 464; O=O 498.)
  d Calculate the enthalpy change in the reaction.
  e Is the reaction exothermic, or endothermic?
  f Where is energy transferred from, and to?

7 Hydrogen peroxide decomposes very slowly to water and oxygen, in this reaction:
$$2H_2O_2 (aq) \longrightarrow 2H_2O (l) + O_2 (g)$$
This is its reaction pathway diagram:

  a i Which represents the enthalpy change (ΔH) for the reaction: A, B, C or D?
   ii Which sign will ΔH have: + or −?
  b Which energy change, A, B, C or D, represents the activation energy?
  c Which energy change, A, B, C or D, is due to bonds being formed, in this reaction?
  d The reaction is exothermic. Explain why, in terms of breaking and making bonds.
  e List as many characteristics as you can for an *endothermic* reaction.

8 a Which are valid reasons today, for drivers to switch to cars with hydrogen-oxygen fuel cells?
   i They cost less than cars with petrol engines.
   ii Hydrogen refill stations are widely available.
   iii The fuel cells do not rely at all on fossil fuels.
   iv No harmful products are formed.
  b Which of the reasons above that are not valid today might be valid in the future? Explain.

9.1 Introducing reaction rates

Objectives: define *rate*; understand that the rate of a reaction is found by measuring how fast a reactant is used up, or how fast a product is formed

Fast and slow

Some chemical reactions are **fast**, and some are **slow**. Compare these:

The precipitation of silver chloride, when you mix solutions of silver nitrate and sodium chloride. This is a very fast reaction.

Concrete setting. This reaction is quite slow. It will take a couple of days for the concrete to fully harden.

Rust forming on an old car. This is usually a very slow reaction. It will take years for the car to rust completely away.

But it is not always enough to know just that a reaction is fast or slow. In factories where they make products from chemicals, they need to know *exactly* how fast a reaction is going, and how long it will take to complete. In other words, they need to know the **rate** of the reaction.

What is rate?

Rate is a measure of how fast or slow something is. Here are some examples.

This plane has just flown 800 kilometres in 1 hour. It flew at a **rate** of 800 km per hour.

This petrol pump can pump out petrol at a **rate** of 50 litres per minute.

This factory can produce bottles of fruit juice at a **rate** of 25 bottles per second.

From these examples you can see that:

Rate is a measure of the change that happens in a single unit of time.

Any suitable unit of time can be used – a second, minute, hour, day ...

THE RATE OF REACTION

Rate of a chemical reaction

When zinc is added to dilute sulfuric acid, they react together. The zinc slowly disappears, and a gas bubbles off.

As the reaction proceeds, there are fewer and fewer bubbles. This is a sign that the reaction is slowing down.

Finally, no more bubbles appear. The reaction is over, because all the acid has been used up. Some zinc remains behind.

The gas that bubbles off is hydrogen. The equation for the reaction is:

zinc + sulfuric acid ⟶ zinc sulfate + hydrogen
$Zn\ (s) + H_2SO_4\ (aq) \longrightarrow ZnSO_4\ (aq) + H_2\ (g)$

Both zinc and sulfuric acid get used up in the reaction. At the same time, zinc sulfate and hydrogen form.

You could measure the rate of the reaction, by measuring:

- the amount of zinc used up per minute *or*
- the amount of sulfuric acid used up per minute *or*
- the amount of zinc sulfate produced per minute *or*
- the amount of hydrogen produced per minute.

For this reaction, it is easiest to measure the amount of hydrogen produced per minute, since it is the only gas that forms. It can be collected as it bubbles off, and its volume can be measured.

In general, to find the rate of a reaction, you should measure:
- **the amount of a reactant used up per unit of time** *or*
- **the amount of a product produced per unit of time.**

1. Here are some reactions that take place in the home. Put them in order of decreasing rate (the fastest one first):
   a. raw egg changing to hard-boiled egg
   b. fruit going rotten
   c. cooking gas burning
   d. bread baking
   e. a metal tin rusting
2. Which of these rates of travel is the slowest?
   5 kilometres per second
   20 kilometres per minute
   60 kilometres per hour
3. Suppose you had to measure the rate at which zinc is used up in the reaction above. Which of these units would be suitable? Explain your choice.
   a. litres per minute
   b. grams per minute
   c. centimetres per minute
4. Iron reacts with sulfuric acid like this:
   $Fe\ (s) + H_2SO_4\ (aq) \longrightarrow FeSO_4\ (aq) + H_2\ (g)$
   a. Write a word equation for this reaction.
   b. Write down four different ways in which the rate of the reaction could be measured, in theory.

9.2 Measuring the rate of a reaction

Objectives: describe how to measure the rate of a reaction that produces a gas; interpret a graph of the results; recall how the rate changes as a reaction proceeds

A reaction that produces a gas

The rate of a reaction is found by measuring the amount of a **reactant** used up per unit of time, or the amount of a **product** produced per unit of time. Look at this reaction:

magnesium + hydrochloric acid ⟶ magnesium chloride + hydrogen
$Mg\,(s)$ + $2HCl\,(aq)$ ⟶ $MgCl_2\,(aq)$ + $H_2\,(g)$

Here hydrogen is the easiest substance to measure, because it is the only gas in the reaction. It bubbles off and can be collected in a **gas syringe**, where its volume is measured.

The experiment to measure its rate

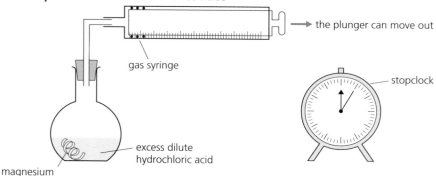

▲ The reactions in fireworks produce carbon dioxide and nitrogen. Difficult to measure the rate!

Clean the magnesium with sandpaper. Put dilute hydrochloric acid in the flask. (The acid is in excess.) Drop the magnesium into the flask. Insert the stopper plus syringe immediately. Start the clock at the same time.

Hydrogen begins to bubble off. It rises up the flask and into the gas syringe, pushing the plunger out:

At the start, no gas has yet been produced or collected. So the plunger is all the way in.

Now the plunger has been pushed out to the 20 cm³ mark. 20 cm³ of gas have been collected.

The volume of gas in the syringe is noted at intervals – for example, every half a minute. How will you know when the reaction is complete?

Typical results

Time / minutes	0	$\tfrac{1}{2}$	1	$1\tfrac{1}{2}$	2	$2\tfrac{1}{2}$	3	$3\tfrac{1}{2}$	4	$4\tfrac{1}{2}$	5	$5\tfrac{1}{2}$	6	$6\tfrac{1}{2}$
Volume of hydrogen / cm³	0	8	14	20	25	29	33	36	38	39	40	40	40	40

You can tell quite a lot from this table. For example, the reaction lasted about five minutes. No more gas was produced after that. But a graph of the results is even more helpful. It is shown on the next page.

THE RATE OF REACTION

A graph of the results

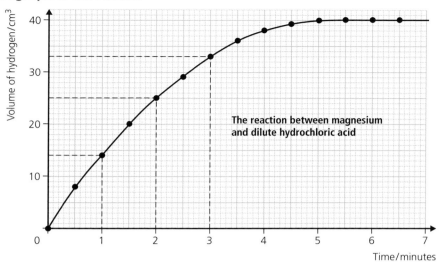

The reaction between magnesium and dilute hydrochloric acid

▲ A graph gives a very clear picture of how the reaction rate changes.

Notice these things about the results:

1. In the first minute, 14 cm³ of hydrogen are produced.
   So the rate for the first minute is 14 cm³ of hydrogen per minute.
   In the second minute, only 11 cm³ are produced. (25 − 14 = 11)
   So the rate for the second minute is 11 cm³ of hydrogen per minute.
   The rate for the third minute is 8 cm³ of hydrogen per minute.
   So the rate decreases as time goes on.
   The rate changes all through the reaction. It is greatest at the start, but decreases as the reaction proceeds.

2. The reaction is fastest in the first minute, and the curve is steepest then. It gets less steep as the reaction gets slower.
   The faster the reaction, the steeper the curve.

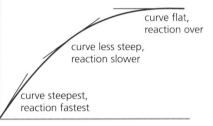

3. After 5 minutes, no more hydrogen is produced, so the volume no longer changes. The reaction is over, and the curve goes flat.
   When the reaction is over, the curve goes flat.

4. Altogether, 40 cm³ of hydrogen are produced in 5 minutes.

   The *average* rate for the reaction $= \dfrac{\text{total volume of hydrogen}}{\text{total time for the reaction}}$

   $= \dfrac{40 \text{ cm}^3}{5 \text{ minutes}}$

   $= \mathbf{8 \text{ cm}^3 \text{ of hydrogen per minute.}}$

Good for gases ...
The method shown in this unit can be used for *any* reaction where a gas is produced.

1 **a** For the experiment in this unit, explain why:
   i the magnesium ribbon is cleaned first
   ii the clock is started the instant the reactants are mixed
   iii the stopper and gas syringe are inserted immediately
   b What is the purpose of the gas syringe?

2 From the graph at the top of this page, how can you tell when the reaction is over?

3 Look again at the graph at the top of the page.
   a How much hydrogen is produced:
      i in the first 2.5 minutes? **ii** in the first 4.5 minutes?
   b How long did it take to collect 20 cm³ of hydrogen?
   c What is the rate of the reaction during:
      i the fourth minute? **ii** the sixth minute?

4 State one possible source of error, in this experiment.

9.3 Changing the rate (part I)

Objectives: describe how the rate of a reaction changes with concentration, and temperature; understand that only one variable must be changed, for a fair test

Ways to change the rate of a reaction

There are several ways to speed up or slow down a reaction. For example, you could change the concentration of a reactant, or the temperature.

1 By changing concentration

Here you will see how rate changes with the **concentration** of a reactant.

The method Repeat the experiment from page 104 twice (A and B below). Keep everything the same each time *except* the concentration of the acid. In B it is *twice as concentrated* as in A. And in both cases the acid is in excess, so all the magnesium is used up.

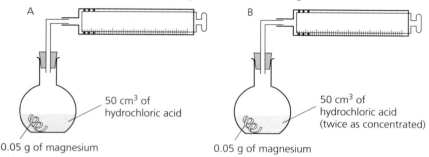

The results Here are both sets of results, shown on the same graph.

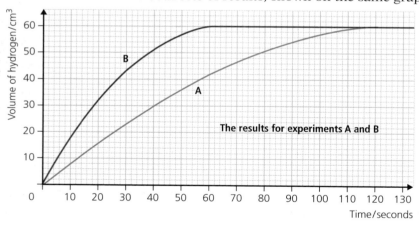

The results for experiments A and B

Notice these things about the results:
1 Curve B is steeper than curve A. So the reaction was faster for B.
2 In B, the reaction lasts for 60 seconds. In A it lasts for 120 seconds.
3 Both reactions produced 60 cm³ of hydrogen. Do you agree?
4 So in B the *average* rate was 1 cm³ of hydrogen per second. (60 ÷ 60)
   In A it was 0.5 cm³ of hydrogen per second. (60 ÷ 120)
   The average rate in B was twice the average rate in A.
   So in this example, doubling the concentration doubled the rate.

A reaction goes faster when the concentration of a reactant is increased.
It follows that you can *slow down* a reaction, by *reducing* concentration.

Variables
- Factors you can change in an experiment are called **variables**. Temperature is an example.
- To find the effect of variable X on variable Y, you keep all other factors unchanged, or **controlled**.
- This lets you focus on how X affects Y.
- X, the variable you choose to change, is called the **independent variable**. Y is the **dependent variable**.

In experiment 1 on the left, the independent variable is *concentration*. The dependent variable is *volume of gas produced*. All other variables, such as the mass of magnesium, are kept the same (controlled).

▲ Bleach reacts with coloured substances, oxidising them to colourless substances. The more concentrated the solution of bleach, the faster this stain will disappear.

THE RATE OF REACTION

2 By changing temperature

Here you will see how rate changes with the **temperature** of the reactants. (So temperature is the *independent variable*. Check the panel on page 106!)

The method Dilute hydrochloric acid and sodium thiosulfate solution react to give a fine yellow precipitate of sulfur. You can follow the rate of the reaction like this:

1. Mark a cross on a piece of paper.
2. Place a beaker containing sodium thiosulfate solution on top of the paper, so that you can see the cross through it from above.
3. Add hydrochloric acid, start a clock at the same time, and quickly measure the temperature of the mixture.
4. The cross grows fainter as the precipitate forms. Stop the clock the moment you can no longer see the cross. Note the time.
5. Now repeat steps 1–4 several times, changing *only* the temperature. You do this by heating the sodium thiosulfate solution to different temperatures, before adding the acid.

▲ The low temperature in the fridge slows down reactions that make food rot.

View from above the beaker:

The cross grows fainter with time

The results This table shows some typical results:

Temperature of mixture / °C	20	30	40	50	60
Time for cross to disappear / seconds	200	125	50	33	24

➡ The higher the temperature, the faster the cross disappears

The cross disappears when enough sulfur has formed to hide it. This took 200 seconds at 20 °C, but only 50 seconds at 40 °C. So the reaction is *four times faster* at 40 °C than at 20 °C.

A reaction goes faster when the temperature is raised. The rate generally doubles for an increase of 10 °C.

That is why food cooks much faster in pressure cookers than in ordinary saucepans. (The temperature in a pressure cooker can reach 125 °C.) And if you want to slow a reaction down, you can lower the temperature.

▲ Oh dear. Oven too hot? Reactions faster than expected?

1. Use the graph on page 106 to answer this question.
   a. How much hydrogen was obtained after 2 minutes:
      i. in experiment A? ii. in experiment B?
   b. How can you tell which reaction was faster?
2. Explain why experiments A and B both gave the same amount of hydrogen.
3. Which is the dependent variable, in experiment 2?
4. Copy and complete: *A reaction goes ...... when the concentration of a ...... is increased. It also goes ...... when the ...... is raised.*
5. Raising the temperature speeds up a reaction. Try to give two (new) examples of how this is used in everyday life.
6. What happens to the rate of a reaction when the temperature is *lowered*? How do we make use of this?

9.4 Changing the rate (part II)

Objectives: describe how the rate of a reaction is affected by changing the surface area of a solid reactant, and the pressure of a gaseous reactant

3 By changing surface area

In many reactions, one reactant is a solid. The reaction between hydrochloric acid and calcium carbonate (marble chips) is an example. Carbon dioxide gas is produced:

$$CaCO_3 \, (s) + 2HCl \, (aq) \longrightarrow CaCl_2 \, (aq) + H_2O \, (l) + CO_2 \, (g)$$

The rate can be measured using the apparatus on the right. The carbon dioxide escapes through the cotton wool. It is a heavy gas, so the flask gets lighter. By weighing the flask at regular intervals, you can follow the rate of the reaction.

The method Place the marble in the flask and add the acid. Quickly plug the flask with cotton wool to stop any liquid splashing out. Then weigh it, starting the clock at the same time. Record the mass at regular intervals until the reaction is complete.

The experiment is repeated twice. Everything is kept exactly the same each time, except the *surface area* of the marble chips.

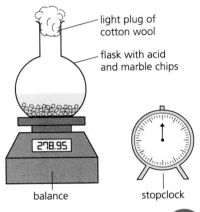

For experiment 1, large chips are used. Their surface area is the total area of exposed surface.

For experiment 2, the same *mass* of marble is used – but the chips are small so the surface area is greater.

The results The results of the two experiments are plotted here:

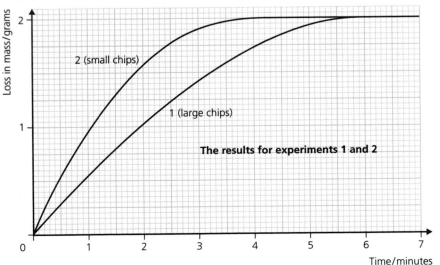

The results for experiments 1 and 2

Look at the graph. Did changing the surface area have any effect?

About the variables

In experiment 3 on the left …

- surface area is the **independent variable**.
- loss of mass is the **dependent variable**.

The other variables, such as mass of the marble and concentration of the acid, must be carefully controlled.

How to draw the graph

First you have to find the *loss in mass* at different times:

loss in mass at a given time = mass at start − mass at that time

Then you plot the values for *loss in mass* against *time*.

THE RATE OF REACTION

Notice these things about the results:
1 Curve 2 is steeper than curve 1. So reaction is faster for the small chips.
2 In both experiments, the final loss in mass is the same: 2.0 grams.
  In other words, 2.0 grams of carbon dioxide are produced each time.
3 The complete reaction takes two minutes longer for the large chips.

These results show that:

A reaction goes faster when the surface area of a solid reactant is increased.

4 By changing pressure

Many reactions involve gases. For example, the reactions of hydrogen with oxygen to make water, and with nitrogen to make ammonia.

As you saw on page 10, you can increase the pressure on a gas by decreasing the volume. Let's take another look at that idea.

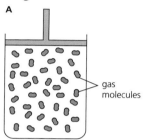

Suppose this container has a volume of 2 dm³, and contains 2 g of a gas. Then the concentration of the gas is **1g/dm³**.

The gas pressure has been doubled by squeezing the gas into a volume of 1 dm³. The concentration of the gas is now **2g/dm³**.

So increasing the pressure of a gas increases its concentration. And the increase in concentration speeds up the reaction rate, just as it does for reactions in solution (page 106).

A reaction involving gases goes faster when the pressure is increased.
It follows that you can slow down the reaction by lowering the pressure.

> ### Explosion!
> Explosions give out a great deal of energy, and produce gases that expand rapidly. They can occur in coal mines, grain silos, wood factories and flour mills, where fine flammable dust (such as coal dust) can collect in the air and catch fire from a spark. Several factors work together, giving a reaction rate that it is out of control.

▲ This old factory produced wood powder for making linoleum and other products. Fine, highly flammable wood dust caught fire. Its large surface area led to a very rapid reaction: an **explosion**.

▲ Down a coal mine. Methane gas is present in mines. It burns in oxygen. Once it reaches a certain concentration, a spark from a drill could cause an explosion. So mines are fitted with methane monitors.

1 Using the graph on page 108, find, for each sample:
  a the mass of carbon dioxide produced in the first minute
  b the average rate of production of the gas per minute
2 State the effect of surface area on the rate of a reaction involving solids.
3 List three variables that must be controlled in the experiment on page 108, to give valid results.
4 The substances in fireworks are powdered. Suggest a reason.
5 Finding the rate by measuring loss of mass would not work for a reaction that produces hydrogen gas. Suggest a reason.
6 Increasing the pressure of a gas increases its concentration. Explain why.
7 What will happen to the rate of a gas reaction, if the gas pressure is *reduced*?
8 Increasing the pressure has no effect on the reaction between two solids. Suggest a reason.

9.5 Explaining rate changes

Objectives: explain what the *collision theory* is; use it to explain the effect of concentration, temperature, surface area, and gas pressure, on reaction rate

The collision theory

Magnesium and dilute hydrochloric acid react together like this:

magnesium + hydrochloric acid ⟶ magnesium chloride + hydrogen

$Mg\ (s)\ +\ 2HCl\ (aq)\ \longrightarrow\ MgCl_2\ (aq)\ +\ H_2\ (g)$

In order for the magnesium and acid particles to react together:
- the particles must collide with each other, and
- the collision must have enough energy to be *successful*. In other words, it must have the activation energy needed for reaction to occur.

This is called the **collision theory**. It is shown by these drawings:

> **Activation energy**
> The activation energy is the minimum energy that colliding particles must have, for reaction to occur. Page 95!

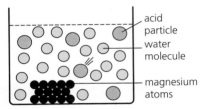
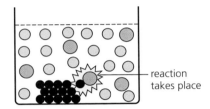
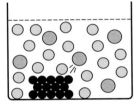

The particles in the liquid move non-stop. To react, an acid particle must collide with a magnesium atom, and bonds must break.

Here the particles have enough energy for a successful collision. Bonds break. Reaction occurs. New bonds form, giving the products.

But here the particles did not have enough energy. The collision was not successful. No bonds were broken. The acid particle just bounced away.

If there are lots of successful collisions in a given minute, then a lot of hydrogen is produced in that minute. In other words, the rate of reaction is high. If there are not many, the rate of reaction is low.

The rate of a reaction depends on how many successful collisions there are in a given unit of time.

Why the rate changes with each factor
Increasing the concentration

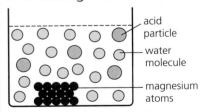
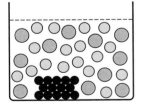
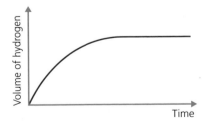

In dilute acid there are fewer acid particles. So there are fewer successful collisions between acid particles and magnesium atoms.

Here the acid is more concentrated: there are more acid particles. So the frequency of successful collisions is higher. The reaction is faster.

But in both cases, the reaction slows down over time – because as the reactant particles get used up, there are fewer particles left to collide.

The reaction rate increases with concentration because there are more particles per unit volume. So there is a higher frequency of collisions – and therefore a higher frequency of successful collisions.

THE RATE OF REACTION

Increasing the temperature
On heating, *all* the particles take in thermal energy. Their energy increases.

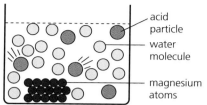 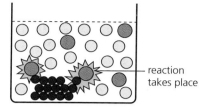 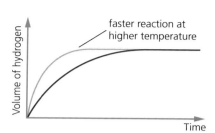

So now the acid particles move faster, and collide more frequently with magnesium particles, and …

… more colliding particles have the activation energy needed for reaction to occur.

So the rate of the reaction increases. It generally doubles for a temperature rise of 10 °C.

The reaction rate increases with temperature for two reasons: there are more collisions, and more have enough energy to be successful.

Increasing the surface area

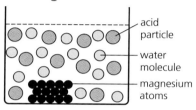 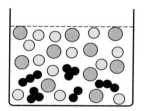

When magnesium ribbon is used, the acid particles can collide only with the atoms in its outer layer.

Many more atoms are exposed in the powdered metal. So successful collisions are more frequent.

The reaction rate increases with surface area because more particles of the solid are exposed, so successful collisions are more frequent.

Frequency and rate
- At any point in a reaction, a % of the collisions will be successful.
- The *frequency* of collisions is the number of collisions per unit of time.
- So if you can increase the frequency of collisions, you will increase the frequency of successful collisions.
- So the reaction rate will rise.

Increasing the gas pressure

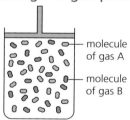 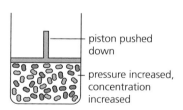 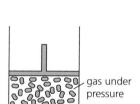

Reaction between gases A and B depends on successful collisions. The pressure can be increased by pushing down the piston.

Now there are more molecules per unit volume. So the frequency of successful collisions increases. So the rate of reaction increases.

The result is the same where a gas reacts with a liquid or solid. (But only gas particles are affected by an increase in pressure.)

The reaction rate increases with gas pressure because there are more gas particles per unit volume, so successful collisions are more frequent.

Q

1. Copy and complete: Two particles can react together only if they …… and the …… has enough …… to be …….
2. What is a *successful* collision?
3. Increasing the gas pressure speeds up the rate of reactions involving gases. Explain why, in terms of the collision theory.
4. Reaction between magnesium and acid speeds up when:
   a the concentration of the acid is doubled. Why?
   b the temperature is raised. Why? (Give *two* reasons.)
   c the acid is stirred. Why?
   d the metal is ground to a powder. Why?

9.6 Catalysts

Objectives: define *catalyst* and *enzyme*; recall that enzymes act as catalysts; state that a catalyst decreases the activation energy (E_a) of a reaction

What is a catalyst?
In Units 9.3 and 9.4, you saw different ways to speed up a reaction. There is another way to speed up some reactions: **use a catalyst**.

A catalyst is a substance that increases the rate of a reaction, and is unchanged at the end of the reaction.

Example: the decomposition of hydrogen peroxide
Hydrogen peroxide is a colourless liquid that breaks down very slowly to water and oxygen:

    hydrogen peroxide ⟶ water + oxygen
    $2H_2O_2\,(l) \longrightarrow 2H_2O\,(l) + O_2\,(g)$

You can show how a catalyst affects the reaction, like this:

1. Pour some hydrogen peroxide solution into three measuring cylinders. The first one is the **control**.
2. Add manganese(IV) oxide to the second, and raw liver to the third.
3. Now use a glowing wooden splint to test the cylinders for oxygen. The splint will burst into flame if there is enough oxygen present.

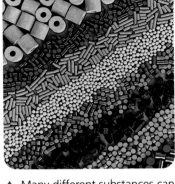

▲ Many different substances can act as catalysts. They are usually made into shapes that offer a very large surface area.

The results

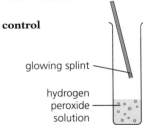
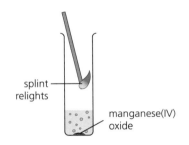
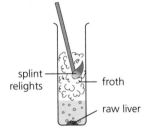

Since hydrogen peroxide breaks down very slowly, there is not enough oxygen to relight the splint.

Manganese(IV) oxide makes the reaction go thousands of times faster. The splint bursts into flame.

Raw liver also speeds it up. The liquid froths as the oxygen bubbles off – and the splint relights.

So manganese(IV) oxide is a catalyst for the reaction: it speeds it up, but is not itself changed. Something in the raw liver acts as a catalyst too. That 'something' is an enzyme called **catalase**.

What are enzymes?
Enzymes are proteins made by cells, to act as biological catalysts. They are found in every living thing.

You have thousands of different catalysts in your body. For example, catalase speeds up the decomposition of hydrogen peroxide in your cells, before it can harm you. Amylase in your saliva speeds up the breakdown of the starch in your food.

Without enzymes, most of the reactions that take place in your body would be far too slow at body temperature. You would die.

> **Removing a catalyst**
> If you *remove* a catalyst before a reaction is complete, the reaction will slow down again!

> **Large, complex …**
> Enzymes are large and complex molecules. They are usually like round blobs in shape.

THE RATE OF REACTION

How do catalysts speed up reactions?
As you saw earlier, a collision between particles must have a minimum amount of energy, for reaction to occur.

This minimum amount of energy is the **activation energy**, E_a. (We met it first on page 110.)

The reaction pathway diagram on the right shows how the activation energy changes when a catalyst is present. It is lower. This is because the catalyst provides another way for particles to react, that needs less energy.

A catalyst lowers the activation energy of a reaction. Now the particles need less energy to react. So more collisions are successful, and the reaction goes faster.

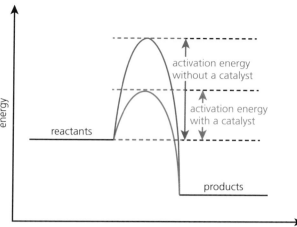

Making use of catalysts
Catalysts are used for many reactions in industry. The product is obtained faster, saving time. Even better, the reaction may go fast enough *at a lower temperature* – which means a lower fuel bill.

So catalysts are very important to the chemical industry. They are often **transition elements** or **oxides** of transition elements. You will see some examples later.

The catalyst is chosen to suit each particular reaction – it may not work for other reactions. Since the catalyst is not changed in the reaction, it can be reused.

▲ A catalyst of platinum and rhodium, in the form of a gauze, being fitted into a tank for making nitric acid.

Enzymes in the wash!
Thousands of enzymes are made by bacteria. We find many uses for them.
- For example, some enzymes help to break down grease and stains on clothing.
- The bacteria which make these enzymes are grown in tanks, in factories.
- The enzymes are removed from the tanks and used in **biological detergents**.
- Enzymes work best in conditions like those inside the bacteria that made them. Higher temperatures destroy them. So the clothes should usually be washed at or below 40°C. That saves fuel too! (Check the instructions on the container.)

1. What is a *catalyst*?
2. Which of these does a catalyst *not* change?
   a the speed of a reaction
   b the products that form
   c the total amount of each product formed
3. a A catalyst can be reused. Explain why.
   b The manganese(IV) oxide was weighed before and after the experiment on page 112. What was discovered? Explain.
4. Explain what an *enzyme* is, and give an example.
5. Why do our bodies need enzymes?
6. When liver is cooked, it has no effect on the rate of decomposition of hydrogen peroxide. Suggest a reason.
7. a Define *activation energy*.
   b Explain why a catalyst speed up a reaction, in terms of activation energy.
8. Catalysts are very important in industry. Explain why.

More about enzymes

Mainly from microbes
Enzymes are proteins made by living things, to act as catalysts for their own reactions. So we can obtain enzymes from plants, and animals, and microbes such as bacteria and fungi. In fact we get most from microbes.

Traditional uses for enzymes
We humans have used enzymes for thousands of years. For example …

- **In making bread** Yeast (a fungus) and sugar are used in making dough for bread. When the dough is left in a warm place, the yeast cells feed on the sugar to obtain energy. This reaction is called **fermentation**. It is catalysed by enzymes in the yeast.

$$C_6H_{12}O_6 \, (aq) \xrightarrow{\text{catalysed by enzymes}} 2C_2H_5OH \, (aq) + 2CO_2 \, (g) + \text{energy}$$
glucose　　　　　　　　　　　　ethanol　　　carbon dioxide

The carbon dioxide makes the dough rise. Later, in the hot oven, the gas expands even more while the dough cooks. So you end up with spongy bread. The heat kills the yeast off.

- **In making yoghurt** To make yoghurt, special bacteria are added to milk. They feed on the lactose (sugar) in it, to obtain energy. Their enzymes catalyse its conversion to lactic acid and other substances, which turn the milk into yoghurt.

Making enzymes the modern way
In making bread and yoghurt, the microbes that make the enzymes are present. But in most modern uses for enzymes, they are not. Instead:

- the bacteria and other microbes are grown in tanks, in a rich broth of nutrients; so they multiply fast
- then they are killed off, and their enzymes are separated and purified
- the enzymes are sold to factories.

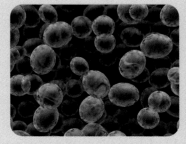

▲ Yeast cells. Cells are living things, but the enzymes they make are not. Enzymes are just chemicals (proteins).

▲ The holes in bread are where carbon dioxide gas expanded.

▲ Anyone home? The tank contains bacteria, busy making enzymes. For example, it could be the enzyme amylase, which catalyses the conversion of starch to sugar.

▲ The amylase is sold to a company that uses it to make a sweet syrup from corn (maize) flour. The syrup is used in biscuits, cakes, soft drinks, and sauces.

THE RATE OF REACTION

Modern uses of enzymes
Enzymes have many different uses. Here are some common ones:
- **In making soft-centred chocolates**   How do they get the runny centres into chocolates? By using the enzyme **invertase**.

  First they make a paste containing sugars, water, flavouring, colouring, and invertase. Then they dip blobs of it into melted chocolate. The chocolate hardens. Inside, the invertase catalyses the breakdown of the sugars to more soluble ones, so the paste goes runny.

  Other enzymes are used in a similar way to 'soften' food, in making tinned food for infants.
- **In making stone-washed denim**   Once, denim was given a worn look by scrubbing it with pumice stone. Now an enzyme does the job.
- **In making biological detergents**   As you saw on page 113, these contain enzymes to catalyse the breakdown of grease and stains.
- **In DNA testing**   Suppose a criminal leaves traces of skin or blood at a crime scene. These contain the criminal's DNA. DNA has a complex structure. An enzyme is used to cut it into bits, for analysis in the lab.

▲ Thanks to invertase …

▲ Thanks to a restriction enzyme …

How do they work?
This shows how an enzyme molecule catalyses the breakdown of a reactant molecule:

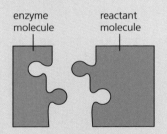
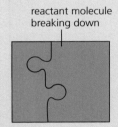
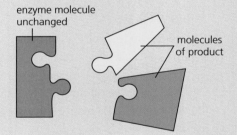

First, the two molecules must fit together like jigsaw pieces. (So the reactant molecule must be the right shape, for the enzyme.)

The 'complex' that forms makes it easy for the reactant molecule to break down. You do not need to provide energy by heating.

When decomposition is complete the molecules of the product move away. Another molecule of the reactant takes their place.

Enzymes are a much more complex shape than the drawing suggests. Even so, this model gives you a good idea of how they work.

The search for extremophiles
Most of the enzymes we use work around 40 °C, and at a pH not far from 7. In other words, in conditions like those in the cells that made them.

But around the world, scientists are searching high and low for microbes that live in very harsh conditions. For example, deep under the ice in Antarctica, or at hot vents in the ocean floor, or in acidic lakes around volcanoes. They call these microbes **extremophiles**.

Why do scientists want them? Because if microbes can cope with harsh conditions, their enzymes will work in the same harsh conditions. So these enzymes may find a great many uses in industry.

▲ Many bacteria live around hot vents in the ocean floor – in water at up to 400 °C.

Checkup on Chapter 9

Revision checklist

Core syllabus content
Make sure you can …
- [ ] explain what *the rate of a reaction* means
- [ ] describe a practical method to investigate:
  - the rate of a reaction that produces a gas, by collecting the gas
  - the effect of concentration on the reaction rate
  - the effect of surface area on the reaction rate (by change in mass)
  - the effect of temperature on the reaction rate
- [ ] give the correct units for the rate of a given reaction (for example, cm^3 per minute, or grams per minute)
- [ ] work out, from the graph for a reaction:
  - how long the reaction lasted
  - how much product was obtained
  - the average rate of the reaction
  - the rate in any given minute
- [ ] describe the effect of these on the rate of reaction:
  - temperature
  - concentration of a solution
  - surface area of a solid reactant
  - pressure of a gas, in reactions involving gases
- [ ] say which of two reactions was faster, by comparing the slopes of their curves on a graph
- [ ] define these terms: *catalyst   enzyme*
- [ ] describe the effect on reaction rate of adding or removing a catalyst, including enzymes

Extended syllabus content
Make sure you can also …
- [ ] describe the collision theory
- [ ] define *activation energy* (E_a)
- [ ] define *a successful collision*, and explain its link to activation energy
- [ ] use the collision theory to explain why the rate of a reaction increases with concentration, temperature (two reasons!), surface area, and gas pressure
- [ ] interpret, and sketch, a reaction pathway diagram showing the effect of a catalyst on activation energy
- [ ] explain why a catalyst speeds up a reaction, in terms of activation energy and collisions

Questions

Core syllabus content

1 The rate of the reaction between magnesium and dilute hydrochloric acid can be measured using this apparatus:

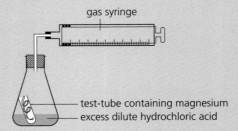

  a What is the purpose of:
   i the test-tube?   **ii** the gas syringe?
  b How would you get the reaction to start?

2 Some magnesium and an *excess* of dilute hydrochloric acid were reacted together. The volume of hydrogen produced was recorded every minute, as shown in the table:

Time / min	0	1	2	3	4	5	6	7
Volume of hydrogen / cm^3	0	14	23	31	38	40	40	40

  a What does an *excess* of acid mean?
  b Plot a graph of the results.
  c What is the *rate of reaction* (in cm^3 of hydrogen per minute) during:
   i the first minute?
   ii the second minute?
   iii the third minute?
  d Why does the rate change during the reaction?
  e How much hydrogen was produced in total?
  f How long does the reaction last?
  g What is the *average rate* of the reaction?
  h How could you slow down the reaction, while keeping the amounts of reactants unchanged?

3 Suggest a reason for each observation below.
  a Hydrogen peroxide decomposes much faster in the presence of the enzyme catalase.
  b The reaction between manganese carbonate and dilute hydrochloric acid speeds up when some concentrated hydrochloric acid is added.
  c Powdered magnesium is used in fireworks, rather than magnesium ribbon.

THE RATE OF REACTION

4 In two separate experiments, two metals A and B were reacted with an excess of dilute hydrochloric acid. The volume of hydrogen was measured every 10 seconds. These graphs show the results:

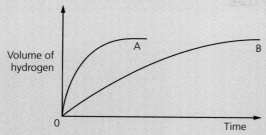

a i Which piece of apparatus can be used to measure the volume of hydrogen produced?
   ii What other measuring equipment is needed?
b Which metal, A or B, reacts faster with hydrochloric acid? Give your evidence.
c Sketch and label the curves that will be obtained for metal B if:
   i more concentrated acid is used (curve X)
   ii the reaction is carried out at a lower temperature (curve Y).

5 Copper(II) oxide catalyses the decomposition of hydrogen peroxide. 0.5 g of the oxide was added to a flask containing 100 cm³ of hydrogen peroxide solution. A gas was released. It was collected, and its volume noted every 10 seconds. This table shows the results:

Time/s	0	10	20	30	40	50	60	70	80	90
Volume/cm³	0	18	30	40	48	53	57	58	58	58

a What is a catalyst?
b Draw a diagram of suitable apparatus for this experiment.
c Name the gas that is formed.
d Write a balanced equation for the decomposition of hydrogen peroxide.
e Plot a graph of the volume of gas (vertical axis) against time (horizontal axis).
f Describe how rate changes during the reaction.
g What happens to the concentration of hydrogen peroxide as the reaction proceeds?
h What chemicals are present in the flask after 90 seconds?
i What mass of copper(II) oxide would be left in the flask at the end of the reaction?
j Sketch on your graph the curve that might be obtained for 1.0 g of copper(II) oxide.
k Name one other catalyt for this reaction.

Extended syllabus content

6 Marble chips (lumps of calcium carbonate) react with hydrochloric acid as follows:
$CaCO_3 (s) + 2HCl (aq) \longrightarrow$
$\quad CaCl_2 (aq) + CO_2 (g) + H_2O (l)$
a What gas is released during this reaction?
b Describe a laboratory method that could be used to investigate the rate of the reaction.
c How will this affect the rate of the reaction?
   i increasing the temperature
   ii adding water to the acid
d Explain each of the effects in **c** in terms of collisions between reacting particles.
e If the lumps of marble are crushed first, will the reaction rate change? Explain your answer.

7 Zinc and iodine solution react like this:
$Zn (s) + I_2 (aq) \longrightarrow ZnI_2 (aq)$
The rate of reaction can be followed by measuring the mass of zinc metal at regular intervals, until all the iodine has been used up.
a What will happen to the mass of the zinc, as the reaction proceeds?
b Which reactant is in excess? Explain your choice.
c The reaction rate slows down with time. Why?
d Sketch a graph showing the mass of zinc on the y axis, and time on the x axis.
e If the experiment is repeated at a higher temperature, the rate of reaction increases. Explain this increase in terms of collisions between particles, and activation energy.
f In industry, catalysts are used for chemical reactions that would otherwise be too slow.
   i What effect does a catalyst have on activation energy?
   ii Explain why this effect leads to a faster reaction.

8 a Ethene reacts with steam to form ethanol, as shown in this equation:
$C_2H_4 (g) + H_2O (g) \longrightarrow C_2H_5OH (l)$
The reaction is carried out at high pressure, in order to increase the reaction rate. Using the collision theory, explain why increasing the pressure leads to a higher rate for this reaction.
b Ethanol reacts with ethanoic acid like this:
$C_2H_5OH (l) + CH_3COOH (l) \longrightarrow$
$\quad CH_3COOC_2H_5 (l) + H_2O (l)$
Increasing the pressure has no effect on the rate of this reaction. Explain why.

10.1 Reversible reactions

Objectives: explain what a *reversible reaction* is, and give examples; understand that a reversible reaction reaches a state of equilibrium, and describe this state

What is a reversible reaction?

The blue crystals above are *hydrated* copper(II) sulfate. On heating, they turn to a white powder. This is *anhydrous* copper(II) sulfate.

The reaction is easy to reverse: add water! The anhydrous copper(II) sulfate gets hot and turns into the blue hydrated compound.

In fact, the anhydrous compound can be used to test for water. If it turns blue when a liquid is added, the liquid must contain water.

So the reaction above is **reversible**.
In a reversible reaction, you can change the direction of the reaction by changing the reaction conditions.

We use the symbol ⇌ instead of a single arrow, to show that a reaction is reversible. So the equation for the reaction above is shown as:

$$CuSO_4 \cdot 5H_2O\ (s) \rightleftharpoons CuSO_4\ (s) + 5H_2O\ (l)$$
    blue   white

The reaction from left to right in the equation is called the **forward reaction**. The reaction from right to left is the **reverse reaction**.

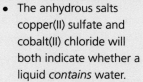

Tests for water
- The anhydrous salts copper(II) sulfate and cobalt(II) chloride will both indicate whether a liquid *contains* water.
- But they will *not tell you* whether it contains other things too!

Another example

These pink crystals are hydrated cobalt(II) chloride. On heating, they turn blue.

The blue crystals are anhydrous cobalt(II) chloride. Add water – and they turn pink again!

So blue cobalt(II) chloride paper is used to test for water. If water is present, the paper turns pink.

The equation for this reversible reaction is:

$$CoCl_2 \cdot 6H_2O\ (s) \rightleftharpoons CoCl_2\ (s) + 6H_2O\ (l)$$
    pink   blue

Reversible reactions and equilibrium

Nitrogen and hydrogen react to make ammonia, in a reversible reaction:

$$N_2 (g) + 3H_2 (g) \rightleftharpoons 2NH_3 (g)$$

Let's look at what happens during the reaction. We show it taking place in a **closed system** where no molecules can escape.

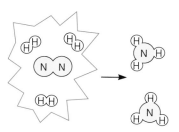 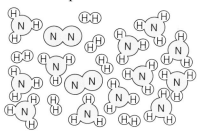 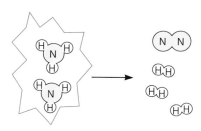

Three molecules of hydrogen and one of nitrogen react to give two molecules of ammonia. If you have the correct mixture of nitrogen and hydrogen in a closed container …

… will it all turn into ammonia? No! Once a certain amount of ammonia is formed, the system reaches a state of **dynamic equilibrium**. From then on …

… every time two ammonia molecules form, another two break down into nitrogen and hydrogen. So the level of ammonia remains unchanged.

What does *dynamic equilibrium* mean?

In the term *dynamic equilibrium*, **equilibrium** means there is no *overall* change. The amounts of nitrogen, hydrogen, and ammonia remain steady.

But **dynamic** means there is continual change at the particle level: ammonia molecules break down while new ones form. The forward and reverse reactions are both taking place, but at the same rate.

The term *dynamic equilibrium* is usually shortened to **equilibrium**.

A reversible reaction in a closed system is at equilibrium when:
- **the forward and reverse reactions take place at the same rate.**
- **the concentrations of reactants and products are no longer changing.**

A challenge for industry

Imagine you run a factory that makes ammonia. You want as high a yield as possible. But you see the problem: the reaction between nitrogen and hydrogen is never complete. Once equilibrium is reached, ammonia molecules break down while new ones form. So what can you do to increase the yield? Find out in the next unit.

▲ Worried about the yield?

1. What is a *reversible reaction*?
2. a What is *anhydrous* copper(II) sulfate?
   b How can it be turned into *hydrated* copper(II) sulfate?
3. a Some blue cobalt(II) chloride paper has turned pink. Why?
   b What will happen if the paper is placed in a warm oven?
4. Look at this equation: $NH_3 (g) + HCl (g) \rightleftharpoons NH_4Cl (s)$
   a How can you tell the reaction is reversible?
   b Write a word equation for the forward reaction.
5. Nitrogen and hydrogen are mixed in a closed system, to make ammonia.
   a Soon, two reactions are going on in the mixture. Give their equations, including the state symbols.
   b For a time, the rate of the forward reaction is greater than the rate of the reverse reaction. Has equilibrium been reached? Explain.
   c The reaction *never* goes to completion. Explain why.

10.2 Shifting the equilibrium

Objectives: state that the yield of product in a reversible reaction can be changed by changing the reaction conditions; explain how this applies in making ammonia

The challenge
Reversible reactions present a challenge to industry, because they *never complete*. Let's look at that reaction between nitrogen and hydrogen again:

$$N_2 (g) + 3H_2 (g) \rightleftharpoons 2NH_3 (g)$$

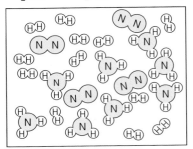
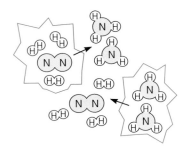
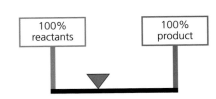

This represents the reaction mixture at equilibrium. The amount of ammonia in it will not increase …

… because for every two molecules of ammonia that form, another two break down.

Here the red triangle represents the equilibrium mixture. It is only part way along the scale. Why?

What can be done?
You want as much ammonia as possible. So how can you increase the yield? This idea will help you:

When a reversible reaction is in equilibrium and you make a change, the system acts to oppose the change, and restore equilibrium. A new equilibrium mixture forms.

A reversible reaction *always* reaches equilibrium, in a closed system. But by changing conditions, you can *shift equilibrium*, so that the mixture contains more product. Let's look at four changes you could make.

1 Change the temperature
Will raising the temperature help you obtain more ammonia? Let's see.

forward reaction:

$$N_2 + 3H_2 \xrightarrow[\text{given out}]{\text{heat}} 2NH_3$$

reverse reaction:

$$2NH_3 \xrightarrow[\text{taken in}]{\text{heat}} N_2 + 3H_2$$

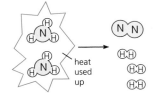
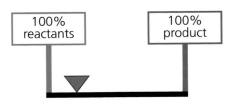

The forward reaction is exothermic – it gives out heat. The reverse reaction is endothermic – it takes heat in. Heating *speeds up both reactions* …

… but it favours the endothermic reaction. More ammonia breaks down (the reverse reaction) in order to use up the heat you added.

So the reaction reaches equilibrium faster – but the new equilibrium mixture has *less* ammonia. The yield has fallen.

What if you lower the temperature? This favors the *exothermic* reaction. More ammonia forms, giving out heat to oppose the change. But if the temperature is *too* low, the reaction takes too long to reach equilibrium.

120

REVERSIBLE REACTIONS AND EQUILIBRIUM

2 Change the pressure

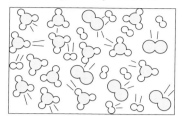

Suppose you increase the pressure by forcing more of the gas mixture into the container. *So there are more molecules present.*

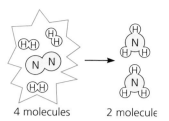

4 molecules   2 molecule

This change favours the forward reaction, because that will *reduce* the number of molecules present. (Look above.)

| 100% reactants | | 100% product |

So the amount of ammonia in the mixture *increases*. Equilibrium shifts to the right. Well done. You are on the right track.

3 Change the concentration
The equilibrium mixture is a balance between nitrogen, hydrogen, and ammonia. If ammonia is removed, more nitrogen and hydrogen will react to form ammonia, to restore equilibrium. So your yield increases!

4 Use a catalyst
A catalyst speeds up the forward and reverse reactions equally. So the reaction reaches equilibrium faster, which saves you time. **But a catalyst does not shift the position of equilibrium.** So the yield of ammonia does not change.

Choosing the optimum conditions
So when manufacturing ammonia, it is best to:
- use high pressure, and remove ammonia, to improve the yield
- use a moderate temperature, and a catalyst, to get a decent rate.

Find out how these ideas are applied in the factory, in the next unit.

What about the reaction rate?
- You know already that the **rate** of a reaction is increased by:
  - increasing the temperature
  - increasing the gas pressure, where a reactant is a gas
  - using a catalyst, where there is a suitable one.
- In a reversible reaction, these 'rules' apply to both the forward *and* reverse reactions. So the overall result is that **equilibrium is reached faster**.
- This is independent of whether the **yield** rises or falls.

Shifting the equilibrium: a summary
Changing concentration
- Adding more reactant or removing product means yield ↑.

Changing temperature
- Forward reaction *exothermic*: temperature ↑ means yield ↓.
- Forward reaction *endothermic*: temperature ↑ means yield ↑.

Changing pressure
(for reactions involving gases)
- Fewer molecules on the right of the equation: pressure ↑ means yield ↑.
- More molecules on the right: pressure ↑ means yield ↓.

Using a catalyst
Speeds up the reaction but has no effect on yield.

1 The yield is never 100%, when making ammonia. Why not?
2 Ammonia is made in factories, using this reaction:
   $N_2 (g) + 3H_2 (g) \rightleftharpoons 3NH_3 (g)$
   Where is the ideal position of equilibrium, for the factory?
   a as near to 100% reactants as possible
   b as near to 100% product as possible
   Explain your answer.
3 Explain why high pressure is used in making ammonia.
4 Which of these does *not* affect the yield of ammonia?
   temperature   pressure   a catalyst
5 Look at this reaction: $NH_3 (g) + HCl (g) \rightleftharpoons NH_4Cl (s)$
   The forward reaction is exothermic. Predict the effect of:
   a adding more solid ammonium chloride to the system
   b heating the reactants   c applying pressure

10.3 The Haber process

Objectives: outline the Haber process, and give its key reaction; describe how the reactants are obtained; state the reaction conditions used, and explain them

Putting it into practice

In Unit 10.2 you looked at ways to increase the yield, in the reversible reaction to make ammonia. Now let's see how those ideas are put into practice in industry. The process used to manufacture ammonia is called the **Haber process**. Follow the numbers in the diagram.

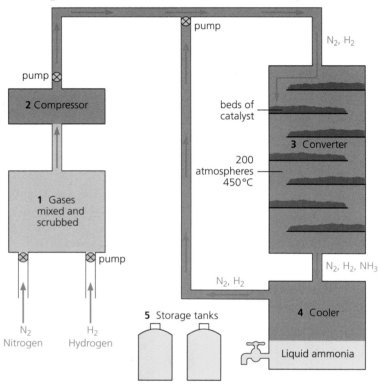

▲ Ammonia factories are often built close to oil refineries, to make use of the hydrogen from cracking.

1. The nitrogen for the Haber process is obtained from air, and the hydrogen by reacting natural gas (methane) with steam, or by cracking hydrocarbons. Check the panel on the right. The two gases are mixed, and **scrubbed** to remove impurities.
2. The mixture is **compressed**. More gas is pumped in until the pressure reaches 200 atmospheres. (Air pressure is around 1 atmosphere.)
3. The compressed gas flows to the **converter**: a round tank with beds of iron at 450 °C, where the key reaction occurs. Iron is the catalyst:

$$N_2 (g) + 3H_2 (g) \rightleftharpoons 2NH_3 (g)$$

But less than 30% of the mixture leaving the converter is ammonia.
4. The mixture is cooled until the ammonia condenses to a liquid. The nitrogen and hydrogen are recycled to the converter for another chance to react. Steps **3** and **4** are continually repeated.
5. The ammonia is run into tanks, and stored as a liquid under pressure.

Obtaining the reactants

nitrogen
Air is nearly 80% nitrogen, and 20% oxygen. The oxygen is removed by burning hydrogen:
$$2H_2 (g) + O_2 (g) \longrightarrow 2H_2O (l)$$
That leaves mainly nitrogen, and a small amount of other gases.

hydrogen
- It is usually made by reacting **natural gas** (methane) with steam:

$$CH_4 (g) + 2H_2O (g) \xrightarrow{catalyst} CO_2 (g) + 4H_2 (g)$$

- It is also made by **cracking hydrocarbons** from petroleum. For example:

$$C_2H_6 (g) \xrightarrow{catalyst} C_2H_4 (g) + H_2 (g)$$
ethane → ethene

REVERSIBLE REACTIONS AND EQUILIBRIUM

Improving the yield of ammonia

Here again is the key reaction:

$$N_2(g) + 3H_2(g) \underset{\text{endothermic}}{\overset{\text{exothermic}}{\rightleftharpoons}} 2NH_3(g)$$

To obtain a high yield of ammonia, the advice in Unit 10.2 was:
- use high pressure and remove ammonia, to improve the yield
- use a moderate temperature and a catalyst, for a decent rate.

The chosen conditions

The temperature and pressure The graph on the right shows that the yield is highest at X, at 350 °C and 400 atmospheres.
But **speed**, **safety**, and **cost** must be taken into account too. So the Haber process uses 450 °C and 200 atmospheres – at Y on the graph. Why? Because:
- at 350 °C, the reaction is too slow. Time costs money, in a factory. Customers are waiting. 450 °C gives a faster rate.
- maintaining a pressure of 400 atmospheres needs powerful pumps, very strong and sturdy pipes and tanks, and a lot of electricity. 200 atmospheres is safer, and saves money.

In fact, the chosen conditions give a low yield in the converter. But the Haber process is designed to cope with this. *The converter is not a closed system*. The gases flow through it continually, reacting on the catalyst. The ammonia is removed, and the unreacted gases are recycled, for another chance to react. So the final yield is almost 100%.

The catalyst Iron speeds up the reaction – but does not change the yield.

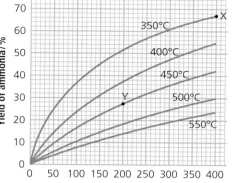

The yield of ammonia at different temperatures and pressures

The conditions: a summary

$$N_2 + 3H_2 \underset{\text{endothermic}}{\overset{\text{exothermic}}{\rightleftharpoons}} 2NH_3$$
4 molecules   2 molecules

to improve the yield:
- a pressure of 200 atmospheres (or 20 000 kilopascals)
- remove ammonia

for a reasonable rate:
- 450 °C
- use a catalyst (iron)

◀ Most of the ammonia from the Haber process is used to make fertilisers, to help crop yields. The rest is used in making dyes, pesticides, and other chemicals.

1. Name the industrial process for making ammonia.
2. State the source of this reactant, for the process in **1**:
   a nitrogen
   b hydrogen
3. Look at the diagram on page 122.
   a Where in the process does the key reaction take place, to produce ammonia?
   b Name the catalyst used.
   c Why are the catalyst beds arranged this way?
   d Why is the ammonia stored as a liquid? Suggest a reason.
4. a 400 atmospheres and 250 °C would give the best yield. Why are these conditions *not* used in the Haber process?
   b State the actual conditions used.
   c What is the yield in the converter, for the chosen conditions? (Use the graph above.)
5. The conditions in the converter do not give a high yield. But two further steps are built into the Haber process, to push the final yield for the process to almost 100%. Describe these steps.

10.4 The Contact process

Objectives: give two sources of sulfur dioxide for the Contact process; outline the process and give the key reactions; state the conditions used, and explain them

Making sulfuric acid
Sulfuric acid is the world's number one chemical. And like ammonia, the key reaction in making it is a **reversible reaction**. The process for its manufacture is called the **Contact process**.

The raw materials
The raw material are **sulfur** *or* **sulfur dioxide**, plus **air** and **water**. Sulfur is mined. Sulfur dioxide is produced when metal sulfide ores (such as lead sulfide) are roasted in air, to obtain the metal.

The process
Let's start with sulfur.

▲ A furnace for burning sulfur to give sulfur dioxide, at a sulfuric acid factory.

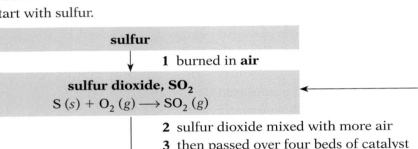

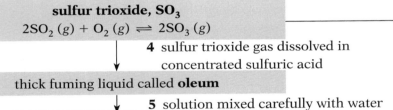

- The key reaction of the process takes place in step **3**. It is **reversible** and **exothermic**. It produces **sulfur trioxide, SO₃**.
- To keep the temperature down to 450 °C in step **3**, heat is removed from the beds of catalyst. Pipes of cold water are coiled around them to carry heat away.
- The sulfur trioxide from step **3** reacts with water to produce **sulfuric acid**. But this is a violent reaction, giving a thick deadly mist of acid. So for safety, it is carried out in two stages: steps **4** and **5**.
- In step **4**, the sulfur trioxide is dissolved in concentrated sulfuric acid, giving a thick fuming liquid. In step **5**, this is then mixed with water.
- The unreacted sulfur dioxide from step **3** is recycled – passed over the beds of catalyst again, for another chance to react.

> **Not a closed system**
> - When a reversible reaction takes place in a closed system, no particles escape.
> - So the reversible reaction reaches equilibrium, and will *never* go to completion.
> - In the Contact process – and the Haber process – the system is not closed. The aim is to *prevent* equilibrium being established!
> - Instead, the product is removed, and unreacted gas(es) are recycled, to get as high a yield as possible.

REVERSIBLE REACTIONS AND EQUILIBRIUM

The conditions used in the Contact process

The key reaction in the Contact process is reversible, as in the Haber process. So the factory faces the same challenge: to increase the yield. But once again, the issues of **speed**, **safety**, and **cost** must be considered.

How to improve the yield
Here again is the key reaction:

$$2SO_2 (g) + O_2 (g) \underset{\text{endothermic}}{\overset{\text{exothermic}}{\rightleftharpoons}} 2SO_3 (g)$$

3 molecules → 2 molecules

Applying what we know already, these will increase the yield:
- a low temperature, since the forward reaction is exothermic
- high pressure, since there are fewer molecules on the right
- remove the product, to lower its concentration so that more will form.

The graph on the right shows how yield falls as the temperature rises.

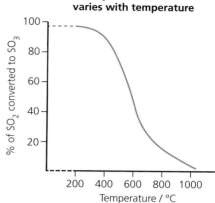

▲ How the yield of sulfur trioxide changes with temperature.

The chosen conditions
In fact the first two conditions above would make the process unprofitable for the manufacturer. So here are the actual conditions used:
- 450 °C. The catalyst is essential, but is inactive below 400 °C, and works better at higher temperatures. So 450 °C is a compromise
- 2 atmospheres (or 200 kilopascal) of pressure. This is low, but the yield is acceptable. Raising the pressure further would cost more
- removal of the sulfur trioxide – safely – by dissolving it in sulfuric acid
- the catalyst of vanadium(V) oxide.

▲ Vanadium(V) oxide, the catalyst.

Note again that the reaction is not carried out in a closed system. Sulfur dioxide flows through the reaction vessel continually, reacting on the catalyst. Sulfur trioxide is continually removed. Unused sulfur dioxide is recycled. The final yield for the process is over 99%.

Sulfuric acid: the world's top chemical
- When you next see a bottle of sulfuric acid in the lab, think about this: it is by far the world's top chemical, in terms of tonnes produced.
- That is because it has so many important uses.
- It is used in making fertilisers, paints, pigments, dyes, plastics, textile fibres, soaps, detergents, and much more. Its main use is for fertilisers.
- It is also the acid used in lead-acid car batteries.

▲ Starting on a journey, thanks to sulfuric acid. It is the acid in the lead-acid battery which powers the starting motor.

1. Sulfuric acid is manufactured using the Contact process.
   a Name the two gases that react, and give their sources.
   b Name the catalyst used.
2. The key reaction in the Contact process is the reaction to convert sulfur dioxide to sulfur trioxide.
   a Write the symbol equation for it.
   b The reverse reaction also occurs on the catalyst beds. Write the symbol equation for it too.
3. The reversible reaction to produce sulfur trioxide is carried out at 450 °C. A lower temperature would give a higher yield, as the graph above shows. Why is 450 °C chosen?
4. These help to raise the yield of sulfur trioxide. Explain how.
   a Several beds of catalyst are used.
   b The sulfur trioxide is removed by dissolving it.
5. a Give one safety precaution taken in the Contact process.
   b Describe one choice that helps to keep costs down.

Checkup on Chapter 10

Revision checklist

Core syllabus content
Make sure you can ...
- [ ] explain what a *reversible reaction* is, with examples
- [ ] write the symbol for a reversible reaction
- [ ] give the equation for the reversible reaction that links:
  - hydrated and anhydrous copper(II) sulfate
  - hydrated and anhydrous cobalt(II) chloride
  and describe how to change the direction of each reaction, using heat and water.
- [ ] explain why the anhydrous compounds copper(II) sulfate and cobalt(II) chloride can be used to test for the *presence* of water (but do not tell you whether a liquid is pure water)

Extended syllabus content
Make sure you can also ...
- [ ] explain that a reversible reaction in a closed container reaches a state of equilibrium
- [ ] decribe what is happening to the reactants and products, at equilibrium
- [ ] explain that the position of equilibrium can be shifted by changing the reaction conditions
- [ ] predict the effect of a change in conditions for a reversible reaction, given the equation for the reaction, and ΔH where required
- [ ] state that a catalyst brings a reaction to equilibrium faster – but has no effect on the yield
- [ ] explain that % yield is important in industry, but that speed, safety and cost are also considered, in choosing the reaction conditions
- [ ] give the equation, with state symbols, for the key reaction in the Haber process
- [ ] state and explain the conditions used in the Haber process
- [ ] give the equations, with state symbols, for the reactions in the Contact process, and indicate the key reaction
- [ ] state and explain the conditions used in the Contact process

Questions

Core syllabus content

1 Hydrated copper(II) sulfate crystals were heated:

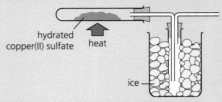

  a What is the ice for?
  b What colour change will occur in the test-tube?
  c The reaction is *reversible*. What does that mean?
  d How would you show that it is reversible?
  e Write the equation for the reversible reaction.

2 Which of these statements are *not* correct for reversible reactions?
  a Forward and reverse reactions are possible.
  b Energy is always taken in, when going from left to right in a reversible reaction.
  c The symbol ⟶ shows that a reaction is reversible.
  d Water is always involved in reversible reactions.
  e If the forward reaction gives out heat, then the reverse reaction will take in heat.
  f Changing the reaction conditions can affect the direction of a reversible reaction.

3 Hydrated cobalt(II) chloride has the formula $CoCl_2 \cdot 6H_2O$. It is heated in a test-tube.

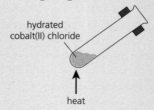

Droplets of liquid appear at the top of the test-tube, and the solid in the test-tube changes colour.
  a i Explain why the droplets of liquid appear.
    ii What colour change occurs in the solid?
    iii Give the name and formula of the compound that remains in the test-tube at the end.
  b The reaction is reversible. Write the chemical equation for it, including the state symbols.
  c What use is made of this reaction in the lab?

Extended syllabus content

4 Many key reactions in industry are reversible. They are *not* carried out in a closed system. The reaction mixture is *not* allowed to reach equilibrium.
  a i What is meant by *a closed system*? (Glossary?)
  ii When a reversible reaction reaches equilibrium in a closed system, the reaction appears to stop. Explain why.
  iii Suggest a reason why industrial processes are not usually allowed to reach equilibrium.
  b In the Haber process, what steps are taken to prevent the reaction from reaching equilibrium?
  c How does using a catalyst affect the relative concentrations of reactants and products, in a reversible reaction?

5 Hydrogen iodide decomposes when heated, into hydrogen and iodine. The reaction is reversible:

$$2HI\,(g) \rightleftharpoons H_2\,(g) + I_2\,(g)$$

Hydrogen iodide and hydrogen are colourless gases. Iodine vapour is purple.
  a What will you expect to see if hydrogen iodide is placed in a closed container, and heated?
  b What observation will indicate to you that equilibrium has been reached?
  c If the equilibrium mixture is heated further, more hydrogen iodide will decompose. Which diagram below, Ⓐ or Ⓑ, represents the reaction pathway for the reaction? Explain your choice.

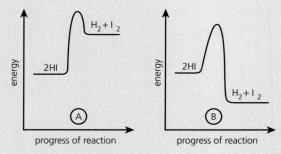

  d What effect will an increase in pressure have on the position of equilibrium? Explain your answer.

6 This ionic equation shows the equilibrium that exists between two different cobalt ions in water:

$$[Co(H_2O)_6]^{2+} + 4Cl^- \rightleftharpoons [CoCl_4]^{2-} + 6H_2O$$
  pink   blue

You have a purple solution containing both ions.
  a How will adding water affect the equilibrium?
  b On heating, the purple solution turns blue. What does that tell you about this reversible reaction?

7 The key reaction in manufacturing sulfuric acid is the conversion of sulfur dioxide to sulfur trioxide. This reversible reaction is carried out at 450 °C and 2 atmospheres pressure, in the presence of vanadium(V) oxide.
  a i Name the industrial process used.
  ii Write the equation for the conversion reaction above, including state symbols.
  b Give two sources for the sulfur dioxide used.
  c i The % of sulfur dioxide converted is much higher at 25 °C than at 450 °C. Why is this?
  ii The process is *not* carried out at 25 °C. Explain why.
  d i The % of sulfur dioxide converted is much higher at 200 atm than at 2 atm. Explain why.
  ii The process is *not* carried out at 200 atm. Why?
  e What effect does vanadium(V) oxide have on:
  i the reaction rate?
  ii the % of sulfur dioxide converted?

8 Hydrogen and bromine react reversibly:
$$H_2\,(g) + Br_2\,(g) \rightleftharpoons 2HBr\,(g)$$
  a Which of these will favour the formation of more hydrogen bromide?
  i add more hydrogen
  ii remove bromine
  iii remove the hydrogen bromide as it forms
  b Pressure has no effect on the position of equilibrium for this reaction. Explain why.
  c However, the pressure *is* likely to be increased, when the reaction is carried out in industry. Suggest a reason for this.

9 Ammonia is made from nitrogen and hydrogen. The reaction is reversible. The enthalpy change for the reaction, ΔH, is -92 kJ/mol.
  a Name a source for the hydrogen used.
  b Write the equation for the reaction.
  c Is the forward reaction *endothermic*, or *exothermic*? Give your evidence.
  d Explain why the yield of ammonia:
  i rises if you increase the pressure
  ii falls if you increase the temperature
  e What effect does increasing this factor have on the rate at which ammonia is made?
  i the pressure
  ii the temperature
  f Why is the reaction carried out at 450 °C, rather than a lower temperature?
  g i Name the catalyst used in this reaction.
  ii What effect does the catalyst have on the % yield of ammonia?

11.1 Acids and bases

Objectives: define *bases* and *alkalis*; name the common acids, alkalis, and indicators used in the lab; state the colour changes for the indicators

Two more groups of chemicals
By now you have met several groups of substances that have similar properties, in chemistry. For example, the metals, non-metals, ionic compounds, and covalent compounds.

This chapter looks at two new groups of compounds, **acids** and **bases**.

Acids

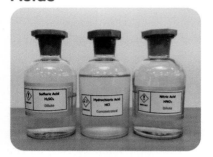  

Acids are **non-metal compounds** in water. These are the main acids you will meet in the lab. They can be **dilute** or **concentrated**.

They must be handled carefully, especially the concentrated acids, because they are **corrosive**. They can eat away metals, cloth, skin.

But some corrode only slowly, even when concentrated. An example is ethanoic acid. It is the acid that gives vinegar its sharp taste.

The formulae of the four acids above are:

sulfuric acid	H_2SO_4 (aq)
hydrochloric acid	HCl (aq)
nitric acid	HNO_3 (aq)
ethanoic acid	CH_3COOH (aq)

There are many other acids too. For example, lemons contain **citric acid**, yoghurt contains **lactic acid**, and fizzy drinks contain **carbonic acid**.

The litmus test for acids
How can you tell if a solution contains an acid? The quickest way is to test it with **litmus**. Litmus is a purple mixture of natural dyes. It can be used as a solution, or adsorbed onto paper. **Acids turn litmus red.**

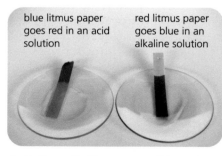

▲ Testing with litmus paper. It's easy!

◀ Far left: fizzy drinks contain carbonic acid, made by dissolving carbon dioxide in water under pressure. When you open the bottle the carbon dioxide fizzes out again.

◀ Left: litmus is made from dyes obtained from certain species of lichen. Lichen grows on rocks and tree trunks.

ACIDS, BASES, AND SALTS

Bases
Bases are oxides and hydroxides of metals.

Examples are: calcium oxide, CaO, magnesium oxide, MgO, sodium hydroxide, NaOH, and magnesium hydroxide Mg(OH)$_2$. As you can see, the hydroxides have **OH** in their formulae.

Bases are generally insoluble in water. But there is an important group of soluble hydroxides: the alkalis.

▲ Three bases. What are their formulae?

Alkalis
Alkalis are soluble bases. They are hydroxides.

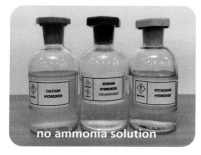

For example, these are pellets of the alkali sodium hydroxide. They dissolve in water, giving an **alkaline** solution.

You will meet these solutions of alkalis in the lab. (Calcium hydroxide is only slightly soluble. Its solution is called **limewater**.)

Alkalis are corrosive, especially the solids and concentrated solutions. They attack skin, and some metals. They can do more harm than acids.

The formulae of the alkalis in the bottles above are:

calcium hydroxide	Ca(OH)$_2$ (aq)
sodium hydroxide	NaOH (aq)
potassium hydroxide	KOH (aq)

The litmus test for alkalis
You can also use litmus to test whether a solution is alkaline. **Solutions of alkalis turn litmus blue.**

Remember:
aci**d** turns litmus re**d**
alkali turns litmus blue

Indicators
Litmus is called an **indicator**, because it indicates whether something is an acid or an alkali.

But it is not the only indicator you will meet in the lab. This table includes two others which are used only in solution (not on test paper).

Look at their colour changes from acid to alkali.

Indicator	Colour in solutions containing ...	
	acids	alkalis
litmus	red	blue
methyl orange	red	yellow
thymolphthalein	colourless	blue

1. What does *corrosive* mean? (Glossary?)
2. Name four laboratory acids, and give their formulae.
3. What is a *base*, in chemistry?
4. a What defines the group of bases called *alkalis*?
   b Give the names and formulae of two alkalis.
5. Why is litmus called an *indicator*?
6. Is this solution *acidic* or *alkaline*?
   a It turns litmus red. b Methyl orange is yellow in it.
   c It turns both litmus and thymolphthalein blue.
7. Give the colour change for the indicator thymolphthalein.

11.2 A closer look at acids and alkalis

Objectives: name the ions responsible for acidity and alkalinity; describe and explain the pH scale; describe universal indicator; define *strong* and *weak* acids

What do all acids have in common?
Let's take hydrochloric acid as an example. It is made by dissolving hydrogen chloride gas in water. Hydrogen chloride is made of molecules. But in water, the molecules break up or **dissociate** into ions:

$$HCl\ (aq) \longrightarrow H^+\ (aq) + Cl^-\ (aq)$$

So hydrochloric acid contains hydrogen ions. And so do all other solutions of acids. The hydrogen ions give them their 'acidity'.
Solutions of acids contain hydrogen ions.

What do all alkalis have in common?
Now let's turn to alkalis, with sodium hydroxide as our example. It is an ionic solid. When it dissolves, all the ions separate:

$$NaOH\ (aq) \longrightarrow Na^+\ (aq) + OH^-\ (aq)$$

So sodium hydroxide solution contains hydroxide ions.
The same is true of all alkaline solutions.
Solutions of alkalis contain hydroxide ions.

> **Remember!**
> - Solutions of acids contain hydrogen ions:
>
>
>
> That is what makes them *acidic*.
> - Solutions of alkalis contain hydroxide ions:
>
>
>
> That is what makes them *alkaline*.

The pH scale
You can say how acidic or alkaline a solution is using a special scale of numbers called the **pH scale**. The numbers go from 0 to 14. Look:

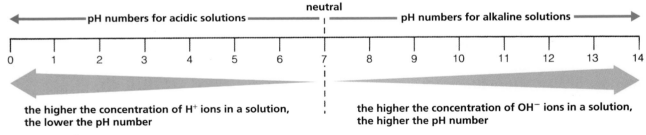

On this scale:

An acidic solution has a pH number less than 7.
An alkaline solution has a pH number greater than 7.
A neutral solution has a pH number of exactly 7.

Pure water is neutral, with a pH of 7. A solution of sugar is neutral too.

◀ Soon to reach his stomach, where the pH is around 2. In our stomachs, a solution of hydrochloric acid activates the enzymes that digest protein. The stomach lining stops it leaking out.

ACIDS, BASES, AND SALTS

Universal indicator

The indicators on page 129 tell you whether a solution is acidic, or alkaline. But **universal indicator** tells you the pH. Its colour changes across the pH scale as shown below. It can be used in solution, or as a strip of test paper:

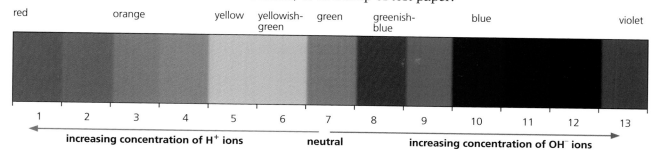

Strong and weak acids

To find the rough pH of a solution, you can use universal indicator. For a precise measurement you use a **pH meter**, as shown on the right.

Solutions of the two acids in the table below were prepared, both with the same concentration. Their pH was measured using a pH meter. You might expect both to have the same pH. But no! Look at the last column:

Acid solution	Concentration	Ions in acid	pH
hydrochloric acid	0.1 mol/dm^3	H^+ and Cl^-	1.08
ethanoic acid	0.1 mol/dm^3	H^+ and CH_3COO^-	2.88

The ethanoic acid solution has a higher pH – which means it has fewer H^+ ions. Why? Compare their equations:

$$HCl\ (aq) \xrightarrow{100\%\ dissociation} H^+\ (aq) + Cl^-\ (aq)$$

In hydrochoric acid solution, all the hydrogen chloride molecules are dissociated into ions. So hydrochloric acid is called a **strong acid**.

Now look at ethanoic acid:

$$CH_3COOH\ (aq) \xrightleftharpoons{less\ than\ 100\%\ dissociation} H^+\ (aq) + CH_3COO^-\ (aq)$$

This time the dissociation is *reversible*. At equilibrium, many ethanoic acid molecules are *not* dissociated. So the ethanoic acid solution has fewer H^+ ions, and a higher pH, than the hydrochloric acid solution. Ethanoic acid is called a **weak** acid.

A strong acid is completely dissociated into ions in solution.
A weak acid is partially dissociated into ions in solution.

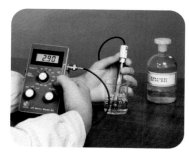

▲ Using a pH meter to measure pH. You dip a probe into the solution. The probe contains two electrodes.

▲ Strong and weak. The car battery contains sulfuric acid, a strong acid. Oranges contain citric acid, a weak acid.

Q

1. All acid solutions have something in common. What is it?
2. All alkaline solutions have something in common. What?
3. A solution turns universal indicator yellow. What is its pH?
4. What can you tell about a solution with this pH value?
   a 9   b 4   c 7   d 1   e 13   f 3
5. Which has a higher concentration of hydroxide ions?
   a a solution with pH 11   b a solution with pH 13
6. Solution A has a higher concentration of H^+ ions than solution B does. Which has a higher pH, A or B?
7. Name the indicator that tells you the pH of a solution.
8. Solutions of acids X and Y have the same concentrations. X has pH 2.5. Y has pH 5.3. Which one is a weak acid?
9. A solution of a strong acid has a pH of 2. How will the pH change when you add more water? Explain your answer.

11.3 The reactions of acids and bases

Objectives: give the key reactions of acids and bases, with equations; define *neutralisation*; explain *why* the neutralisation of an acid by an alkali gives water

When acids react to give salts

Acids react with metals, bases, and carbonates, to give **salts**.
Salts are ionic compounds. Sodium chloride, NaCl, is an example.
The name of the salt depends on the acid you start with:

hydrochloric acid	gives	chlorides
sulfuric acid	gives	sulfates
nitric acid	gives	nitrates

Typical acid reactions

1 **With metals:** acid + metal ⟶ salt + hydrogen

For example:

magnesium + sulfuric acid ⟶ magnesium sulfate + hydrogen

$Mg\,(s) + H_2SO_4\,(aq) \longrightarrow MgSO_4\,(aq) + H_2\,(g)$

The metal **displaces** hydrogen: it drives it from the acid, and takes its place. A solution of the salt magnesium sulfate forms.

2 **With bases:** acid + base ⟶ salt + water

Remember, bases are oxides and hydroxides of metals, and alkalis are soluble bases. All react with acids to give a salt and water.

Example for an acid and alkali:

hydrochloric acid + sodium hydroxide ⟶ sodium chloride + water

$HCl\,(aq) + NaOH\,(aq) \longrightarrow NaCl\,(aq) + H_2O\,(l)$

Example for an acid and insoluble base:

sulfuric acid + copper(II) oxide ⟶ copper(II) sulfate + water

$H_2SO_4\,(aq) + CuO\,(s) \longrightarrow CuSO_4\,(aq) + H_2O\,(l)$

3 **With carbonates:**

acid + carbonate ⟶ salt + water + carbon dioxide

For example:

calcium carbonate + hydrochloric acid ⟶ calcium chloride + water + carbon dioxide

$CaCO_3\,(s) + 2HCl\,(aq) \longrightarrow CaCl_2\,(aq) + H_2O\,(l) + CO_2\,(g)$

Reactions of bases

1 Bases react with acids, as you saw above, giving **only** a salt and water. **That is how you can identify a base.**

2 Bases such as sodium, potassium, and calcium hydroxides react with ammonium salts, driving out ammonia gas. For example:

calcium hydroxide + ammonium chloride ⟶ calcium chloride + water + ammonia

$Ca(OH)_2\,(s) + 2NH_4Cl\,(s) \longrightarrow CaCl_2\,(s) + 2H_2O\,(l) + 2NH_3\,(g)$

This reaction is used to make ammonia in the lab.

▲ Magnesium reacting with dilute sulfuric acid. Hydrogen bubbles off.

▲ In **A**, black copper(II) oxide is reacting with dilute sulfuric acid. The solution turns blue as copper(II) sulfate forms. **B** shows how the final solution will look.

▲ Calcium carbonate reacting with dilute hydrochloric acid. What gas is forming?

Neutralisation

Neutralisation is a reaction with acid that gives *water* as well as a salt.
So the reactions of bases and carbonates with acids are neutralisations.
We say the acid is **neutralised**.

But the reactions of acids with metals are not neutralisations. Why not?

Where does the water come from, in neutralisation?

Neutralisation always produces water. Where does it come form?
Let's look at the reaction between solutions of hydrochloric acid and sodium hydroxide again:

$$HCl\,(aq) + NaOH\,(aq) \longrightarrow NaCl\,(aq) + H_2O\,(l)$$

The hydrochloric acid solution contains H^+ and Cl^- ions. The sodium hydroxide solution contains Na^+ and OH^- ions.

Water is produced when the hydrogen and hydroxide ions combine:

$$\underset{\text{hydrogen ions}}{H^+\,(aq)} + \underset{\text{hydroxide ions}}{OH^-\,(aq)} \longrightarrow \underset{\text{water molecules}}{H_2O\,(l)}$$

The sodium ions and chloride ions are still there, in the solution. If you evaporate all the water, you will get solid sodium chloride.

▲ This soil is too acidic, so the farmer is spreading the base calcium oxide (lime) on it, to neutralise the acidity.

Neutralisation outside the lab

We make use of neutralisation outside the lab – usually to reduce acidity.
For example …

- Bacteria in your mouth feed on sugars in food and sweet drinks, producing acids. These attack the enamel in your teeth, leading to **tooth decay**. But toothpaste contains bases which neutralise the acidity. Brush your teeth!
- Our stomachs produce hydrochloric acid. (A layer of mucus protects the stomach lining from it.) Eating too much of certain foods leads to excess acid, which causes **indigestion**. Medicines for indigestion contain bases, to neutralise the excess acidity.
- Crops grow better in soil with a pH close to 7 (neutral). But sometimes soil is too acidic – because of the rock it came from, or acid rain, or heavy use of fertilisers. So a base – usually **limestone** (calcium carbonate) or **lime** (calcium oxide) or **slaked lime** (calcium hydroxide) is spread on the soil to reduce acidity.

1. Write a word equation for the reaction of dilute sulfuric acid with: **a** zinc **b** sodium carbonate
2. Which reaction in question **1** is *not* a neutralisation?
3. Salts are ionic compounds. Name the salt that forms when calcium oxide reacts with hydrochloric acid, and say which ions it contains.
4. Zinc oxide is a base. Suggest a way to make zinc nitrate from it. Write a word equation for the reaction.
5. In what ways are the reactions of hydrochloric acid with calcium oxide and calcium carbonate:
   a similar?   **b** different?
6. Write the equation showing the reaction between hydrogen ions and hydroxide ions, to give water.
7. **a** *Slaked lime* is the everyday name for solid calcium hydroxide. It helps to control acidity in soil. Explain why.
   b Name one product that will form during the reaction.

11.4 A closer look at neutralisation

Objectives: give the ionic equations for the neutralisation of an acid by soluble and insoluble bases; define acids as *proton donors*, and bases as *proton acceptors*

The neutralisation of an acid by an alkali (a soluble base)

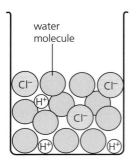

This is a solution of hydrochloric acid. It contains H^+ and Cl^- ions. It will turn litmus red.

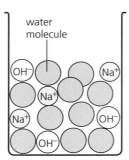

This is a solution of sodium hydroxide. It contains Na^+ and OH^- ions. It will turn litmus blue.

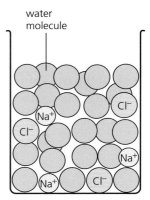

When you mix the two solutions, the OH^- ions and H^+ ions join to form **water molecules**.

The result is a neutral solution of sodium chloride, with no effect on litmus:

$$HCl\,(aq) + NaOH\,(aq) \longrightarrow NaCl\,(aq) + H_2O\,(l)$$

The ionic equation for the reaction

An ionic equation will show what goes on in the neutralisation reaction.

An ionic equation shows just the ions that take part in the reaction.

This is how to write the ionic equation for the reaction above:

1 First, write out the full equation, showing all the ions present.

The drawings above will help you:

$$H^+\,(aq) + Cl^-\,(aq) + Na^+\,(aq) + OH^-\,(aq) \longrightarrow$$
$$Cl^-\,(aq) + Na^+\,(aq) + H_2O\,(l)$$

2 Now cross out any ions that appear on both sides of the equation.

$$H^+\,(aq) + \cancel{Cl^-\,(aq)} + \cancel{Na^+\,(aq)} + OH^+\,(aq) \longrightarrow$$
$$\cancel{Cl^-\,(aq)} + \cancel{Na^+\,(aq)} + H_2O\,(l)$$

The crossed-out ions are present in the solution, but do not take part in the reaction. So they are called **spectator ions**.

3 What is left is the ionic equation for the reaction.

$$H^+\,(aq) + OH^-\,(aq) \longrightarrow H_2O\,(l)$$

So an H^+ ion combines with an OH^- ion to produce a water molecule. This is all that happens during neutralisation.

During neutralisation, H^+ ions combine with OH^- ions to form water molecules.

But an H^+ ion is just a **proton**, as the drawing on the right shows. So, in effect, the acid **donates** (gives) **protons** to the hydroxide ions. The hydroxide ions accept these protons, to form water molecules.

▲ A titration: sodium hydroxide solution was added to hydrochloric acid, from the burette. Neutralisation is complete: the thymolphthalein has turned blue.

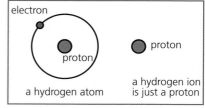

a hydrogen atom

a hydrogen ion is just a proton

134

ACIDS, BASES, AND SALTS

The neutralisation of an acid by an insoluble base

Magnesium oxide is insoluble. It does not produce hydroxide ions. So how does it neutralise an acid? Like this:

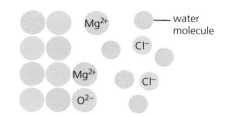

Magnesium oxide is a lattice of magnesium and oxide ions. It is insoluble in water. But when you add dilute hydrochloric acid …

… the acid donates protons to the oxide ions. The oxide ions accept them, forming water molecules. So the lattice breaks down.

The magnesium ions join the chloride ions in solution. If you evaporate the water you will obtain the salt magnesium chloride.

The equation for this neutralisation reaction is:

$2HCl\ (aq) + MgO\ (s) \longrightarrow MgCl_2\ (aq) + H_2O\ (l)$

The ionic equation for it is:

$2H^+\ (aq) + O^{2-}\ (s) \longrightarrow H_2O\ (l)$

Proton donors and acceptors

Now compare the ionic equations for the two neutralisations in this unit:

$H^+\ (aq) + OH^-\ (aq) \longrightarrow H_2O\ (l)$
$2H^+\ (aq) + O^{2-}\ (s) \longrightarrow H_2O\ (l)$

In both:

- the protons are donated by the acids
- ions in the bases accept them, forming water molecules.

So this gives us a new definition for acids and bases:

Acids are proton donors, and bases are proton acceptors.

▲ Help is at hand! Magnesium hydroxide is an insoluble base. Milk of Magnesia is a suspension of magnesium hydroxide. In your stomach, it will neutralise the excess hydrochloric acid that causes indigestion.

1 a What is an *ionic equation*?
   b Hydrochloric acid reacts with a solution of potassium hydroxide. What do you expect the ionic equation for this neutralisation reaction to be? Write it down.
2 What are *spectator ions*? (Glossary?).
3 An H^+ ion is just a proton. Explain why. (Do a drawing?)
4 a Acids act as *proton donors*. What does that mean?
   b Bases act as *proton acceptors*. Explain what that means.
5 A challenge! In a *redox* reaction, oxidation numbers change. (See page 77.) Using the full equation for the MgO / HCl reaction, show that neutralisation is *not* a redox reaction.
6 How to write an ionic equation:
   i Write out the full equation, showing all the ions present.
   ii Cross out ions that appear on both sides of the equation.
   iii What is left is the ionic equation. Rewrite it neatly.
   a Follow steps **i – iii** for the reaction between magnesium oxide and hydrochloric acid, shown on this page.
   b Does your ionic equation match ours? If so, well done!
7 Hydrochloric acid is neutralised by a solution of sodium carbonate. Write the ionic equation for this reaction
8 Give the chemical equation for the reaction that will soon take place in the stomach of the man in the photo above.

11.5 Oxides

Objectives: understand that, in general, metal oxides are bases, and non-metal oxides are acidic; define *amphoteric oxide* and give two examples

What are oxides?
Oxides are compounds containing oxygen and another element. Here we look at different types of oxides, and their behaviour.

Basic oxides
Look how these metals react with oxygen:

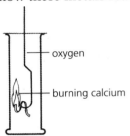

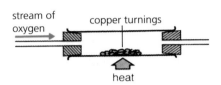

A piece of calcium is heated, then plunged into a gas jar of oxygen. It bursts into flame: intense white with a tinge of red. The white ash that forms is **calcium oxide**:

$Ca\ (s) + O_2\ (g) \rightarrow CaO\ (s)$

Hot iron wool is plunged into a gas jar of oxygen. It glows bright orange, and throws out a shower of sparks. A black solid is left in the gas jar. It is **iron(III) oxide**:

$4Fe\ (s) + 3O_2\ (g) \rightarrow 2Fe_2O_3\ (s)$

Copper is too unreactive to catch fire in oxygen. But when it is heated in a stream of the gas, its surface turns black. The black substance is **copper(II) oxide**:

$2Cu\ (s) + O_2\ (g) \rightarrow 2CuO\ (s)$

The more reactive the metal, the more vigorously it reacts.

The copper(II) oxide produced in the last reaction above is insoluble in water. But it does dissolve in dilute acid:

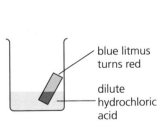
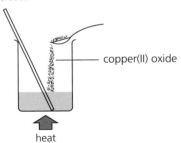
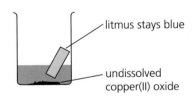

This is dilute hydrochloric acid. It turns blue litmus paper red, like all acids do.

Copper(II) oxide dissolves in it, when it is warmed. But after some time, no more will dissolve.

The resulting liquid has no effect on blue litmus. So the oxide has neutralised the acid.

Copper(II) oxide is called a **basic oxide** since it can neutralise an acid:

$\text{base} + \text{acid} \rightarrow \text{salt} + \text{water}$
$CuO\ (s) + 2HCl\ (aq) \rightarrow CuCl_2\ (aq) + H_2O\ (l)$

Iron(III) oxide and magnesium oxide behave in the same way – they too can neutralise acid, so they are basic oxides.

In general, metals react with oxygen to form basic oxides.
Basic oxides belong to the larger group of compounds called bases.

Acidic oxides
Now look how these non-metals react with oxygen:

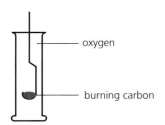 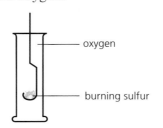 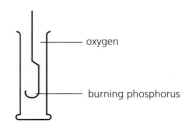

Powdered carbon is heated over a Bunsen burner until red-hot, then plunged into a jar of oxygen. It glows bright red, and the gas **carbon dioxide** is formed:
$$C\ (s) + O_2\ (g) \longrightarrow CO_2\ (g)$$

Sulfur catches fire over a Bunsen burner, and burns with a blue flame. In pure oxygen it burns even brighter. The gas **sulfur dioxide** is formed:
$$S\ (s) + O_2\ (g) \longrightarrow SO_2\ (g)$$

Phosphorus bursts into flame in air or oxygen, without heating. (So it is stored under water!) A white solid, **phosphorus(V) oxide**, is formed:
$$P_4\ (s) + 5O_2\ (g) \longrightarrow P_4O_{10}\ (s)$$

All three oxides will dissolve in water, giving solutions that turn litmus red. So they are called **acidic oxides**.

For example, carbon dioxide forms the weak acid carbonic acid:
$$CO_2\ (g) + H_2O\ (l) \longrightarrow H_2CO_3\ (aq)$$

(The gas is only slightly soluble – but more will dissolve under pressure.)

In general, non-metals react with oxygen to form acidic oxides.

Amphoteric oxides
Aluminium is a metal, so you would expect aluminium oxide to be a base. In fact it is both acidic *and* basic. It acts as a base with hydrochloric acid:
$$Al_2O_3\ (s) + 6HCl\ (aq) \longrightarrow 2AlCl_3\ (aq) + 3H_2O\ (l)$$

But it acts as an acidic oxide with sodium hydroxide, giving a compound called sodium aluminate:
$$Al_2O_3\ (s) + 6NaOH\ (aq) \longrightarrow 2Na_3AlO_3\ (aq) + 3H_2O\ (l)$$

So aluminium oxide is called an **amphoteric oxide**.
An amphoteric oxide will react with both acids and alkalis.
Zinc oxide (ZnO) is another amphoteric oxide.

▲ Zinc oxide: an amphoteric oxide. It will react with both the acid and alkali.

Neutral oxides
- Some oxides of non-metals are neither acidic nor basic. They are **neutral**.
- Neutral oxides do not react with acids or bases.
- Examples are the deadly gas carbon monoxide (CO), and dinitrogen oxide (N_2O, called laughing gas), which is used as an anaesthetic in dentistry.

1. As you saw on page 136, iron burns to give iron(III) oxide. How would you show that this compound is a base?
2. Which metal is more reactive: iron or copper? Give evidence.
3. Copy and complete: Metals usually form .... oxides while non-metals form .... oxides.
4. List carbon, phosphorus and sulfur in order of reactivity, using their reaction with oxygen above.
5. When they escape into the air, the oxides of sulfur and nitrogen cause **acid rain**. Explain why.
6. What is an *amphoteric* oxide? Give two examples.

11.6 Making salts (part I)

Objectives: give the steps in making a salt from an acid, starting with a metal, an insoluble base, or an alkali; explain why titration is needed when an alkali is used

You can make salts by reacting acids with any of these: **metals**, **insoluble bases**, **soluble bases (alkalis)**, and **carbonates**.

Starting with a metal
Zinc sulfate can be made by reacting dilute sulfuric acid with zinc:

$$Zn\ (s) + H_2SO_4\ (aq) \longrightarrow ZnSO_4\ (aq) + H_2\ (g)$$

These are the steps:

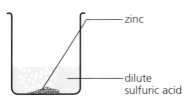
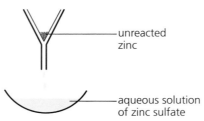

1. Add *excess* zinc to the acid in a beaker. It starts to dissolve, and hydrogen bubbles off. Bubbling stops when all the acid is used up.

2. Some zinc is still left (because the zinc was in excess). Remove it by filtering. This leaves an aqueous solution of zinc sulfate.

3. Heat the solution to evaporate some water, to obtain a **saturated** solution. Leave this to cool. Crystals of zinc sulfate appear.

This method is fine for making salts of magnesium, aluminium, zinc, and iron. But sodium, potassium, and calcium react violently with acids. Lead reacts too slowly, and copper, silver, and gold do not react at all. *So you must choose a different method to make the salts of those metals*.

Starting with an insoluble base or insoluble carbonate
Copper will not react with dilute sulfuric acid. So to make copper(II) sulfate, you could start with either copper(II) oxide, or copper(II) carbonate, which are both insoluble. Here we use copper(II) oxide:

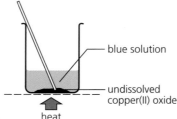
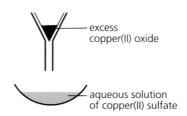
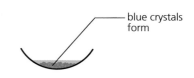

1. Add excess copper(II) oxide to dilute sulfuric acid. It dissolves on warming and the solution turns blue. Add more until no more will dissolve …

2. … which means all the acid has now been used up. Remove the excess solid by filtering. This leaves a blue solution of copper(II) sulfate in water.

3. Heat the solution to obtain a saturated solution. Then leave it to cool. Crystals of **hydrated** copper(II) sulfate form. Their formula is $CuSO_4.5H_2O$.

The steps are the same for copper(II) carbonate. The two equations are:

$$CuO\ (s) + H_2SO_4\ (aq) \longrightarrow CuSO_4\ (aq) + H_2O\ (l)$$
$$CuCO_3\ (s) + H_2SO_4\ (aq) \longrightarrow CuSO_4\ (aq) + CO_2\ (g) + H_2O\ (l)$$

ACIDS, BASES, AND SALTS

Starting with an alkali (soluble base)
You can make sodium chloride safely like this:

NaOH (*aq*) + HCl (*aq*) ⟶ NaCl (*aq*) + H$_2$O (*l*)

Both reactants are soluble, so this time you cannot filter off any excess. So how can you avoid excess reactant getting mixed in with your product? **The answer is: do a titration first.**

In a titration, one reactant is slowly added to the other in the presence of an indicator. The indicator changes colour when the reaction is complete. So now you know the amounts of reactant to use, so that none is left over. **Then you mix these exact amounts, *without* the indicator.**

▲ Thymolphthalein is blue in alkali. As you add acid a little at a time, you get flashes of no colour. A single drop of acid will finally turn the solution colourless. Neutralisation is complete.

The steps in making sodium chloride
We can choose **thymolphthalein** as the indicator. It is blue in alkaline solution, but colourless in neutral and acid solutions. These are the steps:

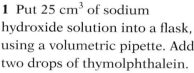
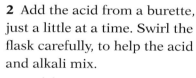
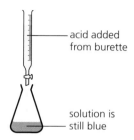

1 Put 25 cm^3 of sodium hydroxide solution into a flask, using a volumetric pipette. Add two drops of thymolphthalein.

2 Add the acid from a burette, just a little at a time. Swirl the flask carefully, to help the acid and alkali mix.

3 The indicator suddenly goes colourless. So the alkali has all been used up. The solution is now neutral. Do not add more acid!

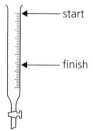
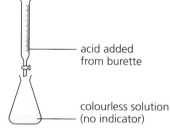
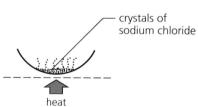

4 Find how much acid you added, using the scale on the burette. This tells you how much acid is needed to neutralise 25 cm^3 of the alkali.

5 Now repeat *without* the indicator. (It would be an impurity.) Put 25 cm^3 of alkali in the flask. Add the correct amount of acid to neutralise it.

6 Finally, heat the solution from the flask to evaporate the water. White crystals of sodium chloride will be left behind.

You could make potassium salts from potassium hydroxide in the same way.

Q

1 What will you start with, to make the salt zinc chloride?
2 You would not make lead salts by reacting lead with acids.
   a Why not?   b Suggest a way to make lead nitrate.
3 Look at step **2** at the top of page 138. The zinc was *in excess*. What does that mean? (Glossary?)

4 You are asked to make the salt potassium chloride, using potassium hydroxide and hydrochloric acid.
   a You will carry out a titration first. Explain why.
   b After the titration, how will you prepare the dry salt?
   c No indicator is added, in your steps in **b**. Why not?

11.7 Making salts (part II)

Objectives: outline the steps in making an insoluble salt by precipitation, and choose suitable starting compounds for a given salt; define *water of crystallisation*

Not all salts are soluble

The salts we looked at so far have all been soluble. You can obtain them as crystals, by evaporating solutions. But some salts are insoluble. Look:

Soluble		Insoluble
all sodium, potassium, and ammonium salts		
all nitrates		
all chlorides, bromides and iodides (halides)	except	the chlorides, bromides, and iodides of silver and lead
sulfates ...	except	calcium, barium, and lead sulfate
sodium, potassium, and ammonium carbonates		but all other carbonates are insoluble

Making insoluble salts by precipitation

Insoluble salts can be made by precipitation.
Precipitation occurs when mixing reactants in solution gives an insoluble product. The insoluble product is called the *precipitate*.

For example, barium sulfate is insoluble. You can make it by mixing solutions of barium chloride and magnesium sulfate:

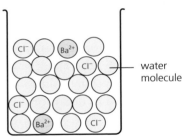
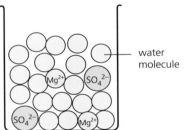

A solution of barium chloride, $BaCl_2$, contains barium ions and chloride ions, as shown here.

A solution of magnesium sulfate, $MgSO_4$, contains magnesium ions and sulfate ions.

When you mix the two solutions, the barium and sulfate ions bond to form solid barium sulfate.

The equation for the reaction is:

$$BaCl_2\ (aq) + MgSO_4\ (aq) \longrightarrow BaSO_4\ (s) + MgCl_2\ (aq)$$

The ionic equation is:

$$Ba^{2+}\ (aq) + SO_4^{2-}\ (aq) \longrightarrow BaSO_4\ (s)$$

The ionic equation does not show the magnesium and chloride ions, because they do not take part in the reaction. They are **spectator ions**.

The steps in making barium sulfate

1. Make up solutions of barium chloride and magnesium sulfate.
2. Mix them. A white precipitate of barium sulfate immediately forms.
3. Filter the mixture. The precipitate is trapped in the filter paper.
4. Rinse the precipitate by running distilled water through it.
5. Then place it in a warm oven to dry.

▲ The precipitation of barium sulfate. It is a very fast reaction.

Choosing the starting compounds
Barium sulfate can also be made from barium nitrate and sodium sulfate, since both are soluble. As long as barium ions and sulfate ions are present, barium sulfate will precipitate.

To precipitate an insoluble salt, you must mix a solution that contains its positive ions with one that contains its negative ions.

Precipitation in lab tests
In Units 19.4 and 19.5, you will learn how to do some clever detective work, using precipitation.

You start with the solution of an unknown compound, and carry out tests to find out what it is. In many of the tests, you check for a precipitate.

For example, if you add silver nitrate solution to the solution, and a pale yellow precipitate forms, the unknown compound must be an iodide (containing the I⁻ ion). You can do other tests to identify the positive ion.

▲ Adding silver nitrate solution to an unknown solution gave this pale yellow precipitate of silver iodide. It proved that the unknown compound was an iodide.

Precipitation in industry
Precipitation has many uses in industry. For example, it is used ...
- to remove harmful substances from waste water, when cleaning it up – such as ions of toxic metals like lead and mercury.
- to make **pigments**, the insoluble compounds used to colour paint and other things. On the right is the pigment Cadmium Yellow (cadmium sulfide) produced by mixing solutions of calcium chloride and sodium sulfide.
- to create a very fine coat of silver bromide, for coating photographic film. The silver bromide breaks down in light, giving the 'negative' image in silver.

Hydrated salts
Many salts have water molecules chemically bonded into their crystals.

These salts are called **hydrated**. The ratio of water molecules to ions is shown in their formulae. Look at the examples on the right.

So $CuSO_4.5H_2O$ tells you that there are five water molecules for each copper ion and sulfate ion. Note where the 5 is placed. A dot separates the water molecules from the rest of the formula.

The water can be driven off by heating, leaving the **anhydrous** salt. But on cooling, water molecules will be taken in again from the surroundings.

The water bonded into hydrated salts is called **water of crystallisation**. It is not usually included when formulae are written.

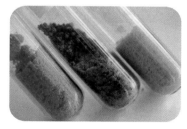

▲ Three hydrated salts. From left to right:
copper(II) sulfate, $CuSO_4.5H_2O$
cobalt(II) sulfate, $CoSO_4.7H_2O$
nickel(II) chloride, $NiCl_2.6H_2O$.

1. Name two families of salts which are all soluble.
2. Name one family of salts which are mostly insoluble.
3. The chemical with formula $CuSO_4.5H_2O$ is called *hydrated*.
   a Explain why.
   b Heating it gives the *anhydrous* form. State its formula.
4. Name four salts you could *not* make by precipitation.
5. Choose two starting compounds to make this insoluble salt:
   a calcium sulfate   b magnesium carbonate
6. Write the ionic equations for your reactions for 5.
7. A sodium salt has the formula $Na_2SO_4.10H_2O$.
   a What can you say about its crystals? *At least* two facts!
   b Define *water of crystallisation*.

11.8 Finding concentration by titration

Objectives: understand that titration is also used to find the concentration of an acid or alkaline solution; follow the four steps to calculate the concentration

How to find concentration by titration

Page 139 showed how the volume of acid needed to neutralise an alkali can be found, by carrying out a **titration**. You add the acid to the alkali a little at a time, and an indicator tells you when the reaction is complete.

If you know the **concentration** of the alkali, you can use your result to find the concentration of the acid (or vice versa).

An example

You are asked to find the concentration of a solution of hydrochloric acid, by titration against a 1 mol/dm³ solution of sodium carbonate.

First, do the titration.

- Measure 25 cm³ of the sodium carbonate solution into a conical flask, using a volumetric pipette. Add a few drops of methyl orange indicator. The indicator goes yellow.
- Pour the acid into a 50 cm³ burette. Record the level.
- Drip the acid slowly into the conical flask. Keep swirling the flask. Stop adding acid when a single drop finally turns the indicator red. Record the new level of acid in the burette.
- Calculate the volume of acid used. For example:
  Initial burette reading: 1.0 cm³
  Final burette reading: 28.8 cm³
  Volume of acid used: 27.8 cm³

So 27.8 cm³ of the acid neutralised 25 cm³ of the alkaline solution.

You can now calculate the concentration of the acid.

Step 1 Calculate the number of moles of sodium carbonate used.

1000 cm³ of the solution contains 1 mole of sodium carbonate, so 25 cm³ contains $\frac{25}{1000} \times 1$ mole or 0.025 mole.

Step 2 From the equation, find the molar ratio of acid to alkali.

$2HCl\ (aq) + Na_2CO_3\ (aq) \longrightarrow 2NaCl\ (aq) + H_2O\ (l) + CO_2\ (g)$
2 moles   1 mole

The ratio is 2 moles of acid to 1 of alkali.

Step 3 Work out the number of moles of acid neutralised.

1 mole of alkali neutralises 2 moles of acid so
0.025 mole of alkali neutralises 2 × 0.025 moles of acid.
0.05 moles of acid were neutralised.

Step 4 Calculate the concentration of the acid.

The volume of acid used was 27.8 cm³ or 0.0278 dm³.

so its concentration = $\frac{\text{number of moles}}{\text{volume in dm}^3} = \frac{0.05}{0.0278} =$ **1.8 mol/dm³**.

Remember!
- Concentration is usually given as moles per dm³ or mol/dm³.
- 1000 cm³ = 1 dm³
- To convert cm³ to dm³ move the decimal point 3 places left. So 250 cm³ = 0.25 dm³.

▲ A final drop of acid has turned the indicator red. This is the **end-point** of the titration. Neutralisation is complete.

Use the calculation triangle

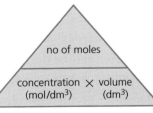

no of moles

concentration × volume
(mol/dm³)   (dm³)

▲ Cover *concentration* with your finger to see how to calculate it.

And also ...
- To find the concentration of an *alkaline* solution, you do things the other way round ... use an acid solution of known concentration.

ACIDS, BASES, AND SALTS

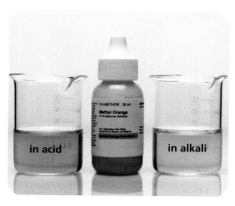

▲ The colour change in methyl orange will help you find concentration …

▲ … *or* you could use a pH meter. How will you know when neutralisation is complete?

> **!** **The standard soution**
> In a titration, the solution with the known concentration is sometimes called the **standard solution**. So you titrate against a standard solution.

Another sample calculation

Vinegar is mainly a solution of ethanoic acid. 25 cm^3 of vinegar were neutralised by 20 cm^3 of sodium hydroxide solution, of concentration 1 mol/dm^3. What is the concentration of ethanoic acid in the vinegar?

Step 1 Calculate the number of moles of sodium hydroxide used.

1000 cm^3 of the sodium hydroxide solution contains 1 mole so
20 cm^3 contains $\frac{20}{1000} \times 1$ mole or 0.02 mole.

Step 2 From the equation, find the molar ratio of acid to alkali.

$$CH_3COOH\ (aq) + NaOH\ (aq) \longrightarrow CH_3COONa\ (aq) + H_2O\ (l)$$
1 mole   1 mole

The ratio is 1 mole of acid to 1 mole of alkali.

Step 3 Work out the number of moles of acid neutralised.

1 mole of alkali neutralises 1 mole of acid so
0.02 mole of alkali neutralise 0.02 mole of acid.

Step 4 Calculate the concentration of the acid. (25 cm^3 = 0.025 dm^3)

Its concentration = $\frac{\text{number of moles}}{\text{volume in dm}^3} = \frac{0.02}{0.025} =$ **0.8 mol/dm^3**.

Note that ethanoic acid is a weak acid: it is only partially dissociated into ions at any given time. But as its H$^+$ ions get used up, more molecules dissociate. This continues until all of the acid has reacted.

▲ Oil and vinegar for putting on salads. The ethanoic acid in vinegar (on the right) gives salad – and chips – a tasty tang.

1. Describe how to identify the end-point of a neutralisation reaction in a titration using an indicator.
2. Solutions of hydrochloric acid and sodium carbonate were prepared, with concentrations of 2 mol/dm^3. What volume of the acid will neutralise 25 cm^3 of the carbonate solution?
3. In a titration, 25 cm^3 of sodium hydroxide solution reacted with 20 cm^3 of sulfuric acid, with concentration 1 mol/dm^3.
   a Name a suitable indicator for this neutralisation reaction.
   b Write the balanced equation for the reaction.
   c Calculate the concentration of the alkaline solution.

Checkup on Chapter 11

Revision checklist
Core syllabus content
Make sure you can ...
- [ ] define a base
- [ ] state that alkalis are soluble bases
- [ ] name the common laboratory acids and alkalis, and give their formulae
- [ ] describe the effects of acids and alkalis on litmus, methyl orange, and thymolphthalein
- [ ] name the ions responsible for acidity and alkalinity
- [ ] describe the pH scale, and state the link between pH numbers and ...
  - acidic, alkaline, and neutral solutions
  - the concentration of H^+ or OH^- ions present
- [ ] describe what universal indicator is, and how its colour changes across pH
- [ ] state what is formed when acids react with:
  metals   bases   carbonates
- [ ] give the typical reactions of bases
- [ ] identify a neutralisation reaction from its equation
- [ ] write the ionic equation showing how H^+ ions and OH^- ions react to form water, during neutralisation
- [ ] explain what *basic oxides* and *acidic oxides* are, and give examples
- [ ] choose suitable reactants for making a salt
- [ ] describe how to make a salt from an acid, using
  - a metal or insoluble base
  - an alkali
- [ ] explain why titration and an indicator are needed, when making a salt from an acid and an alkali
- [ ] give the general rules for solubility of salts
- [ ] explain what *hydrated* and *anhydrous* salts are

Extended syllabus content
Make sure you can also ...
- [ ] explain the difference between strong and weak acids, by showing their equations
- [ ] explain what happens during the neutralisation of any acid by any base, and give the ionic equation
- [ ] define acids and bases in terms of proton transfer
- [ ] define *amphoteric oxide* and give two examples
- [ ] choose two suitable reactants for making an insoluble salt by precipitation
- [ ] calculate the concentration of a solution of acid or alkali, using data from a titration
- [ ] define *water of crystallisation*

Questions
Core syllabus content

1 Rewrite the following, choosing the correct word from each pair in brackets.
Acids are compounds that dissolve in water giving hydrogen ions. Sulfuric acid is an example. It can be neutralised by (acids/bases) to form salts called (nitrates/sulfates). Many (metals/non-metals) react with acids to give (hydrogen/carbon dioxide). Acids react with (chlorides/carbonates) to give (hydrogen/carbon dioxide).
Since they contain ions, solutions of acids are (good/poor) conductors of electricity. They also affect indicators. Litmus turns (red/blue) in acids while thymolphthalein turns (blue/colourless).
The level of acidity of an acid is shown by its (concentration/pH number). The (higher/lower) the number, the more acidic the solution.

2 **A** and **B** are white powders. **A** is insoluble in water, but **B** dissolves. Its solution has a pH of 3. A mixture of **A** and **B** bubbles or effervesces in water, giving off a gas. A clear solution forms.
  a Which of the two powders is an acid?
  b The other powder is a carbonate. Which gas bubbles off in the reaction?
  c Although **A** is insoluble in water, a clear solution forms when the mixture of **A** and **B** is added to water. Explain why.

3 Oxygen reacts with other elements to form oxides. Three examples are: calcium oxide, phosphorus(V) oxide, and copper(II) oxide.
  a Which of these is:
    i   an insoluble base?
    ii a soluble base?
    iii an acidic oxide?
  b When the soluble base is dissolved in water, the solution changes the colour of litmus paper. What colour change will you see?
  c Name the gas given off when the soluble base is heated with ammonium chloride.
  d i   Write a word equation for the reaction between the insoluble base and sulfuric acid.
     ii What is this type of reaction called?
  e Name another acidic oxide.

ACIDS, BASES, AND SALTS

Method of preparation	Reactants	Salt formed	Other products
a acid + alkali	calcium hydroxide and nitric acid	calcium nitrate	water
b acid + metal	zinc and hydrochloric acid	....	....
c acid + alkali	.... and potassium hydroxide	potassium sulfate	water
d acid + carbonate	.... and	sodium chloride	water and ....
e acid + metal	.... and	iron(II) sulfate	....
f acid + ...	nitric acid and sodium hydroxide	....	....
g acid + insoluble base	.... and copper(II) oxide	copper(II) sulfate	....
h acid + ...	.... and	copper(II) sulfate	carbon dioxide and ....

4 The table above is about the preparation of salts.
  i Copy it and fill in the missing details.
  ii Write balanced equations for the eight reactions.

5 The drawings show the preparation of copper(II) ethanoate, a salt of ethanoic acid.

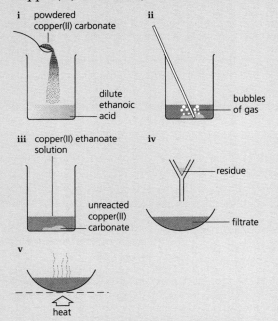

  i powdered copper(II) carbonate / dilute ethanoic acid
  ii bubbles of gas
  iii copper(II) ethanoate solution / unreacted copper(II) carbonate
  iv residue / filtrate
  v heat

  a i Which gas forms in the reaction in stage **ii**?
  ii Write a word equation for this reaction.
  iii How can you tell when it is over?
  b Which reactant above is:
    i present in excess? What is your evidence?
    ii completely used up in the reaction?
  c The copper(II) carbonate is used in powder form, rather than as lumps. Suggest a reason.
  d Name the residue in stage **iv**.
  e Write a list of instructions for carrying out this preparation in the laboratory.
  f Suggest another copper compound to use instead of copper(II) carbonate, to make the salt.

Extended syllabus content

6 Magnesium sulfate ($MgSO_4$) is the chemical name for Epsom salts. It can be made in the laboratory by neutralising the base magnesium oxide (MgO).
  a Which acid should be used to make Epsom salts?
  b Write a balanced equation for the reaction.
  c i The acid is fully dissociated in water. Which term describes this type of acid?
    ii Which ion causes the 'acidity' of the acid?
  d i What is a base?
    ii Write an ionic equation that shows the oxide ion (O^{2-}) acting as a base.

7 a i From the list on page 140, write down two starting compounds that could be used to make the insoluble compound silver chloride.
  ii What is this type of reaction called?
  b i Write the ionic equation for the reaction.
  ii List the spectator ions for the reaction.

8 Washing soda is crystals of hydrated sodium carbonate, $Na_2CO_3.xH_2O$.
The value of x can be found by titration.
In the experiment, a solution containing 2 g of hydrated sodium carbonate neutralised 14 cm^3 of a standard 1 M solution of hydrochloric acid.
  a What does *hydrated* mean?
  b Write a balanced equation for the reaction that took place during the titration.
  c How many moles of HCl were neutralised?
  d i How many moles of sodium carbonate, Na_2CO_3, were in 2 g of the hydrated salt?
    ii What mass of sodium carbonate, Na_2CO_3, is this? (A_r: Na = 23, C = 12, O = 16)
  e i What mass of the hydrated sodium carbonate was water?
    ii How many moles of water is this?
  f How many moles of water are there in 1 mole of $Na_2CO_3.xH_2O$?
  g Write the full formula for washing soda.

12.1 The Periodic Table: an overview

Objectives: give the key points about the Periodic Table; understand that it shows patterns and trends, and can be used to predict the properties of elements

What is the Periodic Table?

Period	Group I	II											III	IV	V	VI	VII	VIII
1							1 H hydrogen 1											2 He helium 4
2	3 Li lithium 7	4 Be beryllium 9											5 B boron 11	6 C carbon 12	7 N nitrogen 14	8 O oxygen 16	9 F fluorine 19	10 Ne neon 20
3	11 Na sodium 23	12 Mg magnesium 24					the transition elements						13 Al aluminium 27	14 Si silicon 28	15 P phosphorus 31	16 S sulfur 32	17 Cl chlorine 35.	18 Ar argon 40
4	19 K potassium 39	20 Ca calcium 40	21 Sc scandium 45	22 Ti titanium 48	23 V vanadium 51	24 Cr chromium 52	25 Mn manganese 55	26 Fe iron 56	27 Co cobalt 59	28 Ni nickel 59	29 Cu copper 64	30 Zn zinc 65	31 Ga gallium 70	32 Ge germanium 73	33 As arsenic 75	34 Se selenium 79	35 Br bromine 80	36 Kr krypton 84
5	37 Rb rubidium 85	38 Sr strontium 88	39 Y yttrium 89	40 Zr zirconium 91	41 Nb niobium 93	42 Mo molybdenum 96	43 Tc technetium –	44 Ru ruthenium 101	45 Rh rhodium 103	46 Pd palladium 106	47 Ag silver 108	48 Cd cadmium 112	49 In indium 115	50 Sn tin 117	51 Sb antimony 122	52 Te tellurium 128	53 I iodine 127	54 Xe xenon 131
6	55 Cs caesium 133	56 Ba barium 137	57–71 lanthanoids	72 Hf hafnium 178	73 Ta tantalum 181	74 W tungsten 184	75 Re rhenium 186	76 Os osmium 190	77 Ir iridium 192	78 Pt platinum 195	79 Au gold 197	80 Hg mercury 201	81 Tl thallium 204	82 Pb lead 207	83 Bi bismuth 209	84 Po polonium –	85 At astatine –	86 Rn radon –
7	87 Fr francium –	88 Ra radium –	89–103 actinoids	104 Rf rutherfordium –	105 Db dubnium –	106 Sg seaborgium –	107 Bh bohrium –	108 Hs hassium –	109 Mt meitnerium –	110 Ds darmstadtium –	111 Rg roentgenium –	112 Cn copernicium –	113 Nh nihonium –	114 Fl flerovium –	115 Mc moscovium –	116 Lv livermorium –	117 Ts tennessine –	118 Og oganesson –

lanthanoids	57 La lanthanum 139	58 Ce cerium 140	59 Pr praseodymium 141	60 Nd neodymium 144	61 Pm promethium –	62 Sm samarium 150	63 Eu europium 152	64 Gd gadolinium 157	65 Tb terbium 159	66 Dy dysprosium 163	67 Ho holmium 165	68 Er erbium 167	69 Tm thulium 169	70 Yb ytterbium 173	71 Lu lutetium 175
actinoids	89 Ac actinium –	90 Th thorium 232	91 Pa protactinium 231	92 U uranium 238	93 Np neptunium –	94 Pu plutonium –	95 Am americium –	96 Cm curium –	97 Bk berkelium –	98 Cf californium –	99 Es einsteinium –	100 Fm fermium –	101 Md mendelevium –	102 No nobelium –	103 Lr lawrencium –

Look at the Periodic Table above. It is the most important document in chemistry! It lists the 118 known elements. Even better, it arranges them in a way that shows the patterns.

- The Periodic Table shows the elements in order of **proton number** (or **atomic number**). The proton number is shown above each symbol. Lithium, with 3 protons, is followed by beryllium, with 4.
- When arranged by proton number, the elements show **periodicity**: elements with similar properties appear at regular intervals.
- Look at the columns numbered I to VIII. All the elements within a numbered column have similar properties. They form a **group**. Lithium is in Group I. Oxygen is in Group VI.
- The rows are called **periods**. They are numbered 1 to 7. Period 1 has only two elements, hydrogen and helium.
- The red zig-zag line separates **metals** from **non-metals**, with the non-metals to its right (except for hydrogen).
- The **transition elements** include the lanthanoids, which belong to Period 6, and the actinoids, in Period 7. (You won't study these.)

Key for Periodic Table

atomic number
symbol
name
relative atomic mass

Proton number

- Remember, the proton number identifies an atom.
- For example, if an atom has 6 protons, it must be a carbon atom.
- The proton number is the same as the number of electrons in that atom.
- The proton number is also called the **atomic number**.

THE PERIODIC TABLE

Key points about the groups and periods
- **The group number is the same as the number of outer shell electrons in the atoms, in Groups I – VII.**
- **The atoms of the Group VIII elements (the noble gases) have full outer shells of electrons. So they are *unreactive*, and *monatomic*.**
- **The period number tells you the number of electron shells.**
  So in the elements of Period 2, the atoms have two electron shells. In Period 3 they have three, and so on.

The metals and non-metals
Look again at the table. The metals are to the left of the zig-zag line. There are far more metals than non-metals. In fact over 80% of the elements are metals.

Hydrogen
Find hydrogen in the table. It sits alone. That is because it has one outer electron, and forms a positive ion (H^+) like the Group I metals – but unlike them it is a gas, and it usually reacts like a non-metal.

The transition elements
The **transition elements** – the block in the middle of the Periodic Table – are all metals. They include several metals you know, like iron, copper, silver, and gold. There is more about them in Unit 12.5.

Patterns and trends in the Periodic Table
The elements in a group behave in a similar way – and also show trends. For example, as you go down Group I, the elements become *more reactive*.

There is a trend across periods too, for Period 2 onwards: a change from metal to non-metal. Look at Period 3, where only sodium, magnesium, and aluminium are metals. The rest are non-metals.

So if you know the position of an element in the Periodic Table, you can predict its properties.

We will explore the trends in Groups I and VII, and across Period 3, in the next three units.

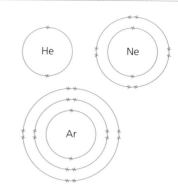

▲ The electronic configuration for three noble gases from Group VIII: helium, neon, and argon. Their outer electron shells are full so the gases are both unreactive and monatomic. (They exist as single atoms.)

▲ Neon and argon are used in advertising signs. Their lack of reactivity makes them safe. They glow when a voltage is applied. The colours they produce can be changed by adding other substances.

Not all natural!
- 94 of the 118 elements in the Periodic Table occur naturally. Metals occur as compounds in rock. Air is mainly nitrogen and oxygen.
- But the rest are **synthetic**. They were made in the lab, usually in tiny amounts. They are radioactive, and their nucleii break down – sometimes within in fractions of a second! Most are in Period 7.

A wealth of compounds
- Everything on Earth is made of the 94 natural elements.
- There are *millions* of compounds – and new ones are made, or discovered, every year.

1. In the Periodic Table, what name is used for:
   a a row of elements?
   b a numbered column of elements?
2. Using *only* the Periodic Table and its key to help you, give *six* facts about each of these elements:
   a nitrogen   b magnesium   c argon
3. In which groups of the Periodic Table are non-metals found?
4. State the link between the group number of an element, and the outer shell electrons in its atoms.
5. The elements of Group VIII are unreactive. Why?
6. Find the element in the Periodic Table that is named after:
   a France   b America   c Einstein   (Try Period 7!)

12.2 Group I: the alkali metals

Objectives: describe physical and chemical properties, and trends, for the elements of Group I; explain why they show similar reactions, and are highly reactive

What are they?
The **alkali metals** is the special name given to the elements in Group I. They are **soft silvery metals**. Their atoms have **1 outer shell electron**.

Only the first three – lithium, sodium and potassium – are safe to keep in the school lab. The others are violently reactive. Francium is radioactive.

the alkali metals

Their physical properties
The alkali metals are *not* typical metals.
- They are softer than most metals. You can cut them with a knife.
- They are also lighter than most other metals – they have **low density**.
- They have low melting points.

The trends in their physical properties
As with any family, the alkali metals are all a little different. Look:

Metal	Density in g/cm³	Melts at /°C
lithium, Li	0.53	181
sodium, Na	0.97	98
potassium, K	0.86	63
rubidium, Rb	1.53	39
caesium, Cs	1.88	29

density increases ↓   melting points decrease ↓

▲ A piece of sodium, cut with a knife.

About density
- **Density** tells you how light or heavy a substance is.
- density = $\dfrac{\text{mass (in grams)}}{\text{volume (in cm}^3\text{)}}$
- 1 cm³ of water has a mass of 1 g. So its density is 1 g/cm³.
- Lithium has the lowest density of all metals.

So there is an overall change for each property, as you go down the group. A gradual change like this is called a **trend**.

Their chemical properties
Their chemical properties show trends too. Look:

1 Reaction with water

Experiment	What you see
The element is placed in a trough of water with some indicator added: (metal; trough of water plus indicator)	• lithium floats around in the water and a gas fizzes • sodium melts into a silver ball and shoots across the water; a gas fizzes • melts, and shoots across the water; a gas catches fire; the potassium catches fire too and burns with a lilac flame • in each case the indicator changes colour, indicating an alkaline solution

increasing reactivity ↓

Note how the reactivity changes. For sodium the reaction is:

sodium + water → sodium hydroxide + hydrogen

The indicator changes colour because sodium hydroxide is an alkali.

The alkali metals react vigorously with water. Hydrogen bubbles off, leaving solutions of their hydroxides, which are alkalis.

▲ Lithium reacting with water.

THE PERIODIC TABLE

2 Reaction with chlorine

If you heat the three metals, and plunge them into gas jars of chlorine, they burst into flame. They burn brightly, forming chlorides. For example:

sodium + chlorine → sodium chloride

Potassium burns the most vigorously of the three.

3 Reaction with oxygen

The heated metals also burst into flame when plunged into gas jars of oxygen. They burn fiercely to form **oxides**. These dissolve in water to give alkaline solutions. Again, potassium burns the most vigorously.

So, from these three reactions in the school lab, we can conclude this:
Reactivity increases as you go down Group I.

Why do they react in a similar way?

All the alkali metals react in a similar way. Why? Because all their atoms have 1 outer shell electron:

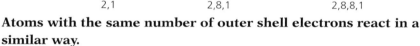

Atoms with the same number of outer shell electrons react in a similar way.

Why are they so reactive?

The alkali metals are the most reactive of all the metals.

Why? Because their atoms need to lose only one electron, to gain a full outer shell. They have a very strong drive to give up this electron to atoms of other elements. That is why they are so reactive.

When the alkali metals react, their atoms become **ions**. The products are **ionic compounds**.

For example, when sodium reacts with chlorine, the ionic compound sodium chloride forms. It is a lattice of Na^+ and Cl^- ions.

The alkali metals form ions with a charge of 1+.
They produce ionic compounds, which are all white solids.
These dissolve in water to give colourless solutions.

▲ Potassium burning in chlorine. Inside the gas jar, countless trillions of atoms are reacting. Potassium atoms are transferring electrons to chlorine atoms, to form solid potassium chloride.

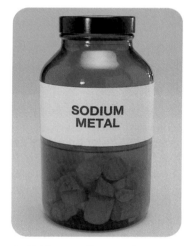

▲ Lithium, sodium, and potassium are stored under oil in the lab, to prevent reaction with oxygen and water.

1 Which best describes the Group I metals:
   a *soft* or *hard*?   b *reactive* or *unreactive*?
2 a The Group I metals show a *trend* in melting points. What does that mean?
   b One measurement in the table on page 148 does not fit its trend. Which one?
3 The Group I metals are called the *alkali metals*. Why?
4 Why do the Group I metals react in a similar way?
5 What makes the Group I metals so reactive?
6 Iron is an important metal in the building industry. But the Group I metals are not used in construction. Why not?
7 a Give the equation for the reaction between potassium and chlorine, with state symbols.
   b What colour will an aqueous solution of the product be?
   c Will this solution conduct electricity? Explain.
8 Rubidium is below potassium, in Group I. Predict how it will react: a with water   b with chlorine
   and name and describe the products that form.

12.3 Group VII: the halogens

Objectives: describe physical and chemical properties, and trends, for the elements of Group VII; explain why they show similar reactions, and are highly reactive

A non-metal group

Group VII is a group of **non-metal elements**. This group is usually called **the halogens**. It includes chlorine, bromine, and iodine, which you will meet in the lab, and fluorine, which is too dangerous. The halogens:

- are all different colours.
- are all poisonous.
- form **diatomic** molecules (containing two atoms). For example, Cl_2.

Trends in their physical properties

As usual, the group shows trends in physical properties. Look at these:

Halogen	At room temperature the element is …	Boiling point /°C
chlorine, Cl_2	a pale yellow-green gas	−35
bromine, Br_2	a red-brown liquid	59
iodine, I_2	a grey-black solid	184

density increases ↓ boiling points increase ↓

▲ Chlorine, bromine, and iodine. Bromine is liquid at room temperature, but readily forms a vapour.

Trends in their chemical properties

The halogens are among the most reactive elements in the Periodic Table. They react with metals to form compounds called **halides**. Look:

Halogen	Reaction with iron wool	The product	Its appearance
chlorine	Hot iron wool glows brightly when chlorine passes over it.	iron(III) chloride, $FeCl_3$	yellow solid
bromine	Hot iron wool glows, but less brightly, when bromine vapour passes over it.	iron(III) bromide, $FeBr_3$	red-brown solid
iodine	Hot iron wool shows a faint red glow when iodine vapour passes over it.	iron(III) iodide, FeI_3	black solid

reactivity decreases ↓

So they all react in a similar way. But look at the trend:
Reactivity decreases as you go down Group VII.
This is the *opposite* of the trend in Group I. (Page 148.)

Why do they react in a similar way?

The halogens react in a similar way because all their atoms have 7 outer shell electrons. Look:

2,7   2,8,7

Atoms with the same number of outer shell electrons react in a similar way.

▲ Iodine is a disinfectant. His skin is being wiped with a solution of iodine in ethanol, before he gives blood.

Why are they so reactive?

Halogen atoms need only 1 electron to reach a full outer shell of 8 electrons. They have a very strong drive to gain this electron from atoms of other elements. That is why they are so reactive.

With metals When halogens react with metals, their atoms accept electrons to form **halide ions**. The products are **ionic compounds**. Iron(III) chloride, $FeCl_3$, is made up of Fe^{3+} and Cl^- ions.

With non-metals When halogens react with non-metals, their atoms *share* electrons, giving molecules with covalent bonds. Chlorine reacts with hydrogen to form molecules of hydrogen chloride, HCl.

How the halogens react with halides

1 Chlorine water is a solution of chlorine in water. When it is added to a colourless solution of potassium bromide, the mixture turns orange as shown in the photo. This reaction is taking place:

$$Cl_2\ (aq) + 2KBr\ (aq) \longrightarrow 2KCl\ (aq) + Br_2\ (aq)$$
pale yellow   colourless   colourless   orange

Bromine has been pushed out of its compound, or **displaced**.

2 When chlorine water is added to a colourless solution of potassium iodide, the mixture turns red-brown, because of this reaction:

$$Cl_2\ (aq) + 2KI\ (aq) \longrightarrow 2KCl\ (aq) + I_2\ (aq)$$
pale yellow   colourless   colourless   red-brown

This time iodine has been displaced.

▲ Chlorine displacing bromine from aqueous potassium bromide.

What happens if you use bromine or iodine instead of chlorine? Look:

If the solution contains ...	when you add chlorine ...	when you add bromine ...	when you add iodine ...
chloride ions (Cl^-)		there is no change	there is no change
bromide ions (Br^-)	bromine is displaced, giving an orange solution		there is no change
iodide ions (I^-)	iodine is displaced giving a red-brown solution	iodine is displaced giving a red-brown solution	

The results in the table confirm this:

A halogen will displace a *less reactive* halogen from a solution of its halide.

1 Describe the appearance of chlorine, bromine, and iodine.
2 a Describe the trend in reactivity in Group VII.
   b Is this trend the same as for Group I?
3 Give two similarities in the products, when chlorine, bromine, and iodine react with iron.
4 Fluorine, F_2, is a pale yellow gas. It reacts instantly with cold iron wool. Where in Group VII would you place it?
   a above chlorine   b below iodine
   Explain your choice.
5 Why are the halogens so reactive?
6 a Write a word equation for the reaction of bromine with potassium iodide. What would you expect to see?
   b Now explain *why* the reaction in **a** occurs.
7 The fifth element in Group VII is astatine (At_2). It is a very rare element. Do you expect it to be:
   a a gas, a liquid, or a solid? Give your reason.
   b coloured or colourless?
   c harmful or harmless?

12.4 More about the trends

Objectives: explain the link between group number and the charge on ions, in the Periodic Table; explain the trends in reactivity; describe changes across Period 3

Here we review the ideas you met already about trends within groups, and add some more. We also look at trends across a period.

The key ideas

1 For Groups I – VII, the group number is the same as the number of outer shell electrons in the atoms.

Group number	I	II	III	IV	V	VI	VII	VIII
Number of outer shell electrons	1	2	3	4	5	6	7	full shell

2 **A full outer shell of electrons makes an atom stable.**
That is why the noble gases (Group VIII) are unreactive.

3 **Atoms achieve full outer shells by taking part in reactions.**
When they lose or gain electrons, ions form. Look:

Group	Number of outer shell electrons	Electrons lost or gained, to reach a full shell	Charge on ion	Examples
I	1	1 lost	1+	Na^+, K^+
II	2	2 lost	2+	Mg^{2+}, Ca^{2+}
III	3	3 lost	3+	Al^{3+}, Ga^{3+}
VI	6	2 gained	2−	O^{2-}, S^{2-}
VII	7	1 gained	1−	Cl^-, Br^-

▲ In a hurry to gain full outer shells. Powdered aluminium is reacting with iodine. The heat given out vapourises the iodine.

4 **The number of electron shells in the atom affects reactivity.**
The nucleus of an atom is positive. It exerts a pull on the outer electrons. But the pull lessens as shells are added, and this affects reactivity. Look:

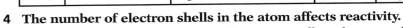

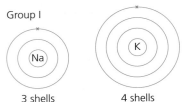

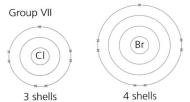

It is easier (takes less energy) for potassium to lose an electron, as its outer shell is further from the nucleus. So it is more reactive than sodium.

It is easier for chlorine to accept an electron, since its outer shell is closer to the nucleus. So it is more reactive than bromine.

You met these trends in reactivity already in Units 12.2 and 12.3.

About reactivity
- If an element reacts readily with other reactants, we describe it as **reactive**.
- A vigorous reaction is a sign that an electron transfer does not need much energy put in.

5 **The number of outer shell electrons affects reactivity too.**
The fewer electrons to lose or gain, the less energy is needed. Look:

Period 3, metal atoms

Period 3, non-metal atoms

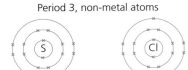

It is easier (takes less energy) to lose 1 outer electron than 2. So sodium is more reactive than magnesium.

It is easier to accept 1 electron than 2, into the outer shell. So chlorine is more reactive than sulfur.

The most reactive …
Of the natural elements:
- fluorine (Group VII) is the most reactive non-metal, and
- caesium (Group I) is the most reactive metal.

This page will help you understand why!

THE PERIODIC TABLE

6 And finally, density generally increases down *all* groups.
As you go down a group, the mass of the atoms increases. Look at the A_r values for three of the Group VII elements on the right. The atoms also get larger – but not by much. The result is that density increases.

Element	A_r	
chlorine	35.5	density increases
bromine	80	
iodine	127	

Trends across Period 3

The main trend *across* a period is the change from metal to non-metal. But it is not the only trend. See how many trends you can spot in Period 3:

Group	I	II	III	IV	V	VI	VII	VIII
element	sodium	magnesium	aluminium	silicon	phosphorus	sulfur	chlorine	argon
number of outer-shell electrons	1	2	3	4	5	6	7	8 (full shell)
element is a ...	metal	metal	metal	metalloid	non-metal	non-metal	non-metal	non-metal
reactivity	high →			low →			high	unreactive
state at room temperarure	solid	solid	solid	solid	solid	solid	gas	gas
boiling point/°C	883	1107	2467	2355	(ignites)	445	−35	−186
oxide is ...	basic		amphoteric				acidic	–
typical compound	NaCl	$MgCl_2$	$AlCl_3$	$SiCl_4$	PH_3	H_2S	HCl	–

The change from metal to non-metal

The change from metal to non-metal is not in fact as clear-cut as shown in the Periodic Table on page 146.

Look at silicon in the table above. It has properties of both a metal *and* a non-metal. For example, it can conduct electricity under certain conditions – but it also forms covalent bonds with oxygen.

So silicon is called a **metalloid**.

The other periods also have metalloids, where the metals give way to non-metals (except for Period 1 which has only two elements).

▲ Silicon is obtained from sand. It is used for the chips in computers and mobiles. It must be 99.9999% pure, for chips!

1. Describe how the number of outer shell electrons changes with group number, across the Periodic Table.
2. Explain why:
   a Group III elements form ions with a charge of 3+.
   b Group VI elements form ions with a charge of 2−.
   c halide ions have a charge of 1−.
3. Atoms of nitrogen, in Group V, form ions in some reactions. They are *nitride* ions. Suggest a formula for them.
4. The boiling points of three of the noble gases are: argon −186 °C, krypton −152 °C, xenon −107 °C.
   a Name the other noble gases, up to Period 6.
   b Use the trend above to suggest boiling points for them.
5. Silicon is a *metalloid*. Explain the term in italics.
6. a Describe three trends in properties across Period 3.
   b Would you expect a similar trend across the numbered groups in Period 4? Explain.

12.5 The transition elements

Objectives: recall that the transition elements are metals; give their typical physical and chemical properties

What are they?
The transition elements are the large block of elements in the middle of the Periodic Table. They are all metals, and include most of the elements we use every day – such as iron, copper, zinc, nickel, and silver.

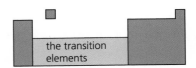

Their physical properties
Here are three of the transition elements:

Iron: the most widely used metal; grey with a metallic lustre (shine). Copper: reddish with a metallic lustre. Nickel: silvery with a metallic lustre.

Here is some data for them, with sodium added for comparison:

Element	Symbol	Density in g/cm³	Melting point/°C
iron	Fe	7.9	1535
copper	Cu	8.9	1083
nickel	Ni	8.9	1455
sodium	Na	0.97	98

Ten transition elements

iron	copper
nickel	zinc
silver	gold
platinum	mercury
chromium	titanium

These are typical properties of transition elements:
- **high density.** Iron is over 8 times heavier than sodium.
- **high melting points.** Look at the three examples in the table. (Mercury is an exception. It is a liquid at room temperature.)
- **hard and strong.** They are not soft like the Group I metals.
- **good conductors of heat and electricity.** Of all the metals, silver is the best conductor of electricity, and copper is next.

Their chemical properties
1 **They are much less reactive than the Group I metals.** For example, copper and nickel do not react with water or catch fire in air, like sodium does. (But iron does rust easily in damp air.)
2 **They do *not* show a clear trend in reactivity.** But those next to each other in the Periodic Table do tend to be similar.
3 **Most transition elements form coloured compounds.** In contrast, the Group I metals form white compounds.
4 **Many transition elements and their compounds acts as catalysts**. A catalyst speeds up a reaction, but remains unchanged itself. Iron is used as a catalyst in making ammonia (page 122).

▲ Compounds of the transition elements are used in pottery glazes, because of their colours.

THE PERIODIC TABLE

5 Most can form ions with different charges. Compare these:

Metal	Charge on ions	Examples
Group I metals	always 1+	sodium: Na^+
Group II metals	always 2+	magnesium: Mg^{2+}
Group III metals	always 3+	aluminium: Al^{3+}
Transition elements	variable	copper: Cu^+, Cu^{2+} iron: Fe^{2+}, Fe^{3+}

Salts of transition elements
- The oxides and hydroxides of all metals are bases.
- So you can make salts of the transition elements by starting with their oxides or hydroxides, and reacting these with acids.

6 Most can form more than one compound, with another element.
That is because they form ions with different charges. Look:

copper(I) oxide, Cu_2O copper(II) oxide, CuO

iron(II) oxide, FeO iron(III) oxide, Fe_2O_3

The Roman numeral tells you how many electrons each metal atom has lost, to form the compound. It is called the **oxidation number**. **The transition elements have variable oxidation numbers.**

Steel
- **Steel** is iron to which other substances have been added, to improve its properties.
- **Stainless steel** contains around 11% chromium, to prevent rust.

▲ Because they are hard and strong, transition elements are used in structures like bridges, buildings, ships, cars and other vehicles, and machinery. Iron is the most used of all – usually as **steel**.

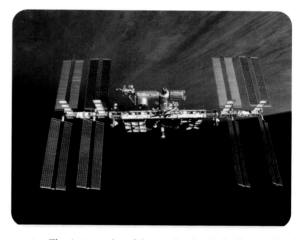

▲ The International Space Station is built mainly of steel, titanium – which is strong, and much lighter than steel – and aluminium (Group III). Some copper is used too.

1. Name five transition elements that you are familiar with.
2. Which best describes the transition elements, overall:
   a soft or hard?
   b high density or low density?
   c high melting point or low melting point?
   d reactive or unreactive, with water at room temperature?
3. Explain why:
   a most paints contain compounds of transition elements
   b the transition element iron is used in building bridges
4. Some transition elements and their compounds are used as catalysts. Define *catalyst*, and give an example. (Glossary?)
5. What is unusual about mercury, compared to the other transition elements?
6. *Most transition elements have more than _____ oxidation number in their compounds.*
   What is the missing word?
7. Write the formulae for these compounds:
   a copper(I) chloride   b iron(II) carbonate
8. Name these compounds: a $CuCl_2$ b NiO c $FeBr_3$

How the Periodic Table developed

Before the Periodic Table
Imagine you find a box of jigsaw pieces. You really want to build that jigsaw. But the lid has only scraps of the picture, and many of the pieces are missing. How frustrating!

That's how chemists felt, 150 years ago. They had found more and more new elements. For example, 24 metals, including lithium, sodium, potassium, calcium, and magnesium, were discovered between 1800 and 1845. They could tell that these fitted a pattern of some kind. They could see fragments of the pattern – but could not work out what the overall pattern was.

And then the Periodic Table was published in 1869, and everything began to make sense.

A really clever summary
The Periodic Table is *the* summary of chemistry. It names all the elements and gives their symbols. It shows the families of elements, and how they relate to each other. It even tells you about the numbers of protons, electrons, and electron shells in their atoms.

Today we take the Periodic Table for granted. But it took hundreds of years, and hard work by hundreds of chemists, to develop. There were some failed attempts along the way, like the 'Law of Octaves'.

The Law of Octaves
By 1863, 56 elements were known. John Newlands, an English chemist, noted that there were many pairs of similar elements. In each pair, the **atomic weights** (or relative atomic masses) differed by a multiple of 8. So he produced a table with the elements in order of increasing atomic weight, and put forward the **Law of Octaves**: *an element behaves like the eighth one that follows it in the table.*

This was the first table to show a repeating or **periodic** pattern of properties. But there were problems. For example, it had copper and sodium in the same group – even though they behave very differently. So it was rejected by other chemists.

Newland's Table of Octaves, presented to the Chemical Society in London in 1865

H	Li	Be	B	C	N	O
F	Na	Mg	Al	Si	P	S
Cl	K	Ca	Cr	Ti	Mn	Fe
Co, Ni	Cu	Zn	Y	In	As	Se
Br	Rb	Sr	Ce, La	Zr	Di, Mo	Ro, Ru
Pd	Ag	Cd	U	Sn	Sb	Te
I	Cs	Ba, V	Ta	W	Nb	Au
Pt, Ir	Tl	Pb	Th	Hg	Bi	Os

▲ 250 years ago, nobody knew of aluminium. Now it is used in most forms of transport: planes, boats, cars, trains …

▲ Lithium was discovered in 1817. It is used in mobile phone batteries. (Mobiles contain about 30 elements, including copper, cobalt, silver, and gold.)

Examples of octaves

element	atomic weight
potassium	39
sodium	− 23 (subtract)
	16 or 2 × 8
calcium	40
magnesium	− 24
	16 or 2 × 8

Now we use **relative atomic mass** instead of **atomic weight**.

◀ Newlands knew of all these, in 1865. How many of them can you name? Find Di (for didymium). This 'element' was later found to be a mixture.

THE PERIODIC TABLE

▲ The Prince of the Periodic Table: Dmitri Mendeleev (1834–1907). Element 101, the synthetic element Mendelevium (Md), is named after him. So is a crater on the moon.

The Periodic Table arrives

Dmitri Ivanovich Mendeleev was born in Russia in 1834, the youngest of at least 14 children. By the age of 32, he was a Professor of Chemistry.

Mendeleev had gathered a lot of data about the known elements. (By then there were 63.) He wanted to find a pattern that made sense of the data, to help his students. So he made a card for each element. He played around with the cards on a table, first putting the elements in order of atomic weight, and then into groups with similar behaviour. The result was the Periodic Table. It was published in 1869.

Mendeleev took a big risk: he left gaps for elements not yet discovered. He even named three: **eka-aluminium**, **eka-boron**, and **eka-silicon**, and predicted their properties. Soon three new elements were found, that matched his predictions – gallium, scandium, and germanium.

Atomic structure and the Periodic Table

Mendeleev put the elements in order of *atomic weight*. But he found some problems. For example, potassium ($A_r = 39$) is lighter than argon ($A_r = 40$), so should come first. But a reactive metal like potassium clearly belongs to Group I, not Group VIII. So he swopped those two around.

In 1911 the proton was discovered. It soon became clear that the **proton number** was the key factor in deciding an element's position in the table, and not A_r. So Mendeleev was right to swop those elements.

It was still not clear why the elements in a group reacted in a similar way. Scientists knew that the numbers of electrons and protons must be equal, since atoms are not charged. And then they discovered that:

- the electrons are arranged in shells
- it is only the outer shell electrons that take part in reactions.

And this explained the similar reactions within a group.

So by 1932, 63 years after it appeared, Mendeleev's table finally made sense. Today's Periodic Table has far more elements. But his table, over 150 years old, is still its foundation.

▲ Mendeleev knew of aluminium, titanium, and molybdenum, which are all used in today's racing bikes.

▲ Mendeleev would recognise all the elements in these health tablets too.

Checkup on Chapter 12

Revision checklist

Core syllabus content
Make sure you can …
- ☐ state that the elements in the Periodic Table are in order of proton number (atomic number)
- ☐ give the meanings of the numbers above and below the symbols in the Periodic Table
- ☐ point out where in the Periodic Table these are:
    Group I   Group VII   Group VIII
    hydrogen   the transition elements
- ☐ state the link between:
    – group number and the number of outer shell electrons
    – period number and the number of electron shells
    – group number and the charge on ions
- ☐ state that there is a change from metal to non-metal, across a period
- ☐ say why elements in a group react in a similar way
- ☐ give the other name for Group I, and name at least three elements in it
- ☐ describe the trends for the Group I elements
- ☐ give the other name for Group VII
- ☐ name chlorine, bromine, and iodine as halogens, and describe their appearance at room temperature
- ☐ describe the trends for the Group VII elements
- ☐ describe how halogens react with solutions of halides, and explain the pattern
- ☐ give the other name for the Group VIII elements, and explain their lack of reactivity
- ☐ give three physical and three chemical properties of the transition elements

Extended syllabus content
Make sure you can also …
- ☐ explain why the elements of Group I and Group VII are so reactive, in terms of electron transfer
- ☐ explain the trends in reactivity in Groups I and VII, in terms of electron shells
- ☐ give more detail about the trends across a period, including:
    – the change from metal to non-metal
    – the change in reactivity
- ☐ state that the transition elements have variable oxidation numbers, explain what this means, and give examples

Questions

Core syllabus content

1 This extract from the Periodic Table shows the symbols for the first 20 elements.

H																	He
Li	Be											B	C	N	O	F	Ne
Na	Mg											Al	Si	P	S	Cl	Ar
K	Ca																

Look at the row from lithium (Li) to neon (Ne).
a What is this row of the Periodic Table called?
b Which element in it is the least reactive? Why? Look at the column of elements from lithium (Li) to potassium (K).
c What is this column of the table called?
d Of the three elements shown in this column, which one is the most reactive?

2 Rubidium is an alkali metal. It lies below potassium in Group I. Here is data for Group I:

Element	Proton number	Melting point/°C	Boiling point/°C	Chemical reactivity
lithium	3	180	1330	quite reactive
sodium	11	98	890	reactive
potassium	19	64	760	very reactive
rubidium	37	?	?	?
caesium	55	29	690	violently reactive

a Describe the trends in melting point, boiling point, and reactivity, as you go down the group.
b Now predict the missing data for rubidium.
c In a rubidium atom:
   i how many electron shells are there?
   ii how many electrons are there?
   iii how many outer shell electrons are there?

3 The elements of Group VIII are called the noble gases. They are all monatomic gases.
a Are the noble gases reactive, or unreactive?
b Name three noble gases, and give their electronic configurations.
c i What is meant by *monatomic*?
   ii Explain why the noble gases, unlike all other gaseous elements, are monatomic.
d *When elements react, they become like noble gases.* Explain what the statement above means.

4 This diagram shows some of the elements in Group VII of the Periodic Table.

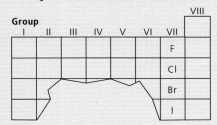

   a What are the elements in this group called?
   b Chlorine reacts explosively with hydrogen. The word equation for the reaction is:
   hydrogen + chlorine → hydrogen chloride
   The reaction requires sunlight, but not heat.
   i How would you expect fluorine to react with hydrogen?
   ii Write the word equation for the reaction.
   c **i** How might bromine react with hydrogen?
   ii Write the symbol equation for that reaction.

5 The Periodic Table is the result of hard work by many scientists, in many countries, over hundreds of years. They helped to develop it by discovering, and investigating, new elements.

The Russian chemist Mendeleev was the first person to produce a table like the one we use today. He put all the elements he knew of into his table. But he realised that gaps should be left for elements not yet discovered. He even predicted the properties of some of these.

Mendeleev published his Periodic Table in 1869. The extract on the right below shows Groups I and VII from his table. Use the modern Periodic Table (page 146) to help you answer these questions.

   a What does Period 2 mean?
   b **i** How does Group I in the modern Periodic Table differ from Group I in Mendeleev's table?
   ii The arrangement in the modern table is more appropriate for Group I. Explain why.
   iii What do we call the Group I elements today?
   c **i** What do we call the Group VII elements?
   ii The element with the symbol Mn is out of place in Group VII. Why?
   iii Where is the element Mn in today's table?
   d Mendeleev left gaps in several places in his table. Why did he do this?
   e There was no group to the right of Group VII, in Mendeleev's table. Suggest a reason for this omission.

Extended syllabus content

6 This question is about **elements** from these families: alkali metals, alkaline earth metals (Group II), transition elements, and halogens.

A a soft, silvery metal; reacts violently with water
B a yellow-green gas at room temperature; reacts readily with other elements, when heated
C a hard solid at room temperature; shows variable oxidation numbers in its compounds
D conducts electricity, and reacts slowly with water; its atoms each give up two electrons
E a red-brown liquid at room temperature; does not conduct electricity
F a hard solid; conducts electricity and rusts easily

   a For each element above, state which of the listed families it belongs to.
   b **i** Comment on the position of elements **A** and **B** within their families.
   ii Describe the outer shell of electrons for each of the elements **A** and **B**.
   c Explain why the arrangement of electrons in their atoms makes some elements very reactive, and others unreactive.
   d Name elements that fit descriptions **A** to **F**.
   e Which of **A** to **F** may be useful as catalysts?

7 Here is data for some elements of Group VI:

Element	Proton number	Melting point / °C	Density g/cm^3
oxygen	6	−219	1.33
sulfur	16	113	2.07
selenium	34	217	4.81
tellurium	52	x	y

   a Describe the trends in: **i** melting point **ii** density for the first three elements of Group VI.
   b Predict the values x and y for tellurium.
   c The trend in properties might not extend to polonium, the next element in this group. Suggest a reason. (Periodic Table page 146.)

An extract from Mendeleev's Periodic Table

	Group I		Group VII	
Period 1	H			
Period 2	Li		F	
Period 3	Na		Cl	
Period 4	K		Mn	
		Cu		Br
Period 5	Rb			
		Ag		I

13.1 Comparing metals and non-metals

Objectives: recall where the metals and non-metals are, in the Periodic Table; give their general physical and chemical properties; give examples of exceptions

Metals and non-metals in the Periodic Table

Over 80% of the elements in the Periodic Table are metals. They lie to the left of the zig-zag line in the outline table below. The non-metals lie to the right.

You know quite a lot about both already. In this unit we compare their properties. But first, a quick reminder!

Group I – the alkali metals; reactive metals including sodium and potassium

Group II includes magnesium and calcium; this group is often called **the alkaline earth metals**

the transition elements – all are metals, and they include most of the metals in everyday use, such as iron, copper, zinc, silver, gold …

Group III includes aluminium, the most abundant metal in Earth's crust

Group VII – the halogens; reactive non-metals, including chlorine and bromine; their compounds are called **halides**

Group VIII – the noble gases; unreactive because their atoms have full outer electron shells

The general properties of metals

Below are the *general* properties of metals. But not all metals have *all* of these properties! (You know already that Group I metals are soft.)

Physical properties
1. They are **strong**. If you press down on them, or drop them, or try to tear them, they won't break – and it is hard to cut them.
2. They are **malleable**; they can be hammered into shape without breaking.
3. They are **ductile**: they can be drawn out into wires.
4. They are **sonorous**: they make a ringing noise when you strike them.
5. They are shiny when polished.
6. They are good conductors of electricity and heat.
7. They have high melting and boiling points. They are all solid at room temperature, except mercury. Look at table ①.
8. They have high **density** – they feel 'heavy'. Look at the green panel.

Chemical properties
1. Most metals react with dilute acids to form **salts**. Hydrogen gas is given off.
2. Metals react with oxygen to form **oxides**, and most of these oxides are **bases** – they neutralise acids, forming salts and water. (See page 136.)
3. Metals form positive ions when they react. The compounds they form are **ionic compounds**.

①

Metal	Melting point / °C	Boiling point / °C
lead	328	1750
zinc	420	980
gold	1065	2710
iron	1540	2890
titanium	1680	3300
mercury	– 39	357

Density: a reminder
It tells you 'how heavy'.

$$\text{density} = \frac{\text{mass (in grams)}}{\text{volume (in cm}^3\text{)}}$$

Compare these:

1 cm^3 of iron, mass 7.86 g
so density = 7.86 g/cm^3

1 cm^3 of lead, mass 11.34 g
so density = 11.34 g/cm^3

160

THE BEHAVIOUR OF METALS

The general properties of non-metals

Non-metals have very different properties from metals.
Again these are *general* properties, and there are exceptions.

Physical properties

1 They have low melting and boiling points. Ten are gases at room temperature, including the oxygen and nitrogen you breathe in. Look at table 2.
2 The solid non-metals are not malleable, nor ductile: they are **brittle**. They break into smaller pieces when hammered.
3 The solid non-metals do not conduct electricity or heat.
4 The solid non-metals look dull rather than shiny.
5 The solid non-metals have low density. They are 'light'.

Chemical properties

1 When non-metals react with oxygen to form oxides, most of the oxides are **acidic oxides.** These dissolve in water to give acidic solutions.
2 When non-metals react with metals they form ions with a negative charge. The compounds are **ionic compounds**.
3 But when non-metals react with each other, they form **covalent compounds**. For example, carbon dioxide, CO_2.

Non-metal	Melting point / °C	Boiling point / °C
oxygen	− 218	− 183
nitrogen	− 210	− 196
argon	− 189	− 186
chlorine	− 101	− 34
phosphorus	44	281
sulfur	119	445

Carbon: exceptions!

- Most non-metals exist as single atoms (the noble gases) or molecules (oxygen, chlorine, ...).
- But carbon is different. Its two forms, diamond and graphite, have giant structures.
- So their properties are not typical.
- Diamond is very hard, with very high melting and boiling points.
- Graphite conducts electricity.

Reactivity

As you saw in Chapter 12, reactivity varies down and across the Periodic Table. The most reactive metals are in Group I. The most reactive non-metals are in Group VII.

In the rest of this chapter, we compare more metals for reactivity – but this time, not by group. We look at metals across the Periodic Table.

Remember, the more easily its atoms give up electrons, the more reactive a metal will be. A reactive metal has **a strong drive** to give up electrons.

◀ We are over 96% non-metal (mainly compounds of oxygen, carbon, nitrogen, and hydrogen). We also contain metal ions essential to life – sodium, potassium, magnesium, calcium, iron, zinc, and more.

1 Metals are *ductile*, *malleable*, and *sonorous*. Explain the terms in italics.
2 10 cm^3 of aluminium weighs 28 g.
10 cm^3 of tin weighs 74 g.
   a Which has a higher density, aluminium or tin?
   b How many times more dense is it than the other metal?
3 Compare the melting points of iron and mercury, table 1.
4 Write sentences that compare the metal and non-metal elements, for these properties:
   a conducting electricity   b conducting heat
   c malleability   d melting and boiling points
5 Which non-metals in table 2 are solid at room temperature?
6 Describe how the oxides of metals and non-metals differ.
7 Carbon is *not* a typical non-metal. Explain. (See the panel!)

161

13.2 Comparing metals for reactivity

Objectives: explain that metals can be put in order of reactivity by comparing their reactions (reactions with silver and gold not done in lab); describe and explain the *displacement* of hydrogen

How to compare?
You can compare different metals for reactivity by reacting them all with the same reactant. If one metal is more reactive than another, it will react faster and more vigorously, and at a lower temperature. So let's see!

1 The reaction of metals with water

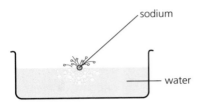

Sodium reacts violently with cold water, whizzing over the surface. Hydrogen gas and a clear solution of sodium hydroxide are formed.

The reaction between calcium and cold water is slower. Hydrogen bubbles off, and a cloudy solution of calcium hydroxide forms.

Magnesium reacts very slowly with cold water, but vigorously with steam: it glows brightly. Hydrogen and solid magnesium oxide form.

You can tell from their behaviour that sodium is the most reactive of the three metals, and magnesium the least.

Compare the equations for the reactions. What pattern do you notice?

$$2Na\ (s) + 2H_2O\ (l) \rightarrow 2NaOH\ (aq) + H_2\ (g)$$
$$Ca\ (s) + 2H_2O\ (l) \rightarrow Ca(OH)_2\ (aq) + H_2\ (g)$$
$$Mg\ (s) + H_2O\ (g) \rightarrow MgO\ (s) + H_2\ (g)$$

Now compare the reactions of those metals with the others in this table:

Metal	Reaction	Order of reactivity	Products
potassium	very violent with cold water; catches fire	most reactive	hydrogen and a solution of potassium hydroxide, KOH
sodium	violent with cold water	↑	hydrogen and a solution of sodium hydroxide, NaOH
calcium	less violent with cold water		hydrogen and calcium hydroxide, $Ca(OH)_2$, which is only slightly soluble
magnesium	very slow with cold water, but vigorous with steam		hydrogen and solid magnesium oxide, MgO
zinc	quite slow with steam		hydrogen and solid zinc oxide, ZnO
iron	slow with steam		hydrogen and solid iron oxide, Fe_2O_3
copper	no reaction	least reactive	

Note the big difference in reactivity. Potassium is the most reactive of these metals, which means it has the strongest drive to give up electrons. Only the first three of the metals produce hydroxides. The others produce insoluble oxides, where they react.

2 The reaction of metals with dilute hydrochloric acid

It is not safe to test sodium and potassium with the dilute acid in the lab, because the reactions are explosively fast. But compare these:

Metal	Reaction with hydrochloric acid	Order of reactivity	Products
magnesium	vigorous	most reactive	hydrogen and a solution of magnesium chloride, $MgCl_2$
zinc	quite slow	↑	hydrogen and a solution of zinc chloride, $ZnCl_2$
iron	slow		hydrogen and a solution of iron(II) chloride, $FeCl_2$
copper silver gold	no reaction, even with concentrated acid	least reactive	

The equation for the reaction with magnesium this time is:

$$Mg\ (s)\ +\ 2HCl\ (aq)\ \longrightarrow\ MgCl_2\ (aq)\ +\ H_2\ (g)$$

Displacement of hydrogen

The equation above shows that magnesium drives out or **displaces** hydrogen from hydrochloric acid. It displaces it from water too.

This tells us that magnesium is *more reactive* than hydrogen, with a stronger drive to exist as a compound.

The same is true for the other metals that react with water and hydrochloric acid. But copper, silver, and gold do not react with either of these. So they are *less reactive* than hydrogen.

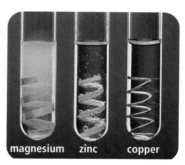

▲ Metals in dilute hydrochloric acid. Does what you see match the table above?

3 The reaction of metals with oxygen

Now look how these metals react with oxygen:

Metal	Reaction on heating in oxygen	Order of reactivity	Products
sodium	burns vigorously with a yellow flame	most reactive	sodium oxide, Na_2O, a white solid
calcium	burns vigorously on strong heating, with an orange-red flame	↑	calcium oxide, CaO, a white solid
magnesium	burns readily with a brilliant white light		magnesium oxide, MgO, a white solid
iron	burns readily with bright sparks, when iron filings are used		iron(II) oxide, FeO, a black solid
copper	reacts slowly at the surface, and a black coat forms	least reactive	copper(II) oxide, CuO, a black solid

Is the overall pattern the same in all three tables?

---

1 Sodium, potassium, and calcium all react with cold water.
   a Which of the three reacts the most vigorously?
   b Which of the three has the strongest drive to give up electrons, and achieve a full outer shell?
   c Describe any pattern you notice in the products.

2 Which of the tested metals react with water to form oxides?

3 Some metals *displace* hydrogen from water and hydrochloric acid. Explain the term in italics.

4 Which element is more reactive? Give your evidence.
   a hydrogen or zinc?   **b** hydrogen or copper?

5 Comment on the products that form when the tested metals were heated in oxygen.

6 Look at the metals in the three tables.
   a Is the order of reactivity the same across the tables?
   b Which metal appears to be:
      the most reactive?   the least reactive?

13.3 Metals in competition

Objectives: explain why carbon can reduce some metal oxides; recall that a more reactive metal will replace a less reactive metal in compounds, and explain why

When metals compete
A reactive metal has a strong drive to give up electrons, and form ions. So let's see what happens when metals compete with each other, and with the non-metal carbon, to form a compound.

1 Competing with carbon

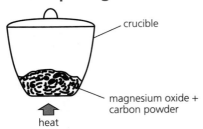

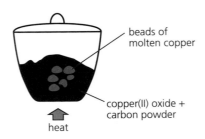

magnesium	more reactive than carbon
aluminium	
carbon	
zinc	
iron	less reactive than carbon
copper	

Magnesium oxide is mixed with powdered carbon and heated. No reaction! So carbon must be *less reactive* than magnesium.

But with copper(II) oxide, there is a reaction. Copper and carbon dioxide gas form. So carbon is *more reactive* than copper.

Oxides of other metals were tested in the same way. Carbon was found to be more reactive than three of the tested metals.

The equation for the reaction with copper(II) oxide is:

$$2CuO\,(s) + C\,(s) \longrightarrow 2Cu\,(s) + CO_2\,(g)$$
copper(II) oxide + carbon ⟶ copper + carbon dioxide

So carbon has reduced the copper(II) oxide to copper, in a redox reaction.

Carbon is more reactive than some metals. It will reduce their oxides to the metal.

2 Competing with other metals, for oxygen

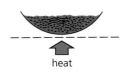

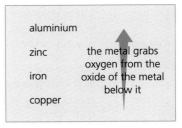

Some powdered iron is heated with copper(II) oxide, CuO. Can the iron grab the oxygen from the copper(II) oxide? Yes!

Once it gets going, the reaction gives out heat. Iron(II) oxide and copper form. So iron has won – it is more reactive than copper.

Other powdered metals were compared in the same way. This shows their order of reactivity. Is it the same as for carbon, above?

The tests confirm that iron, zinc, and aluminium are all more reactive than copper. The equation for the reaction with iron is:

$$Fe\,(s) + CuO\,(s) \longrightarrow FeO\,(s) + Cu\,(s)$$
iron + copper(II) oxide ⟶ iron(II) oxide + copper

The iron is acting as a **reducing agent**, removing oxygen.

A metal will reduce the oxide of a less reactive metal. The reduction always gives out heat – it is exothermic. The metal itself is oxidised.

164

THE BEHAVIOUR OF METALS

3 Competing to form ions in solution

Copper(II) sulfate solution contains blue copper(II) ions, and sulfate ions. An iron nail is placed in it. Will anything happen? Yes!

Copper soon coats the nail. The solution turns green, indicating iron(II) ions. Iron has pushed copper out of the solution.

magnesium	
zinc	
iron	the metal displaces the one below it from solutions of its compounds
copper	
silver	

Other metals and solutions were tested too, with the results above. What do you notice about the order of the metals in this list?

Once again, iron wins against copper. It **displaces** copper from the copper(II) sulfate solution:

$$Fe\ (s)\ +\ CuSO_4\ (aq)\ \longrightarrow\ FeSO_4\ (aq)\ +\ Cu\ (s)$$
iron + copper(II) sulfate ⟶ iron(II) sulfate + copper
(blue) (green)

Other metals displace less reactive metals in the same way.

A metal displaces a less reactive metal from solutions of its compounds.

▲ Zinc has displaced almost all the copper from a copper(II) sulfate solution. What does the solution contain now?

Comparing the drive to form ions

All the reactions in this unit are **redox reactions**: electron transfer takes place. Compare the competitions between iron and copper:

Iron and copper in competition	Competing for oxygen	Competing to form ions in solution
equation	$Fe\ (s)\ +\ CuO\ (s)\ \longrightarrow$ $FeO\ (s)\ +\ Cu\ (s)$	$Fe\ (s)\ +\ CuSO_4\ (aq)\ \longrightarrow$ $FeSO_4\ (aq)\ +\ Cu\ (s)$
the half-equations for electron loss (oxidation) for electron gain (reduction)	$Fe \longrightarrow Fe^{2+} + 2e^-$ $Cu^{2+} + 2e^- \longrightarrow Cu$	$Fe \longrightarrow Fe^{2+} + 2e^-$ $Cu^{2+} + 2e^- \longrightarrow Cu$
the ionic equation (add the half-equations; the electrons on each side of the arrow cancel out)	$Fe + Cu^{2+} \longrightarrow Fe^{2+} + Cu$	$Fe + Cu^{2+} \longrightarrow Fe^{2+} + Cu$

In each case iron gives up electrons to form positive ions. The copper ions accept electrons. So iron has a stronger drive than copper to exist as ions.
The more reactive metal loses electrons more readily, to form positive ions.

1. In the reaction between carbon and copper(II) oxide, which substance is oxidised?
2. a What do you expect to happen when carbon powder is heated with: i calcium oxide? ii zinc oxide?
   b Give a word equation for any reaction that occurs in **a**.
3. When chromium(III) oxide is heated with powdered aluminium, chromium and aluminium oxide form. Which is more reactive, chromium or aluminium?
4. Iron displaces copper from copper(II) sulfate solution. Explain what *displaces* means, in your own words.
5. When copper wire is put into a colourless solution of silver nitrate, crystals of silver form on the wire, and the solution goes blue. Explain these changes.
6. For the reaction described in **5**:
   a write the half equations, to show the electron transfer
   b give the ionic equation for the reaction.

13.4 The reactivity series

Objectives: list the metals of the reactivity series in order; give the key points about the series; explain why aluminium appears to be unreactive

Pulling it all together: the reactivity series

Using experiments in the school lab, and other experiments, we can put the metals in order of reactivity. The list is called **the reactivity series**.

The reactivity series		
potassium, K sodium, Na calcium, Ca magnesium, Mg aluminium, Al	most reactive ↑	metals above the blue line: carbon can't reduce their oxides
carbon zinc, Zn iron, Fe	increasing reactivity	metals above the red line: they displace hydrogen from acids
hydrogen copper, Cu silver, Ag gold, Au	↓ least reactive	

Note where silver and gold appear in the list.
The non-metals carbon and hydrogen are also included, for reference.
This is not a complete list, of course. We could test many other metals.

Points to remember about the reactivity series

1 The reactivity series is really a list of the metals in order of their drive to give up electrons, and form positive ions with full outer shells. The easier this is to do, the more reactive the metal will be.

2 So a metal will react with a compound of a less reactive metal (for example, an oxide, or a salt in solution) by pushing the less reactive metal out of the compound and taking its place.

3 The more reactive the metal, the more **stable** its compounds are. They do not break down easily.

4 The more reactive the metal, the more difficult it is to extract from its ores, since these are stable compounds. For the most reactive metals you need the toughest method of extraction: **electrolysis**.

5 The less reactive the metal, the less readily it form compounds. That is why copper, silver, and gold are found as elements in Earth's crust. The other metals are *always* found as compounds.

Metals we had to wait for
- Because they are easier to extract from their ores, the less reactive metals have been known and used for thousands of years. For example, copper has been in wide use for 6000 years, and iron for 3500 years.
- But the more reactive metals, such as sodium and potassium, had to wait until the invention of electrolysis, over 200 years ago (in 1800) for their discovery.

▲ Copper is used for roofing, since it is unreactive. But over time the surface changes colour due to reactions with the atmosphere. The products that form protect the copper below the surface, so copper roofs last for hundreds of years.

▲ A metal's position in the reactivity series will give you clues about its uses. Only unreactive metals are used in coins.

THE BEHAVIOUR OF METALS

Breaking down metal compounds by heat

- The more reactive a metal is, the more stable its compounds will be. The reverse is also true. Compounds of less reactive metals break down more easily.
- You can break some compounds down by heating. This is called **thermal decomposition**.
- For example, sodium and potassium carbonates do not break down on heating, but the carbonates of other metals decompose to the oxide and carbon dioxide.
- But heat alone will not decompose these oxides further, to give the metals. You would need to heat them with a reducing agent, or use electrolysis.

The photo shows a traditional limestone kiln, where limestone (calcium carbonate) is broken down to lime (calcium oxide). Lime has many uses, including in making cement, neutralising acidity in soil, and to whitewash homes.

The reactivity of aluminium

Look again at the reactivity series. Aluminium is more reactive than iron.

When we use iron, we need to protect it – for example, by painting it or coating it with grease. Otherwise it will corrode or **rust** through reaction with oxygen and water vapour in the air.

But we use aluminium for things like TV aerials, and satellite dishes, and ladders, *without* protecting it. Why?

Because the aluminium has already reacted! It reacts instantly with oxygen, forming a thin coat of aluminium oxide – so thin you cannot see it. This sticks tight to the metal, acting as a barrier to further corrosion.

▲ A layer of aluminium oxide forms instantly on the surface of aluminium, and protects it from further corrosion.

▲ Aluminium is protected from further corrosion by a layer of aluminium oxide. Ideal for scaffolding and ladders!

▲ But aluminium powder is reactive. Here it reduces iron(III) oxide to molten iron, used to weld sections of the rail line together.

Q

1. a List the metals of the reactivity series, in order.
   b Beside each, say where it occurs in the Periodic Table.
   c To which group in the Periodic Table do the most reactive metals belong?
   d Where in the table are the least reactive ones found?
2. Why is magnesium never found as the element, in nature?
3. Iron occurs in Earth's crust mainly as oxides. Suggest a way to reduce these oxides to iron.
4. Aluminium is found in Earth's crust as aluminium oxide.
   a Carbon will not reduce this oxide to aluminium. Why not?
   b Suggest a way to obtain aluminium from its oxide.
5. We need to protect iron from corrosion (rusting) in the atmosphere. We do not need to protect aluminium, even though it is more reactive than iron. Why not? Explain.
6. The process in the photo above is called the *thermit process*. Write the equation for the reaction that is taking place.

13.5 The rusting of iron

Objectives: state that rusting needs oxygen and water; give the word equation for rusting; describe the two approaches to rust prevention, with examples

A big problem
Iron is the world's most widely used metal, by far. But there is a big problem: it corrodes or **rusts** in the atmosphere. So do many steels. Tackling this problem around the world costs billions of dollars a year.

An experiment to investigate rusting
To prevent rusting, we must first understand what causes it. Here is a simple investigation.

1. Stand three identical iron nails in three test-tubes.
2. Then prepare the test-tubes as below, so that:
   - test-tube 1 contains dry air
   - test-tube 2 contains water but no air
   - test-tube 3 has both air and water.
3. Leave the test-tubes to one side for several days.

▲ Rust forms as flakes. Oxygen and moisture can get behind them – so in time the iron rusts right through.

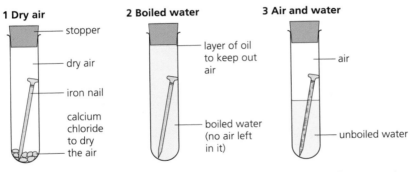

Rusting and salt
- Salty water helps iron rust faster. It conducts, so helps electrons from the iron move more easily.
- This is a problem for ships, and for cars in countries where salt is spread on roads in winter, to lower the freezing point of water.

Result After several days, the nails in test-tubes 1 and 2 show no signs of rusting. But the nail in test-tube 3 has rust on it. So we can conclude …

Rusting requires oxygen and water.

The iron in test-tube 3 has been oxidised by the oxygen in the air:

$$4Fe\,(s) + 3O_2\,(g) + 4H_2O\,(l) \longrightarrow 2Fe_2O_3.2H_2O\,(s)$$

iron + oxygen + water ⟶ hydrated iron(III) oxide (rust)

▲ It was once someone's pride and joy. Now it is rusting away.

▲ Today, steel in car bodies is coated with zinc, to prevent rusting.

THE BEHAVIOUR OF METALS

Ways to prevent rusting

We know that rusting is caused by iron reacting with oxygen and water. So how can we prevent it? There are two approaches.

1 Cover the iron or steel

You can create a barrier to keep out oxygen and water. You could use:

- **paint**. Steel bridges and railings are usually painted.
- **grease**. Tools and machine parts are coated with grease or oil.
- **plastic**. Iron chairs for outdoor use are often coated with plastic.
- **another metal**. Iron is coated with **zinc**, by dipping it into molten zinc, for roofing. Steel is electroplated with zinc, for car bodies. Coating with zinc is given a special name: **galvanising**. For food tins, steel is coated with **tin** by electroplating.

▲ They are painting the outside of this huge ferry, to help prevent rusting.

2 Let another metal corrode instead

During rusting, iron is **oxidised**: it loses electrons.

Magnesium is more reactive than iron, which means it has a stronger drive to lose electrons. So when a bar of magnesium is attached to the side of a steel ship, or the leg of an oil rig, the magnesium will give up electrons in place of the iron:

without magnesium: $Fe \longrightarrow Fe^{3+} + 3e^-$
with magnesium: $Mg \longrightarrow Mg^{2+} + 2e^-$

By giving up electrons, magnesium prevents the iron from giving up electrons and becoming oxidised. The magnesium dissolves. It has been sacrificed to protect the iron. This is called **sacrificial protection**. The magnesium bar must be replaced before it all dissolves.

Zinc is also more reactive than iron, so it could be used too. In fact if the zinc coat on galvanised iron gets damaged, the zinc still protects the iron. So galvanising provides both a barrier *and* sacrificial protection.

◀ Below left: galvanised steel used for roofing. It is often called **corrugated iron**.

▼ Below: blocks of magnesium welded to a boat's hull, for sacrificial protection. The boat is ready for the sea again.

Q

1. Rust is a hydrated oxide.
   a Give its formula. b What does *hydrated* mean?
2. Which two substances are needed for rusting to occur?
3. Suggest a way to prove that it is the *oxygen* in air, not *nitrogen*, that causes rusting.
4. Iron that is covered in grease does not rust. Why not?
5. Rusting can be prevented by *sacrificial protection*.
   a Explain the term in italics.
   b Both magnesium and zinc can be used for it. Explain why.
   c Aluminium cannot be used for it. Why not? (Page 167?)
6. a Name the metal used for *galvanising*,
   b Explain how galvanising prevents rusting in *two* ways.

Checkup on Chapter 13

Revision checklist

Core syllabus content
Make sure you can …
- [ ] explain these terms used about metals:
  malleable   ductile   sonorous   high density
- [ ] give at least three ways in which metals and non-metals differ:
  in physical properties   in chemical properties
- [ ] explain what a *reactive* metal is
- [ ] explain what *the reactivity series* is, and list the metals, plus carbon and hydrogen, in the correct order
- [ ] describe how the metals in the series react with
  - water
  - dilute acids
  - oxygen
  and give word equations where reactions occur
- [ ] explain what *displacement of hydrogen* means
- [ ] explain the position of hydrogen in the reactivity series, in terms of its displacement from water and dilute hydrochloric acid
- [ ] state that more reactive metals have more stable compounds
- [ ] state the conditions required for the rusting of iron and steel
- [ ] give the chemical name and formula for rust
- [ ] state that rusting can be prevented by covering iron and steel with something that acts as a barrier to oxygen and water, and give examples

Extended syllabus content
Make sure you can also …
- [ ] explain that the reactivity of a metal is related to its drive to form positive ions
- [ ] state what the products will be, when:
  - a metal oxide reacts on heating with carbon
  - a metal is heated with the oxide of a less reactive metal
  - a metal is placed in the solution of a compound of a less reactive metal
- [ ] explain why all the reactions of metals are redox reactions
- [ ] describe it, and explain how it prevents rusting:
  - sacrificial protection
  - galvanising (give two ways it prevents rusting)

Questions

Core syllabus content

1

Metal	Density in g/cm^3
aluminium	2.7
calcium	1.6
copper	8.9
gold	19.3
iron	7.9
magnesium	1.7
sodium	0.97

a Define *density*.
b List the metals above in order of increasing density.
c A block of metal has a volume of 20 cm^3 and a mass of 158 g. Identify it, from the table.
d i   Now list the metals in order of reactivity.
   ii The most reactive element has a density of ......?
   iii The least reactive one has a density of ......?
   iv Does there appear to be a link between density and reactivity? If yes, what?
e Using 'lighter' metals for vehicles saves money on fuel and road repairs. Which metal above would be the most suitable for vehicles? Give reasons.

2 The rusting of iron wool gives a way to find the percentage of oxygen in air, using this apparatus:

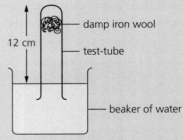

After five days the water level had risen by 2.5 cm, and the iron had rusted to hydrated iron(III) oxide.
a Write a word equation for the reaction.
b The iron wool was dampened with water before being put in the tube. Why?
c Why does the water rise up the tube?
d What result does the experiment give, for the percentage of oxygen in air?
e What sort of result would you get if you doubled the amount of iron wool?
f Explain why the water level would not rise if the iron wool was coated with a layer of grease.

3 For each description below, choose one metal that fits the description. Name the metal. Then write a word equation for the reaction that takes place.
  a A metal that displaces copper from copper(II) sulfate solution.
  b A metal that reacts gently with dilute hydrochloric acid.
  c A metal that floats on water and reacts vigorously with it.
  d A metal that reacts quickly with steam but very slowly with cold water.

4 Look again at the list of metals in **2**. Carbon can be placed between zinc and aluminium.
  a Which two of these pairs will react?
   i carbon + aluminium oxide
   ii carbon + copper(II) oxide
   iii magnesium + carbon dioxide
  b Write a word equation for the two reactions, and underline the substance that is reduced.

Extended syllabus content

5 When magnesium powder is added to copper(II) sulfate solution, a displacement reaction occurs and solid copper forms.

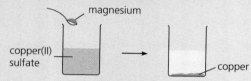

  a Write a word equation for the reaction.
  b Why does the displacement reaction occur?
  c i Write a half-equation to show what happens to the magnesium atoms.
   ii Which type of reaction is this?
  d i Write a half-equation to show what happens to the copper ions.
   ii Which type of reaction is this?
   iii Which metal shows the greater tendency to form positive ions?
  e i Write the ionic equation for the displacement reaction, by adding the half-equations.
   ii Which type of reaction is it?
  f Use the reactivity series of metals to decide whether these will react together:
   i iron + copper(II) sulfate solution
   ii magnesium + calcium nitrate solution
   iii zinc + silver nitrate solution
  g For those that react:
   i describe what you would *see*
   ii write the ionic equations for the reactions.

6 When magnesium and copper(II) oxide are heated together, this redox reaction occurs:
$Mg\,(s) + CuO\,(s) \longrightarrow MgO\,(s) + Cu\,(s)$

  a What does the word *redox* stand for?
  b For the above reaction, name:
   i the reducing agent **ii** the oxidising agent
  c Describe the electron transfer in the reaction.
  d Explain as fully as you can why the *reverse* reaction does not occur.
  e i Name one metal that would remove the oxygen from magnesium oxide.
   ii Does this metal *gain* electrons, or *lose* them, more easily than magnesium does?

7 Chromium is a typical transition element. It reacts with hydrochloric acid in a similar way to iron. It forms a green ion with the formula Cr^{3+}.
  a i What will you see when chromium is added to dilute hydrochloric acid?
   ii Write a balanced equation for the reaction.
  b i What will you see if powdered chromium is added to copper(II) sulfate solution?
   ii Write a balanced equation for this reaction.
  c Chromium can be obtained from its oxide by heating with the metal zinc.
   i Which is more reactive, chromium or zinc?
   ii Write a balanced equation for the reaction.
  d Chromium can also be obtained at the cathode by electrolysis of a solution containing the Cr^{3+} ion.
   Write an ionic equation for the cathode reaction.
  e Another transition metal, titanium, is extracted from its chloride by heating with sodium. What can you say about its reactivity?

8 Zinc is added to a solution of silver nitrate. Which of these equations describe the reaction that occurs?
  a $Ag^+\,(aq) + e^- \longrightarrow Ag\,(s)$
  b $Ag\,(s) \longrightarrow Ag^+\,(aq) + e^-$
  c $Zn^{2+}\,(aq) + 2e^- \longrightarrow Zn\,(s)$
  d $Zn\,(s) \longrightarrow Zn^{2+}\,(aq) + 2e^-$

9 A block of a second metal is attached to an under-water steel pipeline, to protect it from corrosion.
  a i What is the key factor in choosing this second metal?
   ii Name a suitable metal.
   iii Write the half-equation to show what happens to this metal, when attached to the pipeline.
   iv Explain in terms of electrons why this metal protects the steel.
  b What name is given to this type of protection?

14.1 Metal ores and metal extraction

Objectives: state that most metals occur as compounds in rock; define *ore*; explain the link between extraction methods and reactivity; describe extraction as reduction

The composition of Earth's crust

We get our metals from the rock in Earth's crust – the hard outer skin of Earth that we live on.

Rock is mixture of substances. They are mostly **compounds**. But unreactive **elements** such as silver, mercury, platinum, and gold occur **native**, or uncombined.

If you could break the rock in Earth's crust into its elements, you would find it is almost half oxygen. Amazing! This shows its composition:

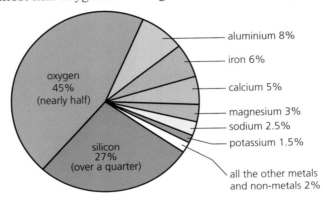

- oxygen 45% (nearly half)
- silicon 27% (over a quarter)
- aluminium 8%
- iron 6%
- calcium 5%
- magnesium 3%
- sodium 2.5%
- potassium 1.5%
- all the other metals and non-metals 2%

▲ We use about nine times more iron than all the other metals put together. It is usually in the form of steel (page 180).

Note that:
- oxygen and silicon make up nearly three-quarters of the crust. They occur together in compounds such as silicon(IV) oxide (**silica**). Oxygen is also found in metal oxides and carbonates.
- just six metals – aluminium to potassium in the pie chart – make up over a quarter of the crust. Aluminium is the most abundant of these, and iron next. All six occur as compounds, because they are reactive.

Metal ores

The rock in an area may contain a large amount of one metal compound, or metal. So it may be worth digging the rock up to extract the metal. The rocks from which we obtain metals are called **ores**. Look at these:

▲ Extracting metals starts with digging! This is the world's largest copper mine, in Utah, USA. It was started in 1906, and is now over 1 km deep and 4 km wide.

Hematite, a common ore of iron. It is mainly iron(III) oxide.

Bauxite, the main aluminium ore. It is mostly aluminium oxide.

Gold. It is unreactive, so is found as seams of the element, in rock.

EXTRACTING AND USING METALS

Extraction

You have a metal ore. And now you have to **extract** the metal from it. How will you do this? It depends on the metal's **reactivity**.

- Where a metal occurs as the element – for example, gold and silver – all you need to do is separate it from sand and other impurities.
- But most metals occur as compounds in their ores. A chemical reaction is needed to reduce them to the metal:

  metal compound $\xrightarrow{\text{reduction}}$ metal

- The compounds of the more reactive metals are very stable, and need electrolysis to reduce them. This is a powerful method. But it is expensive, because it needs a lot of electricity.
- The compounds of the less reactive metals are less stable. They can be reduced to the metal by heating with a suitable reactant.

> **Reduction of metal ores**
> Remember, you can define **reduction** as:
> - loss of oxygen
> $Fe_2O_3 \rightarrow 2Fe$
> - or gain of electrons
> $Al^{3+} + 3e^- \rightarrow Al$
>
> Either way, the ore is **reduced** to the metal.

Extraction and the reactivity series

So the method of extraction is linked to the **reactivity series**. Look:

Metal			Method of extraction from ore		
potassium					
sodium					
calcium			electrolysis		
magnesium	metals more reactive	ores more difficult to decompose		method of extraction more powerful	method of extraction more expensive
aluminium					
carbon					
zinc					
iron			heating in a furnace with carbon or carbon monoxide		
copper					
silver			occur naturally as elements, so reduction is not needed		
gold					

Note that the table includes heating with **carbon monoxide**, as well as carbon. Why? Because when carbon is heated in a furnace, it reacts with the limited supply of oxygen present, to give carbon monoxide gas (CO). This then brings about the reduction.

Iron and aluminium are the two most common metals in Earth's crust, and the two we use most. As the table shows, different methods are used to extract them. We look more closely at these in the next two units.

▲ No need to reduce gold …

1 Which are the two most common elements in Earth's crust?
2 Name the two most common metals in Earth's crust.
3 *The most reactive metals are plentiful in Earth's crust.* True or false? Discuss!
4 What is an *ore*?
5 Gold occurs as the element, in gold ore. Explain why.
6 Name the main ore, and the main compound in it, for:
   a iron   b aluminium
7 Explain why:
   a carbon is not used for the extraction of aluminium
   b electrolysis is not needed for the extraction of iron
   c the change $Fe_2O_3 \rightarrow 2Fe$ is a reduction

173

14.2 Extracting iron

Objectives: describe the blast furnace; recall the raw materials used, and why each is needed; give the key reactions; explain what happens to the waste materials

The blast furnace

Iron is the world's most-used metal. So its extraction is hugely important. It is carried out by reducing iron ore using carbon monoxide, in a **blast furnace**. This is an oven shaped like a chimney:

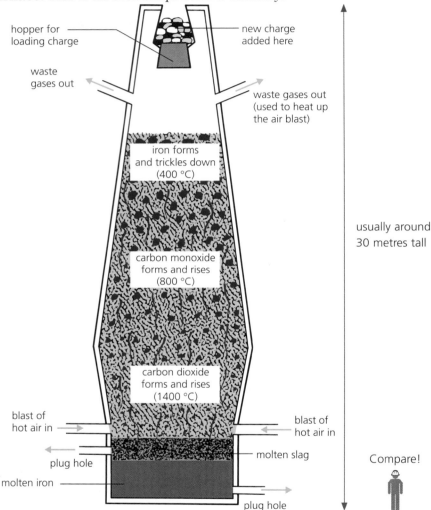

▲ Blast furnaces run non-stop 24 hours a day.

▲ Coke is made by heating coal in an oven in the absence of air. It is almost pure carbon.

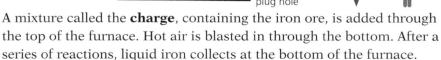

A mixture called the **charge**, containing the iron ore, is added through the top of the furnace. Hot air is blasted in through the bottom. After a series of reactions, liquid iron collects at the bottom of the furnace.

What's in the charge?

The charge contains three things:

1. **Iron ore**. The chief ore of iron is **hematite**. This is mainly iron(III) oxide, Fe_2O_3, mixed with sand and other compounds.
2. **Limestone**. This common rock is mainly calcium carbonate, $CaCO_3$.
3. **Coke**. This is made from coal, and is almost pure carbon.

▲ Mining hematite.

EXTRACTING AND USING METALS

The reactions in the blast furnace

Reactions, products, and waste gases	The symbol equations and further comments
Stage 1: The coke burns to give carbon dioxide and heat The blast of hot air starts the coke burning: carbon + oxygen → carbon dioxide So the carbon is **oxidised** – it gains oxygen.	$C\ (s) + O_2\ (g) \rightarrow CO_2\ (g)$ This is a redox reaction. It is highly exothermic. The heat given out heats the furnace.
Stage 2: Carbon dioxide is reduced to carbon monoxide The carbon dioxide reacts with more coke, like this: carbon + carbon dioxide → carbon monoxide Here the carbon dioxide is **reduced** – it loses oxygen.	$C\ (s) + CO_2\ (g) \rightarrow 2CO\ (g)$ Another redox reaction. It is endothermic – it takes in heat from the furnace. That helps, because stage 3 needs a lower temperature.
Stage 3: Carbon monoxide reduces the iron(III) oxide to iron This is the stage where the extraction of the metal occurs. iron(III) oxide + carbon monoxide → iron + carbon dioxide So the iron(III) oxide is reduced. The iron trickles to the bottom of the furnace.	$Fe_2O_3\ (s) + 3CO\ (g) \rightarrow 2Fe\ (l) + 3CO_2\ (g)$ Another redox reaction. Carbon monoxide is the reducing agent. It is itself oxidised to carbon dioxide.
Stage 4: Impurities are removed as slag There are two reactions. **i** First the limestone from the charge breaks down in the heat of the furnace. This is a **thermal decomposition**: calcium carbonate $\xrightarrow{heat}$ calcium oxide + carbon dioxide **ii** Then the calcium oxide reacts with sand, which is the main impurity in iron ore. Sand is mainly silicon(IV) oxide. The reaction is: calcium oxide + silicon(IV) oxide → calcium silicate The calcium silicate forms **slag** which runs down the furnace and floats on the iron. It is drained off. When it solidifies it is sold, mostly for road building.	$CaCO_3 \xrightarrow{heat} CaO\ (s) + CO_2\ (g)$ $CaO\ (s) + SiO_2\ (s) \rightarrow CaSiO_3\ (s)$ Calcium oxide is a basic oxide. Silicon(IV) oxide is an acidic oxide. Calcium silicate is a salt.
The waste gases: hot **carbon dioxide**, and **nitrogen** from the air, exit from the top of the furnace. Heat is transferred from them to heat the incoming blast of air.	

Where next?

Most of the impurity from the iron ore is removed in the blast furnace – but not all. The molten iron contains carbon, and some sand too.

Some of it is run into moulds to make **cast iron**. This is hard, but its high carbon content makes it brittle. So it is used only for things like canisters for bottled gas, and drain covers, that are not put under much stress.

But most is turned into **steels**. You can find out about steels in Unit 14.5.

▲ A cast-iron drain cover. It was once rock.

1 Name the substances in the charge for the blast furnace.
2 What is the *blast* of the blast furnace? Why is it needed?
3 Give the word equation for the reaction in the furnace that:
 a forms carbon monoxide b gives iron
 c is a thermal decomposition
4 Slag forms in the blast furnace.
 a What is *slag*? b Describe how it forms.
5 Molten iron flows out of the furnace. Is it pure? Explain.
6 Name the oxidising agent in the redox reaction that:
 a heats the furnace b produces the iron

14.3 Extracting aluminium

Objectives: name the main ore of aluminium; outline the extraction process, with equations; explain the use of cryolite, and the other reaction conditions

From rocks to rockets

Aluminium is the most abundant metal in Earth's crust. Its main ore is **bauxite**, which is aluminium oxide mixed with impurities such as sand and iron oxides. These are the steps in obtaining aluminium:

1 First, geologists test rocks and analyse the results, to find out if the bauxite is worth mining. If all is satisfactory, mining begins.

2 Bauxite is red-brown in colour. It is usually near the surface, so is easy to dig up. This is a bauxite mine in Australia.

3 The ore is taken to a bauxite plant, where impurities are removed. The result is white **aluminium oxide**.

4 The aluminium oxide is then broken down by electrolysis, to give aluminium. This shows a cell. An anode is about to be replaced.

5 The aluminium is made into sheets and rolls and blocks, and sold to other industries. It has a great many uses. For example …

6 … it is used to make cans for drinks, food cartons, cooking foil, bikes, TV aerials, electricity cables, cars, trains, and aircraft.

The electrolysis

The electrolysis is carried out in a steel cell. This is lined with carbon, which acts as the cathode. Big blocks of carbon hang down along the cell, and act as anodes. Covers along the cell can be slid open, for access.

Aluminium oxide has a very high melting point, 2072 °C. To save fuel costs it is not melted. It is dissolved in molten **cryolite** or sodium aluminium fluoride. This melts at 1012 °C. It is a good solvent for aluminium oxide, and the solution is a better conductor than aluminium oxide would be.

EXTRACTING AND USING METALS

The electrolytic cell
Here is a diagram of the electrolytic cell.

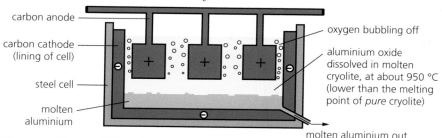

The reactions at the electrodes
Once the aluminium oxide dissolves, its aluminium and oxide ions are free to move. They move to the electrode of opposite charge.

At the cathode (−) The aluminium ions gain electrons:

$$4Al^{3+} (l) + 12e^- \longrightarrow 4Al (l) \quad \text{(reduction)}$$

The aluminium drops to the bottom of the cell as molten metal.

At the anode (+) The oxide ions lose electrons:

$$6O^{2-} (l) \longrightarrow 3O_2 (g) + 12e^- \quad \text{(oxidation)}$$

The oxygen gas bubbles off, and reacts with the anode:

$$C (s) + O_2 (g) \longrightarrow CO_2 (g) \quad \text{(oxidation of carbon)}$$

So the carbon blocks get eaten away, and must be regularly replaced. Along with the cost of electricity, this is further expense.

The overall reaction The aluminium oxide breaks down to aluminium:

aluminium oxide $\longrightarrow$ aluminium + oxygen
$$2Al_2O_3 (l) \longrightarrow 4Al (l) + 3O_2 (g)$$

The molten aluminium is run off at intervals.

▲ Molten aluminium from the tank was run into these moulds, to make blocks.

Amazing aluminium ...
Aluminium has really useful properties.
- It is light, with a much lower density than iron. This is a big advantage for use in vehicles, overhead electrical cables, and planes and other spacecraft.
- It is not as strong as iron – but can be made into strong **alloys**. (Page 180.)
- It is more malleable and ductile than iron. It can be made into complex shapes, and drawn into thick cables like those on the right.
- It conducts electricity – not as well as copper does, but it is lighter *and* cheaper.
- And the big bonus: the thin film of aluminium oxide that forms on its surface. This prevents further corrosion. So the metal does not need to be covered.

1 Name:
  a the main ore of aluminium
  b the compound that remains when this ore is purified
  c the process used to extract the aluminium from it
2 Cryolite is used in the extraction of aluminium. Explain:
  a why it is used   b how it is used

3 a During electrolysis, the oxide ions go to the anode. Why?
  b Write the half-equation for the reaction at the anode.
  c Explain why the anode must be regularly replaced.
4 a Where is the cathode, in the electrolytic cell?
  b Write the half-equation for the reaction at the cathode.
  c How is the aluminium removed from the cell?

14.4 Making use of metals

Objectives: understand that the properties of metals vary; explain that properties dictate how we use a metal; give examples of uses of metals, linked to properties

Thrilling transformations …

We take metals for granted. Think of saucepans, and bikes, and buses, and drinks cans. But next time you look at a metal, remember that it has had a long journey from Earth's crust, and undergone amazing changes.

Look at these examples:

From bauxite (aluminium ore)…

From hematite (iron ore) …

From gold-bearing rock …

… to drinks cans. Aluminium is used for these because it is light, unreactive, non-toxic, and can be rolled into thin sheets.

… to wrought iron gates and lamps. Iron is used for these because it is hard, heavy, and malleable. (But it must be painted to prevent rust.)

… to jewellery. Gold is used for jewellery because it is unreactive, ductile, and lustrous – and it is looked on as a form of wealth.

Properties dictate uses

You know already that metals share general properties. But just as humans differ, so do metals. Some are less dense than others, some less reactive, some better at conducting electricity.

So the way we use a metal depends on its properties, as the three examples above show. We match uses to properties.

The precious metals
- Gold, silver, platinum and palladium are **precious metals**.
- This terms is used for metals which are quite scarce, expensive, and often held as a form of wealth.

EXTRACTING AND USING METALS

▲ A major use of copper is for electrical wiring. Of all the metals, it is the second best electrical conductor. Silver is first.

▲ Rolls of steel coated with zinc. They are off to a car factory, where they will be unrolled and cut into shapes to make car bodies.

Examples of uses for some common metals

Here are uses you met already, and some new ones:

Metal	Used for ...	Properties that make it suitable
aluminium	overhead electricity cables (with a steel core for strength)	a good conductor of electricity, low density (light), ductile, resists corrosion
	cooking foil, food containers, and drinks cans	low density, non-toxic, resists heat and corrosion, can be rolled into thin sheets
	aircraft	low density, resists corrosion
copper	electrical wiring	one of the best conductors of electricity, ductile
	pipes for the water supply inside homes	malleable (so pipes are easy to bend), does not readily corrode, non-toxic
zinc	coating or **galvanising** iron and steel	prevents rusting through sacrificial protection, and also by keeping air and moisture out
	for torch batteries	gives a current when connected to a carbon electrode, in a paste of electrolyte

In the next unit you will see how metals can be even more useful as **alloys**.

What about all the other metals?

There are over 80 naturally occurring metals – and we use them all! Mobile phones alone use at least 30 metals. For example:
- aluminium in the case
- copper, silver, gold, platinum and tungsten in the electrical circuits, and lithium and cobalt in the battery, with lead as a solder
- neodymium, terbium, dysprosium, and tungsten to allow the phone to vibrate
- ytrium, terbium, and dysprosium in the screen display
- indium and tin for the touchscreen function, and lanthanum in the camera lens

So your mobile represents a big chunk of the Periodic Table!

1 Density in g/cm³: iron 7.9 aluminium 2.7 copper 9.0
Using these density values, suggest a reason why:
  a aluminium is not used for anchors for boats, even though it resists corrosion
  b copper is not used in overhead electrical cables, even though it is a better conductor than aluminium

2 Give three reasons why:
  a aluminium is used in cans for soft drinks
  b copper is used for water pipes in homes

3 Which property of zinc leads to its use in the car industry?

4 Silver is a better electrical conductor than copper, but it is not used in wiring homes. Suggest a reason.

14.5 Alloys

Objectives: define *alloy*; explain why alloys are so useful; outline how they are made; interpret diagrams showing alloys; give examples of alloys and their properties

Steels: alloys of iron

Iron is very widely used – but almost never as the pure metal.

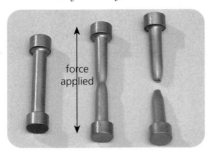

Pure iron is no good for building things, because it is too soft, and stretches quite easily, as you can see from this photo. Even worse, it rusts very easily.

But when a small amount of carbon is added to it – less than 0.5% – the result is **mild steel**. This is hard and strong. It is used for buildings, ships, car bodies, and machinery.

When nickel and chromium are added to iron, the result is **stainless steel**. This is hard and rustproof. It is used for cutlery, and surgical instruments.

So mild steel and stainless steel are **alloys** of iron. There are many other steels too, all with different properties.

An alloy is a mixture of a metal with at least one other element.

The other element can be a metal, or a non-metal.

Alloys make metals more useful

As you can see from above, steels are harder and stronger than iron. So they are more useful. That is why most iron is turned into steels.

Turning a metal into an alloy changes its properties, and makes it more useful.

Brass is another example of an alloy. It is made by adding zinc to copper. it is hard, strong, and shiny. It is used for door locks, keys, knobs, ornaments, and musical instruments such as trumpets and tubas.

▲ This Airbus contains aluminium, aluminium alloys, and titanium. But 53% of it by weight is plastic, reinforced with carbon fibres – strong and very light.

▲ Stainless steel is ideal in restaurant kitchens, because it does not corrode, and is easy to keep clean.

▲ A high-school brass band. Brass is an alloy of copper and zinc. It looks very different from both!

EXTRACTING AND USING METALS

How do alloys work?

 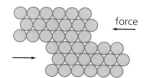 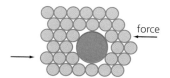

A pure metal is a regular arrangement of metal ions in a sea of electrons. Layers of ions are held in a lattice. (See page 42.)

When pressure is applied, for example by hammering, the layers of ions can slide over each other easily. The metal gets stretched.

But when an alloy is made, new atoms enter the lattice. The layers cannot slide so easily. So the alloy is harder and stronger.

Note these points:

- The new atoms must enter the metal lattice.
- So the main metal must first be melted. Then exactly the right amount of the new element is dissolved in it. The solution cools and hardens.
- The new atoms can be larger, or smaller, than the atoms of the main metal. They will always change its properties.
- It is not only strength that changes. The alloy may be more resistant to corrosion than the main metal. Its ability to conduct electricity and heat will be different. It may look different too.
- And finally, you can add more than one element, in different amounts.

this mixture of metals is not an alloy

an alloy: new atoms in the lattice

▲ Compare the two!

Alloys by design

Today, scientists design new alloys with exactly the right properties for a given use. Exciting work!

Some examples of alloys

A few alloys, such as **brass** and **bronze**, have been made for centuries. Today there are *thousands* of alloys. Here are examples old and new:

Alloy	Made from ... (main metal first)	Special properties	Uses
brass	copper plus zinc	harder than copper, easy to shape, looks attractive, does not corrode	musical instruments, ornaments, door knobs
bronze	copper plus tin	hard, sonorous, resists corrosion	coins, statues, bells, propellers for ships
a typical stainless steel	iron plus chromium, nickel, carbon, and other elements	strong, resists corrosion, looks attractive	cutlery and cooking utensils, construction, equipment in hospitals and factories
a typical aluminium alloy	aluminium plus zinc, magnesium, copper	light but very strong, does not corrode	aircraft, spacecraft

Look at the aluminium alloy. Aircraft need materials that are light but very strong. Pure aluminium is not strong enough. So hundreds of aluminium alloys have been developed for aircraft parts, and more are on the way.

1. Define *alloy*.
2. Iron is usually used in the form of an alloy. Explain why.
3. a Steels are one group of alloys. What is a *steel*?
   b Give two properties of stainless steel which make it more useful than iron.
4. An alloy contains 95% copper and 5% tin.
   a What is the name of this alloy?
   b The copper is melted before tin is added. Why?
   c How does adding tin change the properties of copper?
   d Give one example of a use for this alloy.

Metals, civilisation, and you

No metals, no you

Without metals you probably would not exist. You certainly would not be reading this book. The world would not have nearly 8 billion people.

Our human history has been shaped by the discovery of metals. That is why two of our eras are known as the **Bronze Age**, and the **Iron Age**.

Time past and present
- CE stands for *common era*. It means the period since **1** AD.
- BCE stands for *before common era*. It means the same as BC.

Metals and civilisation

200 000 years ago

The Stone Age
The early humans were hunters and gatherers. We had to kill and chop to get meat and fruit and firewood. We used stone and bone as tools and weapons.

Then, about 10 000 years ago, we began to farm. Farming started in the Middle East. As it spread, great civilisations grew – for example, in the valley of the River Indus, in Asia, around 9000 years ago.

farming starts around here — 8000 BCE

gold known — 6000 BCE

The Bronze Age
By 3500 BCE (over 5000 years ago) copper and tin were known – but not much used, since they are quite soft. Then someone made a discovery: mixing molten copper and tin gives a strong hard metal that can be hammered into different shapes. It was the alloy **bronze**. The Bronze Age had started!

Now a whole new range of tools could be made, for both farming and fighting.

copper and silver known — 4000 BCE

tin and lead known — 3500 BCE

iron began to be used — 1500 BCE

just 7 metals known by 1 CE: gold, copper, silver, lead, tin, iron, and mercury — 1 CE

The Iron Age
Our ancestors had no use for iron – until one day, around 2500 years ago, some got heated up with charcoal (carbon). Probably by accident, in a hot fire. A soggy mess of impure iron collected at the bottom of the fire.

The iron was hammered into shape with a stone. The result was a metal with vast potential. In time it led to the Industrial Revolution. It is still the most widely used metal on Earth.

Industrial Revolution starts around here. 24 metals known — 1750 CE / 1800 CE

65 metals known — 1900 CE

over 90 metals known (some are synthetic) — 2000 CE

The Digital Age
We still depend on iron. But computers now touch every aspect of life. There would be no computing – and no satellites, or TV, or mobile phones – without metals such as selenium, titanium, platinum, cobalt, and old favourites aluminium and copper. There is a flood of new hi-tech uses for metals and their alloys.

Where next?

EXTRACTING AND USING METALS

▲ The lead ore galena (lead sulfide) was used as eye make-up in ancient Egypt. Lead was probably produced by heating galena in an open fire.

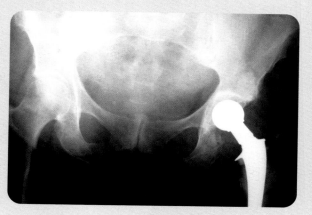

▲ A modern use of metals: this X-ray shows a replacement hip joint made from titanium. Titanium is suitable since it is strong, light, non-toxic, and does not corrode.

The earliest metals

The first metals found were the unreactive ones that exist as elements: gold, copper, and silver. But these were too soft on their own to have many uses, other than for ornaments, jewellery, and coins.

Tin, lead, and iron occur naturally as compounds, so have to be extracted by chemical reactions. It could have happened by accident, at the start. Stones put in a charcoal fire might have been ores. Or the molten metals could have oozed from clay in pottery kilns, where the fires are very hot.

The invention of electrolysis

By 250 years ago, only 24 metals were known: those found naturally as elements, plus others that could be extracted easily in a furnace, using carbon. Nobody had set eyes on sodium or magnesium, for example.

Then in 1800, the first ever electrolysis was carried out (of water). The scientist Humphry Davy heard about it, and tried it out on molten compounds – with amazing results! He discovered potassium and sodium in 1807, and magnesium, calcium, strontium, and barium in 1808.

▲ This gold collar is around 3000 years old. It was found in Ireland.

The discovery of aluminium

Aluminium is the most common metal in Earth's crust. But it was not extracted until 1825, when aluminium chloride and potassium were heated together. (Potassium displaces aluminium from its compounds.)

Only small amounts of aluminium could be extracted this way. So it became more valuable than gold! Then in 1886, the way to extract it by electrolysis, using cryolite, was developed. Aluminium had arrived.

Any more metals to find?

We know about all the natural elements in Earth's crust by now. But scientists are still trying to make new elements. The first synthetic metal was made in 1937. Period 7 of the Periodic Table is now complete, thanks mainly to synthetic metals. The race is on to start Period 8! But the synthetic metals are unstable and give out radiation, and some last for only milliseconds. So … not suitable (yet) for everyday use!

▲ Humphry Davy died when he was 52, of an illness. It was probably caused by harmful vapours from electrolysis.

Checkup on Chapter 14

Revision checklist
Core syllabus content
Make sure you can …
- [ ] name the two most common metals in Earth's crust
- [ ] explain what an *ore* is, and name the main ore of: *iron   aluminium*
- [ ] name a metal that occurs naturally as the element, and explain why it occurs like this
- [ ] explain what *extracting a metal* means, and state that a metal compound is reduced
- [ ] state that the method used to extract a metal depends on the reactivity of the metal
- [ ] explain why iron and aluminium are extracted by different methods
- [ ] for the extraction of iron in the blast furnace:
    - name the raw materials, and explain the purpose of each
    - give word equations for the reactions that take place
    - give uses for the waste products that form
- [ ] state that bauxite is first purified to give aluminium oxide, in the extraction of aluminium
- [ ] state that aluminium is extracted from aluminium oxide by electrolysis
- [ ] give the properties of aluminium that make it suitable for use in: *aircraft   food containers   overhead electrical cables*
- [ ] give the properties of copper that make it suitable for use in electrical wiring
- [ ] explain what alloys are, and draw a diagram to show the structure of an alloy
- [ ] explain why alloys are often used instead of the pure metals
- [ ] say what is in this alloy, and what it is used for: *brass   stainless steel*

Extended syllabus content
Make sure you can also …
- [ ] give the symbol equations for the reactions that take place in the blast furnace
- [ ] outline the extraction of aluminium from aluminium oxide, and
    - explain why cryolite is used
    - explain why the carbon anodes must be replaced
    - give the half-equations for the reactions at the electrodes, and the overall equation

Questions
Core syllabus content

1 This table gives information about the extraction of three different metals from their main ores:

Metal	Formula of main compound in ore	Method of extraction
iron	Fe_2O_3	heating with carbon
aluminium	Al_2O_3	electrolysis
sodium	NaCl	electrolysis

  a Give the chemical name for each compound shown in the table.
  b Which of the metals is obtained from this ore?
     i hematite   ii rock salt   iii bauxite
  c Arrange the three metals in order of reactivity.
  d i How is the most reactive metal extracted?
     ii Give a reason why the same method is not used for the least reactive metal.
  e i How is the least reactive metal extracted?
     ii Explain why this is a reduction of the ore.
  f Which of the methods would you use to extract:
     i potassium?   ii zinc?

2 Aluminium is produced in vast quantities. Give the main reason for its use in the manufacture of:
  a aircraft bodies
  b food containers for takeaway food
  c overhead electrical cables (give two reasons)

3 Iron is extracted from its ore in the blast furnace.
  a Draw a diagram of the blast furnace. Show clearly on your diagram:
     i where air is blasted into the furnace
     ii where the molten iron is removed
     iii where a second liquid is removed
  b i Name the three raw materials used.
     ii What is the purpose of each material?
  c i Name the second liquid that is removed.
     ii When it solidifies, does it have any uses? If so, name one.
  d i Name a waste gas from the furnace.
     ii Does this gas have a use? If so, what?
  e i Write the word equation for the chemical reaction that produces the iron.
     ii Explain why this is a reduction of the iron compound.
     iii What brings about this reduction?

EXTRACTING AND USING METALS

4 A sample of rock contains chalcocite, a low-grade copper ore.
The ore is mainly copper(II) sulfide, CuS. It may also contain small amounts of silver, gold, platinum, iron, cadmium, zinc, and arsenic.

This table shows the yield from 1000 kg of rock.

Mass of rock	Mass of concentrated ore	Mass of pure copper
1000 kg	10 kg	2.5 kg

a What is an *ore*?
b What is a *low-grade* ore? (Glossary?)
c How much waste rock must be removed from 1000 kg of rock, to get the concentrated ore?
d i What % of the concentrated ore is finally extracted as pure copper?
  ii Why might it be economic to extract copper from this ore, even though it is low-grade?
e i Is copper *more* reactive, or *less* reactive, than iron?
  ii What could the concentrated copper ore be reacted with, to obtain the metal?
  iii Name the type of reaction that occurs in **ii**.
f Pure copper is widely used for electrical wiring. Which two properties of copper make it suitable for that use?

5 These diagrams represent two alloys, A and B:

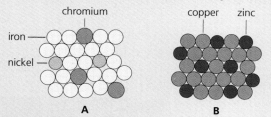

a What is an *alloy*?
b Identify alloys A and B.
c i Which would you choose for cutlery, A or B?
  ii Which properties of your chosen alloy make it suitable for cutlery?
d Alloy A is used to make replacements for worn-out joints in the body. Why would pure iron be unsuitable for a replacement hip joint?
e Alloy B is used for some musical instruments. Give two reasons why it is preferred over copper, for this use.
  f Give one property that is common to the alloys above *and* the pure metals they are made from.

Extended syllabus content

6 Aluminium is the most abundant metal in Earth's crust. Iron is next. Both metals are extracted from their ores in large quantities.

This table summarises the two extraction processes.

The process	Iron	Aluminium
chief ore	hematite	bauxite
formula of main compound in ore	Fe_2O_3	Al_2O_3
energy source used	burning of coke in air (exothermic reaction)	electricity
other substances used in extraction	limestone	carbon (graphite), cryolite
temperature at hottest part of reactor/°C	1900	1000
how the metal separates from the reaction mixture	melts and collects at the bottom	melts and collects at the bottom
other products	carbon dioxide, sulfur dioxide, slag	carbon dioxide

a In each extraction, the metal oxide is *reduced* to the metal. Write a chemical equation for the reaction in which:
  i carbon and carbon dioxide react in the blast furnace to form carbon monoxide, a reducing agent
  ii iron(III) oxide is reduced to iron in the blast furnace
  iii aluminium ions are reduced to aluminium in an electrolytic cell
b Apart from reduction, give one other similarity in the extraction processes for the two metals.
c Explain why each of these substances is used:
  i limestone in the extraction of iron
  ii carbon in the extraction of aluminium
  iii cryolite in the extraction of aluminium
d Write *two* equations to show the role that limestone plays in the blast furnace.
e Aluminium costs over three times as much per tonne as iron, even though it is more abundant in Earth's crust. Suggest two reasons for this.
f Most of the iron produced in the blast furnace is turned into steel, an alloy.
  i Draw two diagrams to show the difference in structure of a pure metal and an alloy.
  ii Explain, in terms of its structure, why steel is harder and stronger than iron.

15.1 Our environment and us

Objectives: explain what our *natural environment* consists of; give several examples of ways in which we mistreat our natural environment

Planet Earth, our home

◀ This famous photo of Earth from space is called *Earthrise*.

Planet Earth, our home. This is the first ever colour photo of Earth, taken from space on December 24, 1968. The grey is the surface of the Moon.

It is estimated that Earth had under 400 million humans in the year 1 CE, and around 3500 million when the photo was taken. Today our number is over 7800 million (or 7.8 billion) … and rising.

Our natural environment
Down here on Earth, our natural environment consists of the air, the climate, the water that flows on Earth's surface and is frozen in icecaps, the rock and soil we live on, and the 1 trillion or more other species that share the planet with us.

Our connection to the environment
We take our natural environment for granted. But in fact we depend on it every second of every day, waking and sleeping. For example:

- **air.** After just a few minutes without the oxygen from air, our bodies may suffer irreversible damage. Deprive us for longer, and we die.
- **water.** Your body is at least 50% water, by mass. It is needed for almost every function in your body, including to carry dissolved nutrients from food to body cells, and to carry waste away.

  You take in water by drinking it, and eating food that contains water.

- **soil.** Crops depend on soil. We depend on crops for food. The animals we rear for meat, milk, and transport depend on plants for food too.

▲ We depend on water outside our bodies too. For washing, cleaning, cooking, watering crops, transport …

CHEMISTRY OF THE ENVIRONMENT

Looking after our natural environment

Although we depend on our natural environment, we do not look after it well. We tend to treat it as a dump. And as our population increases, enviromental problems increase too. Look:

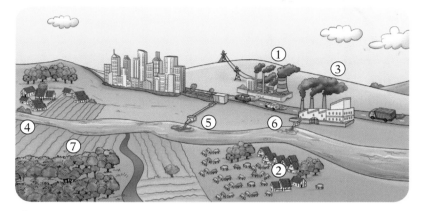

▲ What is in the smoke and fumes?

① Harmful gases from power stations, factories and engines flow into the air. They include carbon dioxide, which causes **global warming**.

② The grazing farm animals we rear give off **methane**, which contributes to global warming too.

③ Tiny particles or **particulates** of substances are pumped into the air from factories and other sources. They may end up in soil and water.

④ Untreated **sewage** may reach the water. It contains harmful **microbes**.

⑤ The liquid from treated sewage is returned to rivers. Substances in it can lead to the suffocation of river life.

⑥ Liquids from factories and mines carry toxic metals and other harmful substances to the river.

⑦ We add **fertilisers** to soil to help crops grow. But they cause problems if they drain into rivers – and overuse of them can harm soil too.

It's chemistry!

When we pollute the natural environment we harm ourselves and other living things. So we must take care.

The waste we dump into the environment is mostly chemical in nature. Chemists can help us to minimise the harm we do. It is a big challenge.

We look in more detail at water, fertilisers, air, and global warming, and also at climate change, in the rest of this chapter.

▲ What is in the effluent?

1. Explain why you are dependent on:
   a the air b the soil c water
2. Treating the natural environment as a dump can harm ourselves too. Give two examples.
3. What do you think is our *main* negative impact on the natural environment? Explain your choice.
4. What do these terms mean?
   You many need to use the glossary.
   a particulates b sewage c fertiliser d microbe
5. Earth's human population is expected to reach around 11 billion by the year 2100. Suggest two ways in which this may put further pressure on the natural environment.

15.2 What is in river water?

Objectives: give examples of *helpful* substances and *harmful* substances found in river water; explain why the water needs treatment before use as a water supply

It is never pure ...

Pure water contains only water molecules. But the water on Earth's surface is never pure. *That is because water is such a good solvent.*

So let's see what you may find in river water. It includes substances you already met in Unit 15.1. The river below could be any river, anywhere ...

Gases dissolved from the air
River water contains dissolved oxygen and nitrogen. And perhaps harmful gases from engines, power stations and factories.

Minerals dissolved from rock and soil
On its journey from source to mouth, a river dissolves many natural substances from the rock and soil it flows through.

Other metal compounds
Metals compounds are also carried to rivers in **effluent** from mines and factories. Some arrive from sewage works too.

Compounds from fertilisers
Nitrates and **phosphates** from fertilisers can get washed off the soil in wet weather, and carried into rivers.

Microbes from untreated sewage
In many places, untreated **sewage** (human waste and used water is) disposed of by letting it soak into soil. From there it may drain into rivers (or wells). Sewage contains many different microbes (bacteria and other microorganisms).

Compounds from sewage works
In some places, sewage and used water from homes is piped to sewage treatment works. Microbes are killed, and the liquid part is cleaned up and put back in the river. But it still contains soluble compounds. For example phosphates from shampoos and detergents.

Plastic waste
Plastic bags and bottles get dumped into the river, or blown in by wind. Some river beds are now lined with plastic bags!

CHEMISTRY OF THE ENVIRONMENT

Helpful or harmful?
How do those substances in water affect us, and other living things?

Helpful
- **Oxygen for aquatic life** Fish and other aquatic life depend on the oxygen dissolved in water. In fish, the water flows over their gills. The oxygen diffuses through their gills and into their blood.
- **Essential minerals for us** Our bodies need ions of around 21 elements, to function. (That's in addition to carbon, nitrogen, oxygen and hydrogen, which make up around 96% of our bodies.)

  For example, calcium is needed for bones and teeth. Magnesium and calcium help to protect us against heart disease. The other essential elements include sodium, potassium, iron, copper, zinc, and fluorine.

  River water contains different amounts of essential minerals, depending on the rock in the river basin. They reach us in drinking water. (We also obtain them from food.)

▲ Scale on shower heads and in kettles is a strong sign that your water contains calcium and magnesium carbonates. They form a scale because their solubility *decreases* with temperature.

Harmful
- **Toxic metals** A number of metals are toxic to humans. They include lead, arsenic, and mercury. Arsenic compounds occur naturally in rock. Compounds of other toxic metals drain in from mines and factories.

  Lead causes brain damage. Arsenic increases the risk of cancer and heart disease. Mercury damages the nervous system.
- **Plastics** Plastics in rivers harm fish and other river animals, which mistake them for food. They cannot digest plastic – but it clogs their guts and they starve to death. See Unit 17.5 for more.
- **Microbes from untreated sewage** They may include the bacteria that cause cholera and typhoid, and the virus that causes hepatitis.
- **Nitrates and phosphates** These promote the growth of tiny water plants called algae, which float on the water. Bacteria feed on the dead algae, using up dissolved oxygen. So fish die. For more, see Unit 15.4.

Using rivers for our water supply
Around the world, most of the water we use and drink is pumped from rivers. As you have seen, river water may contain many different substances. To ensure it is safe to use, it must be cleaned up before it reaches us. Find out how this is done in the next unit.

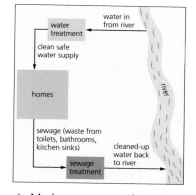

▲ Modern water supply and sanitation. Water from a river is cleaned up and supplied to people's homes. Then the water in the sewage from homes is cleaned up and put back in the river.

1. Rivers have up to about 10 molecules of dissolved oxygen in every million water molecules, at 25 °C. Fish take in this oxygen. How?
2. a Name two groups of compounds that may reach rivers and cause fish to die of oxygen starvation.
   b State the sources of these compounds in **a**.
3. Rivers in limestone areas contain calcium ions. Why?
4. Calcium ions in rivers may end up in our water supply. Explain how these ions benefit us.
5. a River water may also contain compounds of metals that are toxic to humans. Name two of these metals.
   b Name one source of these harmful metal compounds.
6. It is very dangerous to swim in rivers containing untreated sewage. Why?

15.3 Our water supply

Objectives: outline the steps in cleaning up water for a water supply; explain why distilled water is used in the lab; recall the tests to identify water, and pure water

How clean and safe is your water supply?

By 2020, nearly 1 in 10 of the world's population still did not have access to a clean water supply. Over 660 million people drank untreated water from rivers, ponds, and watering holes. Some was contaminated with microbes from human and animal waste. It killed many people.

The rest of us obtained water from **improved sources**, including:

- water piped to homes
- public taps and standpipes, and water delivered by tanker
- wells and boreholes sited in places safe from contamination by sewage
- rainwater
- bottled water.

Note that water from some improved sources may still not be *safe*. It depends on what treatment it has been given before it reaches you.

How to provide a clean safe water supply

These are the general steps in providing a clean safe water supply.

> **1** Find the cleanest source you can, to pump water from.
> For example, a clean stretch of **river**, or an **aquifer**.

↓

> **2** Remove as many solid particles from the water as you can.
> - You could **filter** the water through clean gravel or sand.
> - You could let particles settle to the bottom of a tank. This is called **sedimentation**.
> - You could add something to make fine particles stick together. This is called **flocculation**. You can then skim them off.

↓

> **3** Remove as many unwanted dissolved compounds as you can.
> Filtering the water through **carbon** will remove some substances that cause nasty smells and tastes – for example, hydrogen sulfide which smells like rotten eggs. It can also remove toxic metals.

↓

> **4** Add something to kill the microbes. (Usually chlorine.)

↓

> **5** Store the water safely, ready to pump to taps or tankers.
> It can be kept in a clean covered reservoir.

How clean the final water will be depends on how dirty it is to start with, and what kind of treatment you can afford. So water quality will vary.

Other sources of water !

- Rivers are not the only source of water, for a water supply.
- We pump up **groundwater** held in soil and rocks below the ground, at **hand-dug wells**.
- **Aquifers** are large stores of water held in rock below the ground, and we drill **boreholes** to these.
- Water from below the ground will contain dissolved substances too – but it is usually cleaner and safer than untreated river water.

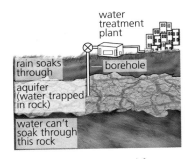

▲ Water being pumped from an aquifer.

▲ Carbon for water treatment. It is made from charcoal, wood or coconut shells. These granules are less than 6 mm long. They have very fine pores, giving a very large surface area to adsorb substances.

A modern treatment plant

This diagram shows a modern water treatment plant. Follow the numbers to see how particles are removed, and microbes killed.

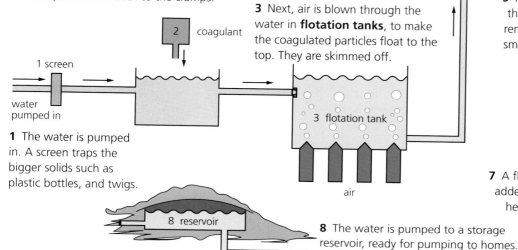

2 A **coagulant** is added, for example iron(III) sulfate. It makes fine suspended particles clump together (flocculate). Some toxic metal compounds will stick to the clumps.

3 Next, air is blown through the water in **flotation tanks**, to make the coagulated particles float to the top. They are skimmed off.

1 The water is pumped in. A screen traps the bigger solids such as plastic bottles, and twigs.

4 The water is passed through a bed of fine sand to filter it.

5 It may also be filtered through **carbon** which removes bad tastes and smells – and some toxic metal ions too.

6 Chlorine is added to kill the bacteria and other microbes.

7 A fluoride compound is added in some plants, to help fight tooth decay.

8 The water is pumped to a storage reservoir, ready for pumping to homes.

This treatment can remove even tiny solid particles. Steps 2 and 5 can remove compounds of toxic metals. Chlorine can kill all the microbes. *But the water will still contain some dissolved substances.*

For example, it may contain calcium and magnesium salts, which are helpful to us. But high levels of some dissolved compounds, such as nitrates, can be dangerous, and especially for babies.

Future filters?

Scientists have developed thin plastic sheets or **membranes** with very tiny holes that let water molecules through, but can trap some ions.

These are not widely used in water treatment – but they may be one day!

Water in the lab

Since tap water contains dissolved substances, it is not used to prepare solutions in the lab. **Distilled water** is used instead. See page 244.

Two tests for water
If a liquid *contains* water, it will …
- turn white anhydrous copper(II) sulfate blue
- turn blue cobalt(II) chloride paper pink.

The test for pure water
If a liquid is *pure* water, it will also …
- boil at 100 °C, and
- freeze at 0 °C.

Q

1 List, as brief bullet points, the five steps in providing a clean safe water supply.
2 Explain these terms:
   a sedimentation   b flocculation   c coagulant
3 Why is chlorine such an important part of water treatment?
4 A fluoride compound may be added to water. Why?
5 Some water can be harmful even after treatment. Why?
6 Tap water is *not* used to make aqueous solutions in the lab.
   a Why is tap water not used?   b What is used instead?
7 You have a liquid, and some blue cobalt(II) chloride paper.
   a How could you prove that the liquid *contains* water?
   b How will you prove that it is *not* pure water?

15.4 Fertilisers

Objectives: name the key elements plants need; define *fertiliser*; explain why fertilisers are needed; explain *NPK fertiliser*; give drawbacks of fertiliser use

What plants need

A plant needs carbon dioxide, light, and water. It also needs several **nutrients**. The main ones are **nitrogen**, **potassium**, and **phosphorus**.

Plants need nitrogen for making chlorophyll, and proteins.

Potassium helps them to produce proteins, and to resist disease.

Phosphorus helps roots to grow, and crops to ripen.

Plants obtain these elements from compounds in the soil, which they take in through their roots as solutions. The most important one is nitrogen. Plants take it in as **nitrate** ions and **ammonium** ions.

Fertilisers

Every crop a farmer grows takes compounds from the soil. Some get replaced naturally. But in the end the soil gets worn out. New crops will not grow well. So the farmer has to add **fertiliser**.

A fertiliser is any substance added to the soil to make it more fertile.

Animal manure is a natural fertiliser. **Synthetic fertilisers** are salts made in factories. Many are ammonium salts, and nitrates. For example, ammonium nitrate, NH_4NO_3, and ammonium phosphate, $(NH_4)_3PO_4$.

An NPK fertiliser is a mixture of salts that provides all three elements plants need: nitrogen, phosphorus, and potassium.

◀ Nutrition for plants: these granules are made of animal manure, a natural fertiliser.

◀ Getting ready to spread fertiliser on fields. This is an NPK fertiliser. The numbers 6.17.15 on the bags tell you that it is 6% nitrogen, 17% phosphorus and 15% potassium by mass.

CHEMISTRY OF THE ENVIRONMENT

◀ Synthetic fertilisers are made by reactions like those below. Then the solutions are evaporated to give solids. This shows fertiliser in storage.

Reactions to make synthetic fertilisers

Many fertilisers are ammonium salts. Ammonia (from the Haber process) is used to make them. Here are two typical reactions.

1 Ammonia reacts with nitric acid to give ammonium nitrate. This fertiliser is an excellent source of nitrogen:

$$NH_3\ (aq)\ +\ HNO_3\ (aq)\ \longrightarrow\ NH_4NO_3\ (aq)$$
ammonia   nitric acid   ammonium nitrate

2 Ammonia reacts with phosphoric acid to give ammonium phosphate:

$$3NH_3\ (aq)\ +\ H_3PO_4\ (aq)\ \longrightarrow\ (NH_4)_3PO_4\ (aq)$$
ammonia   phosphoric acid   ammonium phosphate

The water is then evaporated off to give pellets of fertiliser.

What % is nitrogen?
Ammonium nitrate is rich in nitrogen. What % of it is nitrogen? Find out like this:

Formula: NH_4NO_3
A_r: N = 14, H = 1, O = 16
M_r: (14 × 2) + (4 × 1) + (16 × 3)
     = 80
% of this that is nitrogen
     = $\frac{28}{80}$ × 100%
     = 35%

It's not all good news

Fertilisers help to feed the world. But there are drawbacks.

In the river Fertilisers promote plant growth in water too! They can seep from farmland into rivers. This leads to **eutrophication**, where there is no dissolved oxygen left in the water. It happens because the fertilisers cause excessive growth of algae, which cover the water. When they die, bacteria feed on them, using up all the dissolved oxygen. So fish suffocate.

In the soil Overuse of fertilisers can harm plants. They grow quickly, with shallow roots, making it harder to obtain water.

The soil can be harmed too. Its nutrients become unbalanced and its pH changes, affecting crop yield. It hardens, and may even become useless.

So farmers should use fertilisers sparingly. They should also use them away from river banks – and not spread them in wet weather.

▲ Paddling through the algae. Fertiliser from farms helps algae to grow.

1 What is a *fertiliser*?
2 Why are fertilisers used?
3 An *NPK fertiliser* contains three elements that plants need.
   a Name the three elements.
   b Suggest two salts (an ammonium salt and a nitrate) which will form an NPK fertiliser when mixed together.
4 Fertilisers can harm aquatic life. Explain how.
5 The panel above shows how to work out % composition.
   a Find the % of nitrogen in ammonium sulfate. (A_r: S = 32)
   b Which would provide more nitrogen for the soil: 1 kg of ammonium nitrate or 1 kg of ammonium sulfate?
   c Avoid spreading this fertiliser in wet weather. Why?

15.5 Air, the gas mixture we live in

Objectives: recall the composition of clean dry air; outline the importance of the oxygen in air; describe an experiment to find the % of oxygen in air

What is air?
Air is our everyday name for the layer of gas around us. We live in the **troposphere**, the lowest level of the **atmosphere** that surrounds Earth. The air is at its most dense at Earth's surface, at sea level. It thins out with altitude. So climbers on Mount Everest carry bottled oxygen with them.

What is in air?
This shows the gases that make up clean dry air, at sea level on Earth:

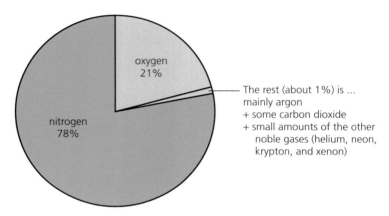

oxygen 21%
nitrogen 78%
The rest (about 1%) is ...
mainly argon
+ some carbon dioxide
+ small amounts of the other noble gases (helium, neon, krypton, and xenon)

But the air around us is not usually clean and dry!

- It contains variable amounts of water vapour.
- As you saw in Unit 15.1, it may also contain harmful substances from human activities. For example, tiny particles of solids, waste gases from factories and other sources, and a growing amount of carbon dioxide. You can read about the harm these do, in later units.

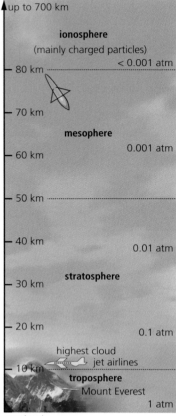

▲ The layers of Earth's atmosphere. Note how the pressure falls with altitude.

▲ We cannot live without oxygen. So deep-sea divers have to bring some with them, in tanks ...

◀ ... and so do astronauts.

194

CHEMISTRY OF THE ENVIRONMENT

Oxygen: the gas we need most

We need most of the gases in air. For example, we depend on plants for food, and they depend on carbon dioxide. And without nitrogen to dilute the oxygen in air, our fuels would burn much too vigorously.

But oxygen is the gas we depend on most. We use it for **respiration**, the process that provides energy. This reaction occurs in all our cells:

glucose + oxygen ⟶ carbon dioxide + water + energy

The energy keeps us warm, and allows us to move, and enables hundreds of reactions to go on in our bodies. (Respiration goes on in the cells of *all* living things – but some organisms use other reactants. For example, bacteria that live deep in soil and water have no available oxygen.)

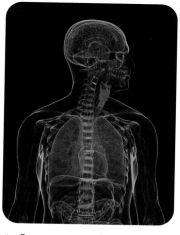

▲ Oxygen enters through your mouth and nose, passes into your lungs, and from there diffuses into your blood.

Measuring the percentage of oxygen in air

How much oxygen is there in air? That is a very important question. Here is a way to find the answer in the lab.

The apparatus A tube of hard glass is connected to two gas syringes A and B. The tube is packed with small pieces of copper wire. At the start, syringe A contains 100 cm³ of air. B is empty.

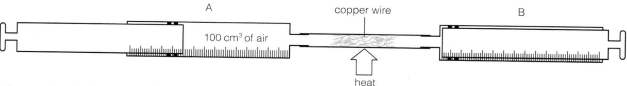

The method These are the steps:

1. Heat the tube containing copper using a Bunsen burner. Then push in A's plunger, as shown above. This forces the air into B. When A is empty, push in B's plunger, forcing the air back to A. Repeat several times. As the air is pushed to and fro, the oxygen in it reacts with the hot copper, turning it black.

2. Stop heating the tube after about 3 minutes, and allow the apparatus to cool. Then push all the gas into one syringe and measure its volume. (It is now less than 100 cm³.)

3. Repeat steps 1 and 2 until the volume of the gas remains steady. This means all the oxygen has been used up. Note the final volume.

The results Starting volume of air: 100 cm³. Final volume of air: 79 cm³. So the volume of oxygen in 100 cm³ air is 21 cm³.

The percentage of oxygen in air is therefore $\frac{21}{100} \times 100 =$ **21%**.

▲ Fish take in the oxygen dissolved in water, through their gills.

Q

1. Suppose you have an average sample of 100 gas particles from clean dry air. How many are likely to be:
   a nitrogen molecules?   b oxygen molecules?
2. Between them, nitrogen and oxygen make up 99% of clean dry air. Name three of the gases in the remaining 1%.
3. Air is not usually clean and dry. Why not?
4. Explain why climbers on Mount Everest carry oxygen.
5. Which is the most *reactive* gas in air? Explain your choice.
6. What happens during *respiration* in humans?
7. a Give the name and formula of the black substance that forms in the experiment above.
   b Suggest a way to turn it back into copper. (Page 72?)

15.6 Air pollution from fossil fuels

Objectives: name the fossil fuels; recall the main pollutants from burning fossil fuels, and the harm they do; describe three ways to help prevent acid rain

The air: a dump for waste gases

Everyone likes clean fresh air. But every year we pump billions of tonnes of harmful gases into the air. Most come from burning **fossil fuels**.

The fossil fuels

These are **coal**, **petroleum**, and **natural gas**. They are all based on carbon. Natural gas is mainly methane, CH_4. Coal and petroleum are mixtures of many compounds. Most are **hydrocarbons** – they contain only carbon and hydrogen. But some contain other elements too, such as sulfur.

We use fossil fuels for generating electricity, for transport, and for heating. But burning them causes some big problems.

▲ Don't breathe in!

The main pollutants from fossil fuels

Pollutant	How is it formed?	What harm does it do?
carbon dioxide, CO_2 colourless heavy gas, sharp smell when concentrated; soluble in water	forms in the **complete combustion** of fuels that contain carbon (coal, gas, petrol and other fuels obtained from petroleum, and wood) in a plentiful supply of air	carbon dioxide occurs naturally in the air; but increasing the amount leads to an increase in global warming, which in turn leads to climate change
carbon monoxide, CO colourless gas, insoluble, no smell	forms in the **incomplete combustion** of fuels that contain carbon, when there is a limited supply of air – for example, in car engines	poisonous even in low concentrations; it reacts with the haemoglobin in blood, and prevents it from carrying oxygen around the body – so you die from lack of oxygen
particulates tiny solid particles (may appear as dark smoke)	form in the incomplete combustion of fuels that contain carbon; for example, particles of soot from diesel engines	linked to respiratory problems, and cancer
sulfur dioxide, SO_2 an acidic gas with a sharp smell	forms when sulfur compounds in the fossil fuels burn; power stations are the main source of this pollutant	dissolves in rain to form **acid rain**; this attacks stonework in buildings – especially limestone and marble, which are forms of calcium carbonate; it acidifies rivers and lakes, killing fish and other river life; it kills trees and insects.
nitrogen oxides, NO and NO_2 acidic gases	form when the nitrogen and oxygen in air react together inside hot motor engines and furnaces	cause respiratory problems, dissolve in rain to give acid rain; form a brown haze called a **photochemical smog** in the air, when UV light from the sun reacts with them

▶ Older coal-fired power stations have been a major source of pollution. Coal can contain a great deal of sulfur, and it also forms soot.

CHEMISTRY OF THE ENVIRONMENT

More about acid rain
Acid rain was first identified in 1963. A scientist discovered that rainwater at a site in the USA was about 100 times more acidic than expected. And then people noticed that forests and lakes were dying – not only in North America but across Europe. Acid rain was the culprit.

Scientists traced the problem to coal-burning power stations that pumped out sulfur dioxide and nitrogen oxides.

Tackling the acid rain problem
The acid rain problem has been tackled in these ways:

- **using low-sulfur fuels** Coal, gas, and fuels from petroleum are treated to remove as much sulfur as possible before they are burned.
- **flue gas desulfurisation** In this process, flue (waste) gases from power stations are treated with a slurry of calcium oxide (CaO, lime). This reacts with sulfur dioxide to give calcium sulfite ($CaSO_3$).
- **catalytic converters** These are fitted into the exhaust system in cars. They convert nitrogen oxides to nitrogen. But that is not all: they also deal with carbon monoxide, and unburnt hydrocarbons from petrol.

▲ A forest in (Europe) killed by acid rain.

How do catalytic converters work?
The converter usually has two compartments, marked **A** and **B** below:

In **A**, oxides of nitrogen are removed by **reduction**.
For example:
$$2NO\,(g) \longrightarrow N_2\,(g) + O_2\,(g)$$
$$NO\,(g) + 2CO\,(g) \longrightarrow N_2\,(g) + 2CO_2\,(g)$$
Note that the second reaction removes carbon monoxide too. The gases then flow into **B**.

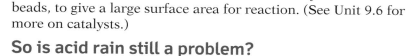

In **B**, the remaining carbon monoxide, and unburnt hydrocarbons, are removed by **oxidation**. This uses the oxygen from **A**:
$$2CO\,(g) + O_2\,(g) \longrightarrow 2CO_2\,(g)$$
$$\text{hydrocarbons}\,(g) + O_2\,(g) \longrightarrow xCO_2\,(g) + yH_2O\,(g)$$
The gases flow out the exhaust pipe and into the air. Now only the carbon dioxide is harmful!

The catalysts are transition elements: platinum, rhodium and palladium. They are coated onto a ceramic honeycomb or ceramic beads, to give a large surface area for reaction. (See Unit 9.6 for more on catalysts.)

So is acid rain still a problem?
Thanks to the steps above, acid rain is no longer a problem in many countries. But it is a problem in countries with weak laws to control emissions. And their acidic gases can flow across borders in the wind.

1. Look at the pollutants in the table on page 196.
   a. Which come from petrol burned in car engines?
   b. Which come from the air itself? How do these form?
   c. Which result in acid rain?
2. Name the brown haze that forms over some cities in sunlight, and the pollutants which give rise to it.
3. Natural gas (methane) is a fossil fuel. Give the word equation for the reaction that occurs when it burns in plenty of air.
4. If methane burns in a poor supply of air, carbon monoxide will form. Explain why this gas is harmful.
5. a. Write a word equation for the reaction that takes place during *flue gas desulfurisation*.
   b. Outline two other ways to help prevent acid rain.
6. a. Name the typical catalysts used in catalytic convertors.
   b. Give the equation for a reaction that turns nitrogen monoxide (NO) from the car engine into a harmless gas.

15.7 Two greenhouse gases

Objectives: name and state the sources of the two main greenhouse gases; explain why they cause global warming; give some predicted impacts of climate change

Carbon dioxide and methane: greenhouse gases

Through our activities, we humans are increasing the amount of **carbon dioxide** in the air. We are adding **methane** too.

Carbon dioxide (CO_2) and methane (CH_4) are called **greenhouse gases**. They are causing **global warming**: Earth is warming up.

The sources of these gases

Carbon dioxide

Carbon dioxide occurs naturally in air. But we add more when we burn fuels that contain carbon, in a plentiful supply of oxygen.

Those fuels are mainly fossil fuels – coal, natural gas, and petrol, kerosene, diesel and the other liquid fuels we obtain from petroleum. The burning of trees and vegetation contributes too.

Methane

- Natural gas is mainly methane. Some escapes into the air at gas wells.
- Methane forms when microbes break down carbon-based material in the absence of oxygen, for example in boglands, and rice paddy fields.
- But the main source is the grazing animals we rear – cattle, sheep, goats, buffalo. Microbes in their stomachs break down grass, and methane is one product. The gas exits from both ends of the animal.

There is far more carbon dioxide in the air than methane. But methane is about 28 times more powerful than carbon dioxide at warming Earth.

▲ Some methane manufacturers eating hay. It is estimated that there are around 1 billion cattle on Earth (and around 8 billion humans).

How do greenhouse gases work?

Greenhouse gases keep Earth warm by preventing heat escaping from the atmosphere to space.

1. The sun sends out energy as light and UV rays.
2. These warm Earth, which reflects some of the energy away again, as heat (thermal energy).
3. Some of this heat escapes from the atmosphere.
4. But some is absorbed by greenhouse gases in the atmosphere. So the air, and Earth, stay warm.

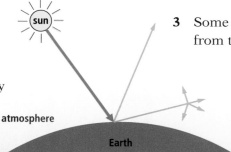

We need greenhouse gases. Without them, we would freeze to death at night, when the sun is not shining.

But the increasing amounts of greenhouse gases in the air are causing global warming: Earth's average temperature is rising. This in turn is leading to **climate change**.

CHEMISTRY OF THE ENVIRONMENT

Climate change

Earth's temperature affects rainfall, and cloud cover, and wind patterns. So as Earth warms up, climates around the world change too.

Scientists try to predict what will happen, using computer models. It is difficult, because the links between weather, clouds, and the ocean are not fully understood. But they have predicted that:

- some dry places will become even drier, with more frequent drought; other places will become wetter
- melting land-ice in the Arctic and Antarctica will cause sea levels to rise, so low-lying countries will be at greater risk of flooding
- glaciers in high mountain ranges will shrink away, reducing the annual supply of meltwater that feeds many of the world's great rivers
- storms, floods, and wildfires will be more frequent and severe
- farmers may no longer be able to grow their usual crops
- more people will become climate refugees, moving away to survive
- species that cannot adapt to the changing climate will become extinct
- harmful insects, and diseases, will spread to new countries that have grown warm enough for them.

Now climate change has caught up with many of the predictions. It is well underway, and speeding up.

▲ Climate change: drought kills crops.

▲ Flooding will be more frequent in some places …

▲ … and heatwaves more frequent in others, causing wildfires.

1 a What is a *greenhouse gas*? (Glossary?)
  b We need greenhouse gases. Why?
  c Why are greenhouse gases now a problem?
2 a Name the two main greenhouse gases we add to the air through our activities.
  b Give the main source for each gas you named in **a**.
3 Define *global warming*.
4 Explain *why* greenhouse gases are able to make Earth warmer.
5 *Globla irmgwan* leads to *alitmec neahgc*. Unjumble!
6 Give two reasons why climate change is likely to affect: **a** food supplies **b** water supplies in many countries.
7 *Climate change will have no impact on me*. Discuss!

15.8 Tackling climate change

Objectives: list non-carbon ways to obtain electricity and power vehicles; explain why trees help us tackle climate change; give a way to reduce methane emissions

What can we do?

The two main greenhouse gases we humans add to the air are **carbon dioxide** and **methane**. So our challenge is to cut the levels of both. Below are ways to do this.

By using non-carbon fuels

Our two main uses of fossil fuels are:

- **to generate electricity** in power stations; coal and gas are the main fuels for this. The diagram on the right shows how it works.
- **in transport of all kinds**: cars, trucks, planes, ships. Liquid fuels obtained from petroleum are burned in their engines.

So reducing carbon dioxide means switching to non-carbon ways to generate electricity and power transport. Let's look at examples.

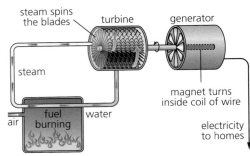

▲ In a power station, energy of motion is converted to electricity when a magnet turns inside a coil of wire (or vice versa). Steam is used to spin the turbine blades, and the turbine turns the magnet. The fuel may be coal, gas, petroleum, or biomass (wood chips or similar).

Non-carbon electricity

We can use:

- **wind**, **waves**, **tides**, and **fast flowing rivers**. All can spin the blades of turbines. Electricity generated using rivers is called **hydroelectricity**.
- **nuclear power**. Nuclei of unstable isotopes such as uranium-235 are forced to break up, giving out an enormous amount of energy. This is used to heat water to make steam, to spin the turbine blades in nuclear power stations.
- **solar power**. We convert energy from sunlight into electrical energy using **photovoltaic cells**, or **PV cells**. This can be stored in batteries.

Non-carbon power for transport

We can power vehicles using:

- **batteries** which are charged using non-carbon electricity.
- **hydrogen–oxygen fuel cells**, which produce only water. You can find out about these in Unit 8.4.

> **! Renewable energy**
> - The energy of the wind is called **renewable energy**.
> - That's because it does not get used up. The wind keeps on blowing!
> - The energy of the tides, waves, rivers, and sunlight is renewable too.

▲ Solar power for the village: all you need is PV cells and sunlight.

▲ A wind farm on farmland.

CHEMISTRY OF THE ENVIRONMENT

By planting trees

It is estimated that over 2000 trees are destroyed per minute around the world, to clear land for a different use – for example, farming or roads.

Deforestation increases carbon dioxide emissions. First, dead trees no longer take in carbon dioxide. And second, they may be burned or left to rot, so their stored carbon is released as carbon dioxide.

Now we must do the opposite: plant more trees, to take in carbon dioxide. This occurs during **photosynthesis**.

In photosynthesis, carbon dioxide and water react in plant leaves, to give glucose and oxygen. Chlorophyll, a green pigment in leaves, is a catalyst for the reaction. Light from the sun provides the energy:

$$\text{carbon dioxide} + \text{water} \xrightarrow{\text{light}}_{\text{chlorophyll}} \text{glucose} + \text{oxygen}$$

$$6CO_2\,(g) + 6H_2O\,(l) \xrightarrow{\text{light}}_{\text{chlorophyll}} C_6H_{12}O_6\,(s) + 6O_2\,(g)$$

▲ A young tree, busy helping us to lower carbon dioxide levels – and giving out oxygen at the same time.

Trees then use the glucose, and nitrates and other compounds from soil, to make the other carbon-based compounds they need.

By reducing methane emissions

As you saw earlier, the rearing of grazing livestock – cattle, sheep, goats, buffalo – is the main source of methane.

To reduce methane, farmers need to rear fewer of these animals – and we need to change our eating habits. That means eating less meat and dairy products from those animals. We can eat fish and chicken instead, or even better, increase the proportion of plant-based foods in our diets.

It's urgent!

Climate change is the major challenge facing the world today.

Scientists say we must ugently cut carbon dioxide emissions in particular, to stop climate change spiralling out of control.

Each of us can do something. Even if it is only turning off an unwanted light, or walking or cycling instead of going by car, or eating meat a little less often. And if billions of people do it …

Carbon capture

- Scientists are also trying out ways to *capture* carbon dioxide and store it underground.
- It could be captured from a power station by absorbing it into a suitable solvent …
- … and then releasing it from the solvent, compressing it, and sending it by pipe or tanker to an underground storage site.
- The storage site could be an empty oil or gas well, for example.

1 In a power station, water is heated to make steam. The steam then spins the blades of a turbine, which makes a magnet turn inside a coil of wire. This gives electricity.
What spins the turbine blades:
   a in windfarms?   b for hydroelectricity?

2 Look at the section *Non-carbon electricity* on page 200. Only one of the methods it mentions does not need a turbine. Which one?

3 Wind is a form of renewable energy.
Define *renewable energy*.

4 Name two other forms of renewable energy used to generate electricity. Explain why each is renewable.

5 Explain why planting millions of trees can help to reduce the levels of carbon dioxide in the air.

6 Write the chemical equation for photosynthesis.

7 Why does this help to limit the rise in greenhouse gases?
   a choosing to buy a battery-powered car
   b choosing to eat fish or chicken instead of meat from grazing animals

8 *I can do nothing about climate change*. Discuss!

Checkup on Chapter 15

Revision checklist

Core syllabus content
Make sure you can ...
- [ ] explain what our *natural environment* is
- [ ] state that water from natural sources (like rivers) may contain many substances, and give examples
- [ ] state that some of the substances in water
  - are beneficial to us or other living things
  - can be harmful to us or other living things and give examples
- [ ] describe the steps in the treatment of water to give a clean safe water supply
- [ ] explain why distilled water is used in the lab
- [ ] give two tests to show that a liquid contains water
- [ ] state how to prove that a liquid is pure water
- [ ] name the three main elements plants need for growth (in addition to carbon dioxide and water)
- [ ] explain what fertilisers are, and why they are needed
- [ ] explain what *NPK fertilisers* are
- [ ] describe the harm fertilisers can do in rivers
- [ ] state the composition of clean dry air
- [ ] describe a way to find the % of oxygen in air
- [ ] name the fossil fuels
- [ ] – name five air pollutants that arise from burning fossil fuels, including in vehicle engines
  - describe the harm each does
- [ ] describe how acid rain forms, and the harm it does
- [ ] give three ways to help prevent acid rain
- [ ] explain what greenhouse gases are, and
  - name the two main ones our activities produce
  - give the main sources of these
- [ ] explain the terms *global warming* and *climate change*
- [ ] give ways to generate electricity for homes using renewable energy
- [ ] give non-carbon ways to power vehicles
- [ ] give a way to reduce emissions of methane

Extended syllabus content
Make sure you can also ...
- [ ] explain how catalytic converters work, and
  - name the metals used as catalysts
  - give the chemical reactions that take place to remove nitrogen monoxide and carbon monoxide
- [ ] give the chemical reaction for photosynthesis

Questions

Core syllabus content

1 Copy and complete:
  a Clean, dry air contains approximately 78% .... and 21 % .... The remaining 1% is made up of .... and ....
  b In some areas the air contains particulates, and carbon monoxide.
    i Name a likely source of these pollutants.
    ii How can particulates affect a person's health?
  c Analysis of the air in industrial areas shows high levels of the gas sulfur dioxide.
    i What is the probable source of this gas?
    ii What is the environmental impact of this gas?
    iii Describe two strategies to lower the concentration of sulfur dioxide in the air.

2 In many places, natural or artificial lakes are used as sources of water for the water supply. Rainfall helps to keep them filled.
  a Why might it be easier to clean up water from these sources than from rivers?
  b The diagram shows one stage in the treatment of water to make it ready for piping to homes:

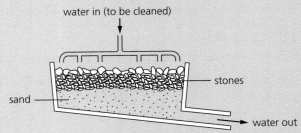

  i Name the process being carried out here.
  ii Which kind of impurities will this process remove from the water?
  c Water is also filtered through carbon.
    i Carbon can remove toxic metal compounds. Name two metals which are toxic to humans.
    ii After filtering through carbon, the water also tastes and smells better. Explain why.
  d The water then needs further treatment to kill microbes.
    i Why must microbes be killed?
    ii Name the chemical used to kill microbes.
  e At the end, the water may still contain dissolved metal ions. Describe the beneficial effect of two different metal ions in drinking water.

CHEMISTRY OF THE ENVIRONMENT

3 Anhydrous copper(II) sulfate and anhydrous cobalt(II) chloride are used in lab tests for water.

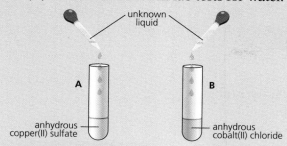

  a If the unknown liquid contains water, what colour change will be seen:
     i in test-tube A?   ii in test-tube B?
  b i How could a sample of water be tested to check whether it was pure or not?
     ii Give the expected result in **i**, for pure water.
  c Why is distilled water used in preference to tap water in the chemistry laboratory?

4 Farmers use chemicals to improve their crop yields.
  a Explain what a farmer means by *crop yield*.
  b What general name is given to the chemical compounds that farmers spread on fields?
  c One of the chemicals has the formula $Ca_2(PO_4)_3$.
     i Name this chemical.
     ii Name the element a plant needs for growth, which this chemical supplies.
  d Another chemical has the formula NH_4NO_3.
     i Name this chemical.
     ii Identify the key element that it supplies.
  e The chemicals in **c** and **d** can have harmful effects in local rivers and streams.
     i   How might they reach a local river?
     ii Name the ion in each chemical which is responsible for the harmful effects.
     iii The chemicals lead indirectly to a change in the amount of oxygen dissolved in the water. How does the amount change?
     iv Give one harmful effect of the change in **iii**.

5 Methane is one of the gases causing global warming.
  a i Explain the term *global warming*.
     ii Global warming leads to climate change. Give any two predicted impacts of climate change.
  b Give one way in which the amount of methane entering the atmosphere could be reduced.
  c Name the other main gas that contributes to global warming.
  d Suggest a reason why the issue of climate change needs to be addressed by politicians.

6 Catalytic converters are fitted in modern cars:

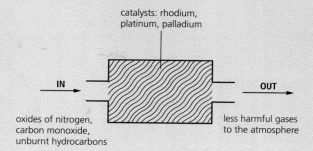

  a i   Where in the car is the catalytic converter?
     ii What is the purpose of a catalytic converter?
     iii Which type of elements are rhodium, platinum, and palladium?
  b i   Oxides of nitrogen enter the converter. How and where are they formed?
     ii Oxides of nitrogen contribute to acid rain. Describe one other adverse effect when oxides of nitrogen are released into the air.

Extended syllabus content
7 A student drew this diagram for global warming:

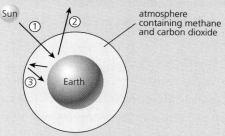

  a Suggest a term the student could use instead of *methane and carbon dioxide*.
  b What does each arrow 1, 2 and 3 represent?
  c As greenhouse gas levels rise, how will this change?
     i the relative amounts of thermal energy (heat) that are transferred, for arrows 2 and 3
     ii Earth's average temperature

8 In the catalytic converters in modern cars, carbon monoxide and oxides of nitrogen from the engine are converted to other substances.
  a i Why is carbon monoxide removed?
     ii Give one harmful effect of nitrogen dioxide.
  b What is meant by a *catalytic* reaction?
  c In one reaction in a catalytic converter, nitrogen monoxide (NO) reacts with carbon monoxide, to form nitrogen and carbon dioxide. Write a balanced equation for this reaction.

16.1 Petroleum: a fossil fuel

Objectives: name the fossil fuels; define *hydrocarbon*; describe petroleum as a mixture of hydrocarbons, and explain its importance; explain *displayed formula*

The fossil fuels
The **fossil fuels** are **petroleum** (or crude oil), **coal**, and **natural gas**. They are called fossil fuels because they are the remains of plants and animals that lived millions of years ago.

Petroleum formed from the remains of dead organisms that fell to the ocean floor, and were buried under thick sediment. High pressures slowly converted them to petroleum, over millions of years.

Natural gas is mainly **methane**. It is often found with petroleum, and is formed in the same way. But high temperatures and pressures caused some compounds to break down to gas.

Coal is the remains of lush vegetation that grew in ancient swamps. The dead vegetation was buried under thick sediment. Pressure and heat slowly converted it to coal, over millions of years.

What is in petroleum?
Petroleum is a smelly mixture of hundreds of different compounds. They are **organic compounds**, which means they contain carbon atoms, usually in chains or rings, and usually with hydogen attached.

In fact most are **hydrocarbons** – they contain *only* carbon and hydrogen. Look at these examples:

This is a molecule of pentane, C_5H_{12}. It has a **straight chain** of 5 carbon atoms.

This is a molecule of 3-methyl-pentane, C_6H_{14}. Here 6 carbon atoms form a **branched chain**.

This is a molecule of cyclohexane, C_6H_{12}. Here a chain of 6 carbon atoms forms a **ring**.

Drawings like these are called **displayed formulae**, because they show all the atoms and bonds. Each short line represents **a covalent bond**.

Note how the carbon atoms are bonded to each other. Carbon can form long chains of atoms. Compounds in petroleum have from 1 to 70 carbon atoms. The hydrogen atoms are bonded to the carbon atoms.

Carbon atoms form the spines of the molecules in organic compounds.

Why on land?
- Petroleum and natural gas are found in the ocean floor, where they form – and also on land!
- That's because some ocean floor has been uplifted, giving land.

ORGANIC CHEMISTRY

Why is petroleum so important?
Of all the fossil fuels, petroleum is the most important. The world has depended on it for decades, because it has so many uses. Billions of litres of petroleum are used up around the world every day.

Over half of the petroleum pumped from oil wells is used for transport. It provides fuel for cars, trucks, planes, and ships. The fuel is burned in their engines.

Much of the rest is burned for heat, in factories, homes, and power stations. In power stations, the heat is used to turn water to steam, to drive turbines to make electricity.

Around 16% is used as a **feedstock** of chemicals to make many other things: plastics, detergents, beauty products, medical drugs, fabrics, and more.

Before it can be used for these different purposes, petroleum must be separated into groups of compounds. The next unit tells you about that.

Will petroleum always be so important?
We humans have known about petroleum – and the other fossil fuels – since ancient times. They were burned for heat and light long before electricity was discovered, or the internal combustion engine invented.

But as you know, all three fossil fuels are linked to **climate change**. Carbon dioxide forms when they burn, and it is a **greenhouse gas**.

So the demand for all three fossil fuels is expected to fall, as the world tackles climate change, and we develop cleaner sources of energy.

The compounds you will study
In this chapter you will study four families of organic compounds. The first family, the **alkanes**, are obtained directly from petroleum. The alkanes then become the starting point for the other three families: **alkenes**, **alcohols**, and **carboxylic acids**.

▲ The invention of the internal combustion engine triggered the demand for petroleum. Mass production of cars began with the Model T Ford, in 1908.

Carbon and you ...
- Carbon is special in its ability to form chains and rings of carbon atoms.
- So there are *millions* of organic compounds. They are the main compounds in every living thing. That includes you!

1 The other name for petroleum is ... ?
2 Why is petroleum called a *fossil fuel*?
3 Name the other two fossil fuels.
4 Which two elements are found in hydrocarbons?
5 What type of bonding is found in hydrocarbons?
6 What is in it?   a natural gas   b petroleum
7 Explain the term *displayed formula*, and draw an example.
8 Explain why the world depends so much on petroleum.
9 Explain why the demand for fossil fuels is likely to fall away. Use the term *greenhouse gas* in your answer. (Page 198?)

16.2 Refining petroleum

Objectives: describe how petroleum is refined into fractions; describe the trends the fractions show; name the fractions, and give a use for each

What does *refining* mean?

Petroleum contains *hundreds* of different hydrocarbons. But a big mixture like this is not very useful. So the first step is to separate the compounds into groups with carbon chains of a similar length. This is called **refining** the petroleum.

The key ideas behind refining

These are the key ideas behind refining:

- As chain length increases, the boiling points of compounds rise.
- If you heat petroleum enough, compounds will **boil** to form gases. If these rise up a tall tower, they will cool as they rise, and **condense** at different heights in the tower, depending on their boiling points.
- So you can collect the groups of compounds in trays at different heights in the tower. The compounds in each tray will have carbon chains that are quite close in length (within a range).
- Each group of compounds can then be used in a different way.

So the compounds have been separated into different groups or **fractions**. The process is called **fractional distillation**. Look:

▲ A fractionating tower in a petroleum refinery. It can operate non-stop for years.

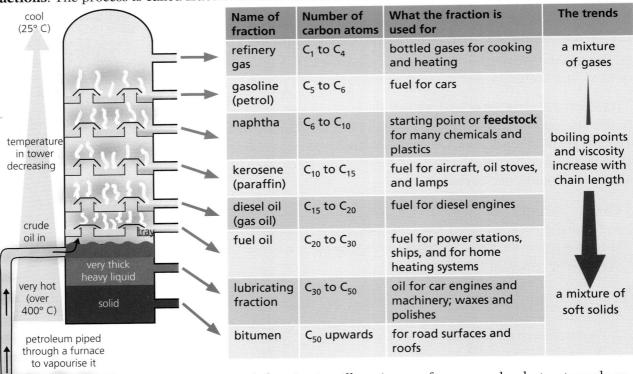

Name of fraction	Number of carbon atoms	What the fraction is used for	The trends
refinery gas	C_1 to C_4	bottled gases for cooking and heating	a mixture of gases
gasoline (petrol)	C_5 to C_6	fuel for cars	boiling points and viscosity increase with chain length
naphtha	C_6 to C_{10}	starting point or **feedstock** for many chemicals and plastics	
kerosene (paraffin)	C_{10} to C_{15}	fuel for aircraft, oil stoves, and lamps	
diesel oil (gas oil)	C_{15} to C_{20}	fuel for diesel engines	
fuel oil	C_{20} to C_{30}	fuel for power stations, ships, and for home heating systems	
lubricating fraction	C_{30} to C_{50}	oil for car engines and machinery; waxes and polishes	a mixture of soft solids
bitumen	C_{50} upwards	for road surfaces and roofs	

Each fraction is still a mixture of compounds – but not nearly as big a mixture as petroleum is. Note how many fractions are used as fuels!

The trends the fractions show

Look at the last column of the table on page 206. **Viscosity** is a measure of a fluid's resistance to flow. A thick sticky liquid has high viscosity.

The viscosity of the fractions is linked to boiling points – and to other properties too. Compare these fractions of petroleum collected in a lab:

chain length, boiling points, and viscosity *increase* →

A boiling point range 25–100 °C   **B** boiling point range 100–150 °C   **C** boiling point range 150–200 °C

This liquid is thin and runny – it has low viscosity. It catches fire easily – it is **flammable**. The high flame shows that it evaporates easily too – it is **volatile**.

This one is less runny, or more **viscous**. It catches fire quite easily. The flame burns lower than in A, showing that it is less volatile than A.

This one is even more viscous. The flame height shows that it is less volatile than B. It does not catch fire so readily or burn so easily, so it is less flammable too.

volatility and flammability *decrease* →

Properties dictate uses

The properties of each petroleum fraction dictate its uses. Most of the fractions – but not all – are used as **fuels**. For example:

- the gasoline (petrol) fraction is thin and runny, volatile, and highly flammable. So it is suitable for burning in car engines, where a fuel must catch fire with just a spark.
- the fuel oil fraction is thicker and much less volatile. It suits furnaces, for example, in ships, where a volatile fraction might cause explosions.
- the bitumen fraction is a soft solid that does not burn until over 200 °C. So it is no good as a fuel. But when melted and mixed with stone chips, it binds them together. This flexible mixture is used to surface roads.

▲ Filling the tank with petrol.

Where next for the fractions?

The fractions undergo further treatment before use. For example:

- the gas fraction is separated into its component gases – methane, ethane, propane, and butane
- the liquid fuel fractions may contain sulfur. This must be removed, because sulfur burns to give sulfur dioxide gas, which causes acid rain.
- part of a fraction may be **cracked**. Molecules are broken into smaller molecules. You can read about cracking in Unit 16.5.

▲ A new road surface: a mixture of bitumen and small stone chips.

Q

1. Two opposite physical changes take place during the fractional distillation of petroleum. Which are they?
2. A group of compounds collected during fractional distillation is called a _____. What is the missing word?
3. Define the term: **a** volatile **b** viscous **c** flammable
4. List four ways in which the properties of fractions differ.
5. Name the petroleum fraction that: **a** is used for petrol **b** has the smallest molecules **c** is the most viscous **d** has molecules with 20 to 30 carbon atoms
6. Give several reasons why petrol is a suitable fuel for cars.
7. Explain why bitumen: **a** is not used as a fuel **b** is used in road building

16.3 Four families of organic compounds

Objectives: explain how the compounds are named; describe three ways to show formulae; define *functional group*; give characteristics of a homologous series

Four families

The rest of this chapter is about four families of organic compounds: **alkanes**, **alkenes**, **alcohols**, and **carboxylic acids**. We obtain the alkanes from petroleum, and they are the starting point for the other families.

Naming the compounds in these families

Study these two tables, to see how the compounds are named:

The family	The names of its members end in ...	Example
the alkanes	-ane	ethane, C_2H_6
the alkenes	-ene	ethene, C_2H_4
the alcohols	-ol	ethanol, C_2H_5OH
the carboxylic acids	-oic acid	ethanoic acid, CH_3COOH

▲ The alkane **methane** occurs as natural gas. It is piped into many homes as fuel.

If the name of the compound begins with ...	... its molecules have this many carbon atoms ...	Example from the alkane family
meth-	1	methane, CH_4
eth-	2	ethane, C_2H_6
prop-	3	propane, C_3H_8
but-	4	butane, C_4H_{10}
pent-	5	pentane, C_5H_{12}

So the name of the organic compound tells you both the family it belongs to, and how many carbon atoms it has.

Showing their formulae

You can show the formula of an organic compound in three ways. Make sure you understand the formulae in this table:

▲ The alcohol **ethanol** is used in marker pens. It evaporates quickly, so the ink dries.

Type of formula	Propane	Propene	Propan-1-ol	Propanoic acid
displayed formula shows all the bonds between atoms, both single and double	H H H | | | H—C—C—C—H | | | H H H	H H \ / C=C—C—H / | | H H H	H H H | | | H—C—C—C—O—H | | | H H H	H H O | | ‖ H—C—C—C | | \ H H O—H
structural formula shows each unit in the structure, but no bonds except for C=C bonds	$CH_3CH_2CH_3$	$CH_2=CHCH_3$	$CH_3CH_2CH_2OH$	CH_3CH_2COOH
molecular formula shows the numbers of atoms, but with – OH and – COOH separate	C_3H_8	C_3H_6	C_2H_5OH	C_2H_5COOH

ORGANIC CHEMISTRY

Comparing the four families
This table shows one member from each family. Compare them.

Family	A member	Displayed formula	Comments
the alkanes	ethane, C_2H_6	H H H–C–C–H H H	• The alkanes contain only carbon and hydrogen, so they are **hydrocarbons**. • The bonds between their carbon atoms are all **single covalent bonds**.
the alkenes	ethene, C_2H_4	H H C=C H H	• The alkenes are **hydrocarbons**. • All alkenes contain **carbon–carbon double covalent bonds**. • The C=C bond is called their **functional group**.
the alcohols	ethanol, C_2H_5OH	H H H–C–C–O–H H H	• The alcohols are **not** hydrocarbons. • They are like the alkanes, but with an **OH group**. • The OH group is their functional group.
the carboxylic acids	ethanoic acid, CH_3COOH	H O H–C–C H O–H	• The carboxylic acids are **not** hydrocarbons. • All carboxylic acids contain the **COOH group**. • The COOH group is their functional group.

The flowchart on the right shows how the four families are linked.

Functional groups
A functional group is an atom or group of atoms in a molecule that dictates how the compound will react.

So all the alkenes react similarly because they all have the C=C group. All the alcohols react similarly because they all have the –OH group.

Homologous series
Each family forms a **homologous series**.
A homologous series is a family of similar compounds that have the same functional group, so have similar chemical properties.

In a homologous series:
- all the compounds fit the same **general formula**.
  For the alkanes the general formula is C_nH_{2n+2}, where n is the number of carbon atoms. For methane n is 1, giving the formula CH_4. For ethane n is 2, giving C_2H_6. Does propane fit the general formula?
- the chain length increases by one – CH_2 – unit at a time.
- the compounds show a trend in physical properties, as the chain gets longer. **Melting and boiling points** rise. **Viscosity** increases – they flow less easily. **Flammability** decreases – they burn less easily.

How the four families are linked

alkanes
obtained from petroleum, where they occur naturally (and natural gas is methane)

↓

alkenes
made from alkanes by cracking

↓

alcohols
made from alkenes by reaction with water

↓

carboxylic acids
made by oxidising alcohols

1. Name four families of organic compounds.
2. *Propan-1-ol* is an organic compound.
   a How many carbon atoms does it contain?
   b Which family does it belong to?
3. Name the alkane with this
   formula: a CH_4 b C_4H_{10}
4. Now draw the displayed formula for each compound in **3**.
5. Define *functional group*, and give two examples.
6. Give the four characteristics of a homologous series.
7. The straight-chain alkane *hexane* has 6 carbon atoms. Show its formula in three ways: molecular, structural, and displayed.
8. Give the molecular formula for the alkane with:
   a 20 carbon atoms   b 32 carbon atoms

16.4 The alkanes

Objectives: name the first four alkanes, and draw their displayed formulae; give the characteristics of the alkanes; define *structural isomer* and give an example

Alkanes: a reminder
This is what you have learned about the **alkanes** so far:

- The alkanes are the simplest family of organic compounds. Their names all end in *-ane*.
- They occur naturally in petroleum.
- They are **hydrocarbons**: they contain only carbon and hydrogen.
- Their carbon–carbon bonds are all **single covalent bonds**.
- They form a homologous series, with the general formula C_nH_{2n+2}.

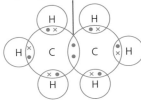

▲ The bonding in ethane. Each bond is a single covalent bond.

The first four alkanes
The first four alkanes are obtained in the **refinery gas fraction** in the petroleum refinery (page 206). Methane also occurs as natural gas:

Compound	methane	ethane	propane	butane
Molecular formula	CH_4	C_2H_6	C_3H_8	C_4H_{10}
Displayed formula	H–C(H)(H)–H	H–C(H)(H)–C(H)(H)–H	H–C(H)(H)–C(H)(H)–C(H)(H)–H	H–C(H)(H)–C(H)(H)–C(H)(H)–C(H)(H)–H
Number of carbon atoms in the chain	1	2	3	4
Boiling point/°C	−164	−87	−42	−0.5

→ boiling points increase with chain length →

All four are gases at room temperature. But boiling points increase with chain length. So the next twelve alkanes are liquids, and the rest are solids.

Chemical properties of the alkanes

1. Because all their carbon–carbon bonds are single covalent bonds, the alkanes are called **saturated**. Look at their displayed formulae.
2. Generally, the alkanes are quite unreactive.
3. The shorter-chain alkanes burn very well in a good supply of oxygen, forming carbon dioxide and water vapour, and giving out plenty of heat. So they are used as **fuels**. This shows the combustion of methane:

   $CH_4(g) + 2O_2(g) \longrightarrow CO_2(g) + 2H_2O(l)$ + heat energy

4. If there is not enough oxygen, the alkanes undergo **incomplete combustion**, giving poisonous carbon monoxide. For example:

   $2CH_4(g) + 3O_2(g) \longrightarrow 2CO(g) + 4H_2O(l)$ + less heat energy

▲ Cylinders of LPG (a mixture of propane and butane) for delivery to homes as fuel.

5. Alkanes also react with chlorine in sunlight. For example:

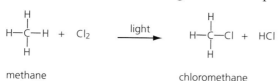

methane → chloromethane

A chlorine atom replaces a hydrogen atom, so this is called a **substitution reaction**.

In a substitution reaction, one atom or group of atoms is replaced by another atom or group of atoms.

If there is enough chlorine, all four hydrogen atoms will be replaced. The reaction can be explosive in sunlight. But it will not occur in the dark, because it is a **photochemical reaction**. The energy from light provides the **activation energy** (E_a) for the reaction.

Structural isomers

butane
boiling point 0 °C

2-methylpropane
boiling point −10 °C

Compare these alkane molecules. Both have the formula, C_4H_{10}. But they have different structures. The first has a straight chain. In the second, the chain is branched. They are **structural isomers**.

Structural isomers are compounds with the same molecular formula, but different structural formulae.

Note that the compound with the branched chain has a lower boiling point. Molecules with branched chains cannot get so close together. So it takes less energy to separate them to form a gas.

The more carbon atoms in a compound, the more structural isomers it will have. There are 75 structural isomers with the formula $C_{10}H_{22}$!

Cracking the alkanes

As you saw, the alkanes are quite unreactive. But they can be turned into reactive compounds, the **alkenes**, by a process called **cracking**. You will learn about cracking in the next unit.

Chlorine and methane

The hydrogen atoms can be replaced one by one:

chloromethane	CH_3Cl
dichloromethane	CH_2Cl_2
trichloromethane	$CHCl_3$
tetrachloromethane	CCl_4

All four are used as solvents. But they can cause health problems, so are used less and less.

▲ Branched-chain hydrocarbons burn less easily than their straight-chain isomers. So they are used in petrol to control combustion – and especially for racing cars.

Structural isomers?

- **a** and **b** below are *not* structural isomers. Both have the structural formula CH_3CH_2Br.
- **b** and **c** are not structural isomers either. Why not?

a, b, c structures shown

Q

1. Describe the bonding in ethane. You could do a drawing.
2. Why are alkanes such as methane and butane used as fuels? Give two reasons.
3. Butane burns in air in a similar way to methane. Write a word equation for its combustion.
4. The reaction of chlorine with methane is called a *substitution* reaction. Why?
5. What special condition is needed for the reaction in **4**?
6. Ethane reacts with chlorine in a substitution reaction.
   a. Draw the displayed formulae for the structural isomers when *two* chlorine atoms are replaced. (See panel above!)
   b. Write the molecular formula for each compound in **a**.
7. The compound C_5H_{12} has *three* structural isomers.
   a. Draw the displayed formula for each isomer.
   b. Their boiling points are 9.5, 28, and 36 °C. Match these to your drawings, and explain your choice.

16.5 Cracking alkanes

Objectives: define *cracking*; give two reasons for its importance; describe the manufacture of alkenes by cracking; describe how ethane is cracked to ethene

What is cracking?

Cracking is a way to make more use of petroleum.
Cracking breaks molecules into smaller ones.

The process is usually carried out by spraying liquid hydrocarbons through a very hot catalyst (a porous mineral called a **zeolite**). The liquid vapourises on the catalyst and the hydrocarbon molecules break up.

Because it uses a catalyst, this is called **catalytic cracking**.

Example: cracking an alkane

Look how this alkane molecule from the naphtha fraction breaks up:

decane, $C_{10}H_{22}$ from naphtha fraction

↓ 540 °C, catalyst

pentane, C_5H_{12} suitable for petrol + propene, C_3H_6 + ethene, C_2H_4

▲ Decane is one of the hydrocarbons in **white spirit**, a solvent used to thin oil-based paint, and clean paint brushes.

So the decane molecule has broken into three smaller molecules. Two of the products, propene and ethene, have **carbon–carbon double covalent bonds**. So they belong to a different family, the **alkenes**.

The decane molecule can break up in other ways too. There are several possibilities. Look at this one:

decane

↓ 540 °C, catalyst

hexane, C_6H_{14} suitable for petrol + ethene, C_2H_4 + ethene, C_2H_4

So several products form when decane is cracked. The naphtha fraction contains many other molecules too. So cracking part of the naphtha fraction leads to a big mix of products. They can be separated later.

Cracking can be carried out on other fractions from petroleum too. It depends on which groups of products are needed.

▲ Before a fraction goes for cracking, sulfur is removed. Refineries are a major source of sulfur, shown above. It can be sold to make sulfuric acid (page 124).

ORGANIC CHEMISTRY

Why crack?
Cracking is really important in the petroleum industry. Why?
- **It helps you make the best use of petroleum, to meet demand.**

    Suppose you have too much of the naphtha fraction (C_6–C_{10}) and not enough of the gasoline (petrol) fraction (C_5–C_6). Your customers want more petrol. So you can crack some of the naphtha fraction to get the more flammable shorter-chain compounds suitable for petrol.

- **It also gives alkenes, with a carbon–carbon double covalent bond.**

    Alkenes are not found in petroleum, but they are extremely useful. The double bond makes them **reactive**. So they are the starting point for a huge range of compounds, including plastics and medical drugs.

Cracking ethane
Ethane has very short molecules – but even they can be cracked, to give ethene and hydrogen. Look:

$$H_3C-CH_3 \xrightarrow{\text{steam}, >800\,°C} H_2C=CH_2 + H_2$$

ethane → ethene + hydrogen

▲ How many plastic items can you see? They all began with the cracking of alkanes to give alkenes.

The two gases are heated in a furnace to a very high temperature.

The hydrogen from this reaction is used to make ammonia (page 122). Ethene has a very different future. As you will see in Chapter 17, it is used to make the plastic **polythene**.

▲ Cracking is carried out at the petroleum refinery. The products are separated as needed, and then stored ready for customers.

▲ Delivery from the refinery to a petrol station, where the petrol will be pumped into underground tanks.

1. What happens during cracking?
2. The cracking of decane is described as *catalytic* cracking.
    a. Explain why.
    b. What other reaction condition is needed for this process?
3. What is *always* produced in a cracking reaction?
4. Explain why cracking is so important.
5. The cracking of ethane is especially important in industry. Why? Give two reasons.
6. a. A straight-chain hydrocarbon has the molecular formula C_5H_{12}. Draw its displayed formula, showing all bonds.
    b. Now show *one* of the possible reactions, when this compound is cracked.

16.6 The alkenes

Objectives: name the first three alkenes, and draw their displayed formulae; recall how alkenes are made; give addition reactions of ethene; give the test for unsaturation

The alkene family
- The **alkenes** are hydrocarbons.
- They form a homologous series, with the general formula C_nH_{2n}.
- They all contain the C=C double bond. It is their functional group.

Here are the first three family members. They are all flammable gases:

▲ A model of the ethene molecule.

Compound	ethene	propene	but-1-ene
Molecular formula	C_2H_4	C_3H_6	C_4H_8
Displayed formula	H₂C=CH₂	CH₂=CH–CH₃	CH₂=CH–CH₂–CH₃
Number of carbon atoms	2	3	4
Boiling point/°C	−102	−47	−6.5

Because of their C=C double bonds, the alkenes are called **unsaturated**. (Alkanes have only C–C single carbon bonds, so are **saturated**.)

Key points about the alkenes

1 The alkenes do not occur naturally in petroleum. They are made from alkanes by **cracking**, as you saw in the last unit. For example:

ethane —(steam, >800 °C)→ ethene + H_2
hydrogen

> **! Structural formulae?**
> Are these structural formulae correct, for the three compounds in the table? Check!
> ethene $CH_2=CH_2$
> propene $CH_2=CHCH_3$
> but-1-ene $CH_2=CHCH_2CH_3$

2 They are much more reactive than alkanes, because the double bond can break, to add on other atoms. For example, ethene can add on hydrogen again, with nickel as catalyst, to form ethane:

$CH_2=CH_2$ (g) + H_2(g) —(heat, pressure, nickel as catalyst)→ CH_3-CH_3 (g)

ethene → ethane

It also adds on steam, in the presence of an acid catalyst, to form ethanol, an alcohol:

$CH_2=CH_2$ (g) + H_2O(g) ⇌ (heat, pressure, acid catalyst) CH_3-CH_2-OH (l)

ethene → ethanol

These reactions are called **addition reactions**. Can you see why? **An addition reaction turns an unsaturated alkene into a saturated compound. There is only one product.**

▲ Ethene occurs naturally in fruit, as the hormone that causes ripening. Fruit produces ethene even after it is picked, and this finally leads to decay. The ethene in fruit is usually called **ethylene**.

ORGANIC CHEMISTRY

A test for unsaturation

You can use **bromine water** to test whether a hydrocarbon is unsaturated. It is an orange solution of bromine in water. If a C=C bond is present, a reaction takes place, and the colour disappears. For example:

$$CH_2=CH_2 \text{ (g)} + Br_2 \text{ (aq)} \longrightarrow CH_2Br-CH_2Br \text{ (l)}$$

ethene (colourless) + (orange) → 1,2–dibromoethane (colourless)

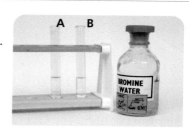

▲ **A** and **B** contained liquid hydrocarbons that do not mix with water. Bromine water was added to both test-tubes. Which one had the unsaturated compound?

This is an example of an addition reaction. The bromine adds on. Note that the same reaction will occur using pure bromine, a liquid.

Structural isomers in the alkene family

In alkenes, the double bonds can be in different positions in the chain.

Compare the compounds below. Both have the formula C_4H_8. But the double bond is in a different position in each molecule. So they are structural isomers:

but-1-ene   but-2-ene

What do the numbers tell you? Count the carbon atoms from the left!

▲ Bananas for export are picked before they are ripe. When they reach their destination they are exposed to ethene gas (behind those yellow doors) for final ripening before sale to shoppers.

▲ Alkenes are among the most important organic compounds, because of their reactivity. Most plastics are made from alkenes. Poly(ethene) is one of the plastics used for toys.

1. Name the simplest alkene, and draw its displayed formula.
2. The table on page 214 shows the first three alkenes. The next one in the series is pent-1-ene. What can you say about:
   a its boiling point?   b its functional group?
3. Alkenes are much more reactive than alkanes. Why?
4. a Propene is *unsaturated*. What does that mean?
   b Name a *saturated* compound with three carbon atoms.
5. Describe a test to check whether a liquid is an alkene.
6. Write the equation for the reaction of ethene with bromine water, using molecular formulae.
7. How would you turn propene into:
   a an alcohol?   b an alkane?
8. The reactions in **7** are a____ reactions. Give the missing word.
9. a Define the term *structural isomer*.
   b Give the structural formulae for the structural isomers but-1-ene and but-2-ene, in that order.

16.7 The alcohols

Objectives: name the first four alcohols, and draw their displayed formulae; state the properties and uses of ethanol; give, and compare, two ways to make ethanol

What are alcohols?
Alcohols contain the **OH** functional group. It is called the **hydroxyl group**.
Look at the first four alcohols.
They are all liquid at room temperature:

Alcohol	methanol	ethanol	propan-1-ol	butan-1-ol
Molecular formula	CH_3OH	C_2H_5OH	C_3H_7OH	C_4H_9OH
Displayed formula	H–C(H)(H)–O–H	H–C(H)(H)–C(H)(H)–O–H	H–C(H)(H)–C(H)(H)–C(H)(H)–O–H	H–C(H)(H)–C(H)(H)–C(H)(H)–C(H)(H)–O–H
Number of carbon atoms	1	2	3	4
Boiling point / °C	65	78	87	117

Note that:
- they form a homologous series, with the general formula $C_nH_{2n+1}OH$.
- their OH functional group means they all react in a similar way.
- two of the names above have -1- in. This tells you that the OH group is attached to the carbon atom at one end of the chain. (Either end!)

Ethanol, an important alcohol
- Ethanol is a good solvent. It dissolves many substances that are insoluble in water.
- It also evaporates easily – it is **volatile**. That makes it a suitable solvent for use in things like marker pens, perfumes, and aftershaves.
- It burns readily in oxygen, and the reaction gives out plenty of heat. So ethanol is used as a fuel. See the next page for more.
- It can kill viruses and bacteria, so is used in hand sanitisers and wipes.

propan-2-ol   butan-2-ol

▲ All alcohols with more than two carbon atoms have structural isomers, because the –OH group can bond to carbon atoms *within* the chain. Compare these with the corresponding compounds in the table.

Two ways to make ethanol
Ethanol is made in two ways, one chemical (linked to the petroleum industry) and one biological.

1 **By reacting ethene with steam**
   This is an **addition reaction**. You already met it on page 214. Look:

   ethene + H_2O (g) $\rightleftharpoons$ ethanol

   300 °C, 60–70 atm
   an acid catalyst (phosphoric acid)

   Note that this reaction is *reversible*, and exothermic. High pressure and a low temperature would give the best yield. But in practice the reaction is carried out at 570 °C, to give a decent rate.

   A catalyst is also used. So the reaction is called a **catalytic addition**.

▲ Ethanol can kill viruses and bacteria. So it is used in hand sanitisers and similar disinfectant products.

ORGANIC CHEMISTRY

2 By fermentation – the biological way

It is made from glucose (a sugar) using **yeast**, *in the absence of oxygen*. Yeast is a mass of living cells. The enzymes in it catalyse this reaction:

$$C_6H_{12}O_6 \ (aq) \xrightarrow[25-35°C]{\text{enzymes in yeast}} 2C_2H_5OH \ (aq) + 2CO_2 \ (g) + \text{energy}$$

glucose   ethanol   carbon dioxide

The process is called **fermentation**, and it is exothermic.

You can start with anything that contains sugars, starch, or cellulose, because all these break down to glucose. For example, maize, or wood.

The yeast stops working when the % of ethanol reaches a certain level, or if the mixture gets too warm. The ethanol is then removed from the reaction mixture using **fractional distillation** (page 206).

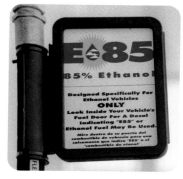

▲ Ethanol from fermentation is mixed with petrol in many countries, to help the environment. The fuel advertised above is 85% ethanol and 15% petrol.

Comparing the two ways to make ethanol

Both methods have advantages, and disadvantages. Compare them:

By reaction of ethene with steam	By fermentation
Advantages	Advantages
• The reaction is fast. • You can run it non-stop. Just keep removing the ethanol. • It gives pure ethanol. No need for fractional distillation.	• It uses renewable resources. You can grow more maize, or more sugarcane. • You can use waste plant material too. • The reaction takes place close to room temperature.
Disadvantages	Disadvantages
• Ethene is made from ethane, obtained from petroleum. Petroleum is not a renewable resource. When it is used up, it's gone. • A lot of electricity is needed: to raise the pressure, boil water to make steam, and keep the reaction at 570 °C. Expensive! • The reaction is reversible, and at 570 °C the yield is not high. So you must keep recycling the unreacted ethene and steam.	• A lot of land is needed to grow enough plant material. • Fermentation is slow, and has to be carried out a batch at a time. • The yeast stops working when the % of ethanol reaches a certain level – even if there's still glucose left. • The ethanol is not pure. It has to be separated by fractional distillation. That needs heat – an extra cost.

Ethanol as a fuel

Ethanol burns well in oxygen. The reaction gives out plenty of heat:

$$C_2H_5OH \ (l) + 3O_2 \ (g) \longrightarrow 2CO_2 \ (g) + 3H_2O \ (l) + \text{heat}$$

So it is increasingly used as a fuel for cars and other vehicles – usually mixed with petrol. Demand has grown because:

- it can be made from plant material, a renewable resource.

- many countries have no petroleum of their own, and must buy petrol from other countries; it costs a lot, so ethanol is an attractive option.

- ethanol forms carbon dioxide when it burns. But when crops are grown to make it, they take in carbon dioxide. So overall, ethanol from plant material has less impact on the environment than petrol does.

▲ Corn (maize) is grown in many countries for fuel. Fuel from plant material, using yeast or bacteria, is called **biofuel**.

Q

1. Draw the displayed formula for ethanol.
2. Give the word equation for the reaction of ethanol as a fuel.
3. In Brazil, sugarcane is used to make ethanol.
   a. Name the process used, and the catalyst.
   b. Now describe how to make ethanol using steam.
4. a. Compare the two methods of making ethanol in industry, giving two advantages and two disadvantages for each.
   b. Which method uses enzymes as a catalyst?
5. Give the structural formulae for the two isomers with molecular formula C_3H_7OH.

16.8 The carboxylic acids

Objectives: name, and draw displayed formulae for, the first four carboxylic acids; recall two ways to make ethanoic acid; give the reactions of ethanoic acid

The carboxylic acid family
Now we look at the family of organic acids: **the carboxylic acids**.
Here are the first four members of the family:

Name of acid	methanoic	ethanoic	propanoic	butanoic
Molecular formula	HCOOH	CH_3COOH	C_2H_5COOH	C_3H_7COOH
Displayed formula	H–C(=O)(O–H)	CH_3–C(=O)(O–H)	C_2H_5–C(=O)(O–H)	C_3H_7–C(=O)(O–H)
Number of carbon atoms	1	2	3	4
Boiling point/°C	101 °C	118 °C	141 °C	164 °C

- The family forms a homologous series with the general formula $C_nH_{2n+1}COOH$. Check: do the four formulae in the table match this?
- The functional group **COOH** is called the **carboxyl group**.

We focus on ethanoic acid in this unit. It is a liquid at room temperature, and we use it in the lab in aqueous solution. Other carboxylic acids behave in a similar way, because they all contain the carboxyl group.

Two ways to make ethanoic acid
Ethanoic acid is made by oxidising ethanol:

$CH_3CH_2OH \xrightarrow{[O]} CH_3COOH$

▲ Carboxylic acids form compounds called **esters**. The apple smells of natural esters. The shampoo contains synthetic esters that smell like those in apples.

The oxidation can be carried out in two ways.

1 **Using an oxidising agent – the chemical way**
   Ethanol is warmed with the powerful oxidising agent potassium manganate(VII), in the presence of acid. The manganate(VII) ions are themselves reduced to Mn^{2+} ions, with a colour change.
   The acid provides the H^+ ions for the reaction:

   $MnO_4^- + 8H^+ + 5e^- \longrightarrow Mn^{2+} + 4H_2O$
   purple   colourless

2 **By fermentation – the biological way**
   When ethanol is left standing in air, bacteria bring about its oxidation to ethanoic acid. This method is called **acid fermentation**.

   Acid fermentation is used to make vinegar (a dilute solution of ethanoic acid). Vinegar starts as foods such as apples or rice. They are fermented to give ethanol. This is then oxidised further to ethanoic acid.

▲ Organic chemistry at the dinner table. Vinegar (left) is mainly ethanoic acid and water. Olive oil (right) is made of **esters**.

ORGANIC CHEMISTRY

Ethanoic acid: typical acid reactions

1 A solution of ethanoic acid in water turns litmus red.
2 It reacts with metals, bases, and carbonates, to form **salts**. The hydrogen of the carboxyl group is replaced by a metal ion. Compare:

$2CH_3COOH$ (aq) + Mg (s) ⟶ $(CH_3COO)_2Mg$ (aq) + H_2 (g)
ethanoic acid   magnesium   magnesium ethanoate   hydrogen

CH_3COOH (aq) + NaOH (aq) ⟶ CH_3COONa (aq) + H_2O (l)
ethanoic acid   sodium hydroxide   sodium ethanoate   water

$2CH_3COOH$ (aq) + K_2CO_3 (s) ⟶ $2CH_3COOK$ (aq) + CO_2 (g) + H_2O (l)
ethanoic acid   potassium carbonate   potassium ethanoate   carbon dioxide   water

The ions in ethanoic acid
- Ethanoic acid dissociates reversibly:
  CH_3COOH (aq) ⇌ CH_3COO^- (aq) + H^+ (aq)
- At any given time, only some of the molecules have dissociated.
- So it is called a *weak* acid.
- The displayed formula for the ethanoate ion is:

$$\begin{array}{c} H \\ | \\ H-C-C \\ | \\ H \end{array} \begin{array}{c} O \\ \diagup \\ \diagdown \\ O^- \end{array}$$

Esters

Carboxylic acids react with alcohols, to give compounds called **esters**:

$$\text{carboxylic acid + alcohol} \underset{}{\overset{\text{acid catalyst}}{\rightleftharpoons}} \text{ester + water}$$

In the example below, the alcohol molecule has been reversed, and the displayed formulae simplified a little, to help you see what is going on:

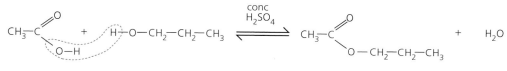

ethanoic acid   propanol   propyl ethanoate (an ester)   water

In making the ester:
- two molecules join to make a larger molecule, with the loss of a small molecule – in this case water. This is called a **condensation reaction**.
- the reaction is reversible, and sulfuric acid acts a catalyst.
- the part of the molecule that comes from the alcohol goes *last* in the formula – but *first* in the name. (Compare with sodium ethanoate!)

Esters are big business!
- Many esters have attractive smells and tastes. For example, propyl ethanoate (above) smells like pears, and ethyl propanoate like pineapples.
- So esters are added to shampoos and soaps and cosmetics for their smells, and to ice cream and other foods as flavours.
- The world trade in esters is worth around 1 billion dollars a year ... and growing.

Two more esters

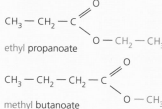

ethyl propanoate

methyl butanoate

■ 'acid' part of name
■ 'alcohol' part of name

▲ The second table on page 208 will help you check the names of esters.

1 What is the functional group of the carboxylic acids?
2 Show the displayed formula for ethanoic acid.
3 Copy and complete.
    carboxylic acid + metal ⟶ _____ + _____
    carboxylic acid + base ⟶ _____ + _____
    carboxylic acid + carbonate ⟶ _____ + _____
                                       + _____
4 Is ethanol *oxidised*, or *reduced*, to make ethanoic acid?
5 Name a catalyst used in making esters.
6 Suggest starting compounds for making this ester:
    a ethyl ethanoate   b methyl propanoate
7 Name the ester with this structural formula:
    a CH_3COOCH_3 b $HCOOCH_2CH_3$
8 Now draw the displayed formula for each ester in **7**.

Checkup on Chapter 16

Revision checklist

Core syllabus content
Make sure you can ...
- [ ] name the fossil fuels and say how they were formed
- [ ] explain what a *hydrocarbon* is
- [ ] explain why petroleum has to be refined, and:
  - describe the refining process
  - name the different fractions
  - say what each fraction is used for
- [ ] explain what *cracking* is, and why it is so useful
- [ ] give the equation for the cracking of ethane
- [ ] draw the displayed formula for
  methane ethane ethene ethanol ethanoic acid
- [ ] say what family a compound belongs to, from its formula, or its name
- [ ] give the functional groups for the alkenes, alcohols, and carboxylic acids
- [ ] give the reactions of alkanes, and state in general, they are unreactive
- [ ] explain the terms *saturated* and *unsaturated*
- [ ] give a test to identify an unsaturated hydrocarbon
- [ ] describe the two ways to make ethanol
- [ ] give two uses for ethanol (as solvent, and fuel)
- [ ] give the equation for the combustion of ethanol
- [ ] give examples of reactions to show that ethanoic acid is a typical acid

Extended syllabus content
Make sure you can also ...
- [ ] name, and draw the displayed formulae for, the four simplest members of each family: alkanes, alkenes, alcohols, and carboxylic acids
- [ ] give the general properties of a homologous series
- [ ] explain what *structural isomers* are, with examples
- [ ] describe the substitution reactions of alkanes with chlorine as *photochemical reactions*, and explain this term
- [ ] describe the addition reactions of alkenes with hydrogen, steam, and bromine
- [ ] compare the two ways of making ethanol
- [ ] describe the two ways to make ethanoic acid
- [ ] explain why ethanoic acid is a weak acid
- [ ] describe the reaction of a carboxylic acid with an alcohol, and name and draw the displayed formula for the ester that forms (for acids and alcohols with up to four carbon atoms each)

Questions

Core syllabus content

1 Petroleum is separated into fractions, like this:

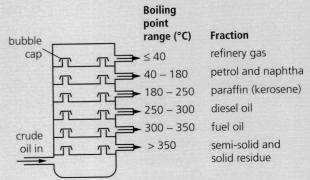

Boiling point range (°C)	Fraction
≤ 40	refinery gas
40 – 180	petrol and naphtha
180 – 250	paraffin (kerosene)
250 – 300	diesel oil
300 – 350	fuel oil
> 350	semi-solid and solid residue

a **i** What is this process called?
  ii It uses the fact that different compounds have different .... What is missing?
b **i** Is naphtha just one compound, or a group of compounds? Explain.
  ii Using the terms *boil* and *condense*, explain how naphtha is produced.
c Give one use for each fraction obtained.
d A hydrocarbon has a boiling point of 260 °C.
  i Are its carbon chains shorter, or longer, than those found in naphtha?
  ii Is it more viscous, or less viscous, than the compounds found in naphtha?

2

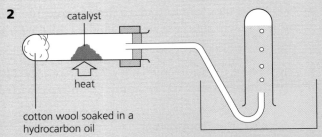

An alkane hydrocarbon can be cracked in the lab using the apparatus above.
a What is *cracking*?
b Which two conditions are needed, to crack the hydrocarbon?
c Give two reasons why cracking is so important in the refining industry.
d **i** Bromine water is added to one of the test-tubes of gas. What change will you see?
  ii Explain the change observed in **i**.
e Ethane, C_2H_6, can be cracked to give ethene, C_2H_4, and hydrogen. Write an equation for this.

3 Answer these questions about the alkanes.
a Which two elements do alkanes contain?

b Which alkane is the main compound in natural gas?
c After butane, the next two alkanes in the series are *pentane* and *hexane*. How many carbon atoms are there in a molecule of:
  i pentane?   **ii** hexane?
d Will pentane react with bromine water? Explain.
e Alkanes burn in a good supply of oxygen. Name the gases formed when they burn.
f Write the word equation for the complete combustion of pentane in oxygen.
g Name a harmful substance formed during *incomplete* combustion of pentane in air.
h Pentane reacts with chlorine in sunlight. Which word describes the reaction that occurs?
  oxidation   combustion   substitution   cracking

4 Alkenes are unsaturated hydrocarbons. They form a homologous series with the general formula C_nH_{2n}.
  a What is a *homologous series*?
  b Explain what the term *unsaturated* means.
  c What is the functional group for this series?
  d i Give the formula and name for the member of the series with two carbon atoms.
    ii Draw its displayed formula.
  e Alkenes react with steam under certain conditions.
    i Name the group of organic compounds formed in this reaction.
    ii What conditions are required for the reaction? Name three.
    iii Which of these terms describes the reaction?
      catalytic oxidation   catalytic cracking
      catalytic refining   catalytic addition
  f i Name the homologous series which has the general formula $C_nH_{2n+1}OH$.
    ii Give the name and molecular formula for the member with two carbon atoms.
    iii Now draw the displayed formula for the compound in **ii**.

Extended syllabus content
5 But-1-ene takes part in addition reactions.
  a Define *addition reaction*.
  b Give the structural formula of the product that forms when but-1-ene reacts with:
    i bromine   **ii** hydrogen   **iii** steam
  c i Draw the displayed formula for but-2-ene.
    ii Will but-2-ene react like but-1-ene does, in addition reactions? Explain your answer.

6 Propan-1-ol is a member of a homologous series.
  a Give four characteristics of a homologous series.
  b i Which homologous series is propan-1-ol in?
    ii What is the functional group in this series?
  c i Write the structural formula for propan-1-ol.
    ii Give the structural formula and name for the isomer of propan-1-ol.
  d Name the type of organic compound that is obtained when propan-1-ol is reacted with acidified aqueous potassium manganate(VII).

7 Ethanoic acid is a member of a homologous series.
  a i Name this series.
    ii Give the general formula for the series.
  b What is the functional group for this series?
  c Ethanoic acid is made from ethanol in two ways.
    i What type of chemical change takes place?
    ii Which method uses room temperature?
  d Ethanoic acid is a *weak* acid. Explain what this means, using an equation to help you. (Page 219!)
  e Ethanoic acid reacts with carbonates.
    i What would you *see* during this reaction?
    ii Write a balanced equation for the reaction with sodium carbonate.
  f i Name the member of the series for which $n = 3$, and draw its displayed formula.
    ii Give the symbol equation for the reaction between this compound and sodium hydroxide.

8 Ethanoic acid reacts with ethanol in the presence of concentrated sulfuric acid.
  a Name the organic product formed.
  b Which type of compound is it?
  c Name the other product of the reaction.
  d What is the function of the sulfuric acid?
  e The reaction is *reversible*. What does this mean?
  f Write an equation for the reaction, using displayed formulae.

9 Hex-1-ene is an unsaturated hydrocarbon. It melts at $-140°C$ and boils at $63°C$. Its empirical formula is CH_2. Its relative molecular mass is 84.
  a i To which family does hex-1-ene belong?
    ii What is its molecular formula?
  b i Hex-1-ene reacts with bromine water. Write a symbol equation to show this reaction.
    ii What is this type of reaction called?
    iii What would you *see* during the reaction?
  c Another member of the family has a boiling point of $30°C$. What can you deduce about it?

17.1 Introducing polymers

Objectives: define the term *polymer*; describe the polymerisation reaction; give examples of natural and synthetic polymers; name the two types of polymerisation

What are polymers?
Polymers are very large molecules, built up from small molecules.
For example, look what happens when ethene molecules join:

This test-tube contains ethene gas. When ethene is heated to 50 °C, at a few atmospheres pressure and over a special catalyst …

… it turns into a liquid that cools to a waxy white solid. This is found to contain very long molecules, made by the ethene molecules joining.

And it is really useful. It is used to make tables and chairs, buckets, bowls, squeezy bottles, crates, water pipes …

The reaction that took place is:

ethene molecules (monomers)

$$\text{H}_2\text{C}=\text{CH}_2 \quad \text{H}_2\text{C}=\text{CH}_2 \quad \text{H}_2\text{C}=\text{CH}_2 \quad \text{H}_2\text{C}=\text{CH}_2 \quad \text{H}_2\text{C}=\text{CH}_2 \quad \text{H}_2\text{C}=\text{CH}_2$$

↓ polymerisation

part of a poly(ethene) molecule (a polymer)

$$-\text{CH}_2-\text{CH}_2-\text{CH}_2-\text{CH}_2-\text{CH}_2-\text{CH}_2-\text{CH}_2-\text{CH}_2-\text{CH}_2-\text{CH}_2-\text{CH}_2-\text{CH}_2-$$

> **It's polythene!**
> - In everyday life, poly(ethene) is known as **polythene**.
> - Do you use anything made of polythene?

The drawing shows six ethene molecules joining together. In fact *thousands* of ethene molecules join to give very large molecules.

- The ethene molecules are called **monomers**.
- The product is called a **polymer**. The polymer made from ethene is called **poly(ethene)**. *Poly-* means *many*.
- The reaction is called a **polymerisation**.

In a polymerisation reaction, thousands of small molecules join to make a large molecule. The small molecules are called monomers.

Synthetic polymers
Poly(ethene) is a **synthetic polymer**. *Synthetic* means it is made in a factory. Other synthetic polymers include nylon, polyester, polystyrene, and acrylic. Hair gels and shower gels contain water-soluble polymers. We are surrounded by synthetic polymers!

▲ The hair gel did it! Hair gel is a thick solution of a water-soluble polymer. The water evaporates, and the gel stiffens.

Natural polymers

Poly(ethene) was invented in 1935. But for millions of year, nature has been busy making **natural polymers**. Look at these examples:

Starch is a polymer made by plants. The starch molecules are built from molecules of **glucose**, a sugar. We eat plenty of starch in rice, bread, and potatoes.

Plants also use glucose to make another polymer called **cellulose**. Cotton T-shirts and denim jeans are almost pure cellulose, made by the cotton plant.

Your skin, hair, nails, bones, and muscles are mostly made of polymers called **proteins**. Your body builds these up from monomers called **amino acids**.

The wood in trees is about 50% cellulose. Paper is made from wood pulp, so this book is mainly cellulose. The polymer in your hair and nails, and in wool and silk, and animal horns and claws, is called **keratin**. The polymer in your skin and bones is called **collagen**.

So – you contain polymers, you eat polymers, you wear polymers, and you use polymers. Polymers play an enormous part in your life!

The reactions that produce polymers

All polymers, natural and synthetic, are large molecules, formed by small molecules joined together.

But polymers are not all formed in the same way. There are two quite different types of reaction: **addition polymerisation**, and **condensation polymerisation**. You can find out more about these in the next two units. Look for the differences!

▲ Wood: over 50% cellulose. This wood may end up as paper.

1 What is:
   a a polymer?   b polymerisation?   c a monomer?
2 *Polymers are scarce compounds.* True or false? Give evidence to support your answer.
3 Explain the term *poly*, in the name *poly(ethene)*.
4 What is the difference between a *natural polymer* and a *synthetic polymer*?
5 Name the natural polymer found in:
   a your hair   b this book
6 Name three other items you own, or use, that are made of:
   a natural polymers   b synthetic polymers
7 There are two types of polymerisation reaction. Name them.
8 When carbon burns in oxygen, their atoms bond to form carbon dioxide. Explain why this is *not* a polymerisation.

17.2 Addition polymerisation

Objectives: recall how poly(ethene) is formed; explain why it does not have an exact formula; name other addition polymers; identify a monomer, from a chain

Another look at the polymerisation of ethene
Here again is the reaction that produces poly(ethene):

ethene molecules (monomers)

$$H_2C=CH_2 \quad H_2C=CH_2 \quad H_2C=CH_2 \quad H_2C=CH_2 \quad H_2C=CH_2 \quad H_2C=CH_2$$

↓ polymerisation

part of a poly(ethene) molecule (a polymer)

$$-CH_2-CH_2-CH_2-CH_2-CH_2-CH_2-CH_2-CH_2-CH_2-CH_2-CH_2-CH_2-$$

The reaction can be shown in a short form like this:

$$n \begin{pmatrix} H & H \\ C=C \\ H & H \end{pmatrix} \xrightarrow{\text{heat, pressure, a catalyst}} \begin{pmatrix} H & H \\ -C-C- \\ H & H \end{pmatrix}_n$$

where n stands for a large number. It could be many thousands. The catalyst for the reaction is usually a mixture of titanium and aluminium compounds.

It's an addition reaction
The reaction above takes place because the C=C bond in the ethene molecules breaks, allowing the molecules to *add on* to each other. So it is called **addition polymerisation**.

In addition polymerisation, double bonds in molecules break, and the molecules add on to each other.

The monomer
The small starting molecules in a polymerisation are called **monomers**. In the reaction above, ethene is the monomer.

For addition polymerisation to take place, the monomers must have C=C double bonds.

Why poly(ethene) has only a general formula
During polymerisation, *the molecules that form have different chain lengths*. So we cannot write an exact formula for poly(ethene). Instead we use the general formula with brackets, shown above.

By changing the reaction conditions, the *average* chain length can be changed. But the chains will never be all the same length.

The relative molecular mass (M_r) of an ethene molecule is 28. The average M_r for a sample of poly(ethene) was found to be over 500 000. This means that over 17 000 ethene molecules had joined, on average. (500 000 / 28) So the *average* chain length was over 34 000 carbon atoms!

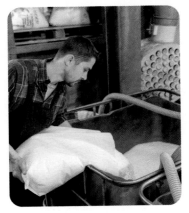

▲ Poly(ethene) is made and sold to factories as pellets. The pellets he is pouring out will be melted and turned into something useful.

▲ To make a bottle, pellets of poly(ethene) are melted. A molten tube of polymer is fed into a bottle-shaped mould. A powerful jet of air blasts it against the sides of the mould. When the mould is opened, out comes a bottle!

POLYMERS

Drawing the polymer chain

If you know the formula for the monomer in addition polymerisation, it is easy to draw the structure of the polymer. Check out these examples:

The monomer	The structure of the polymer	The equation for the reaction
chloroethene (vinyl chloride) — $CH_2=CHCl$	poly(chloroethene) or polyvinyl chloride (PVC)	n stands for a large number!
tetrafluoroethene — $CF_2=CF_2$	poly(tetrafluoroethene) or Teflon	
phenylethene (styrene) — $C_6H_5CH=CH_2$	poly(phenylethene) or polystyrene	

Identifying the monomer

You can also do things the other way round. Given the structure of the addition polymer, you can work out what the monomer was. Like this:

- identify the repeating unit. (It has two carbon atoms side by side, in the main chain.) You could draw brackets around it.
- then draw the unit, but put a double bond between the two carbon atoms. That is the monomer.

For example:

This shows part of a molecule of **poly(propene)**. The unit within brackets is the repeating unit.

So this is the monomer that was used. It is the alkene **propene**. Note the C=C double bond.

▲ PVC is used for hoses and water pipes, and as an insulating cover for electrical wiring. It can be made flexible or rigid.

1. Why is it called *addition* polymerisation?
2. Could methane (CH_4) be used as a monomer for addition polymerisation? Explain your answer.
3. Draw a diagram to show the polymerisation of ethene.
4. It is not possible to give an exact formula for the polymer molecules in poly(ethene). Why not?
5. Draw a diagram to show the polymerisation of:
   a chloroethene   b phenylethene
6. A polymer has the general formula shown on the right. Draw the monomer that was used to make it.

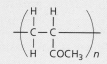

225

17.3 Condensation polymerisation

Objectives: outline *condensation polymerisation*; describe how a *polyamide* and a *polyester* are made; draw the repeating unit for each, and identify the linkage

Condensation polymerisation is different!
Addition polymerisation uses a single monomer, with a C=C bond. The only product is the polymer. **Condensation polymerisation** is different.
- It does not depend on C=C bonds.
- Two different monomers are used. Each has *two functional groups*.
- They join at their functional groups, by getting rid of or **eliminating** a small molecule. So there are *two* products: the polymer and another.

Let's look at two examples.

1 Making a polyamide
Two different monomers react together to make a **polyamide**. Let's call them **A** and **B**. Only their functional groups take part in the reaction. So we can show their carbon chains as blocks, for simplicity:

Monomer **A** has a **carboxyl** group (COOH) at each end. So it is a **dicarboxylic acid**. (*di* means *two*.)

Monomer **B** has an **amine** group (NH_2) at each end. So it is a **diamine**.

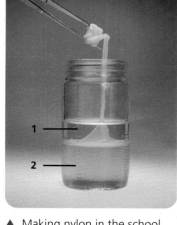

▲ Making nylon in the school lab. Layer **1** is a solution in hexane of a dicarboxylic acid compound. (It reacts the same as the acid.) Layer **2** is a diamine in water. Nylon forms where the solutions meet. You can pull it up as a rope.

The reaction between them is shown below. Some bonds are drawn vertically, to make the reaction easier to follow:

… and another **A** reacts here … and so on

water forms   then another **B** reacts here …

So the carbon atom of the carboxyl group bonds to the nitrogen atom of the amine group. A water molecule is eliminated. The reaction continues at the other ends of A and B, with thousands of monomer molecules joining. Here is part of the polymer molecule:

the **amide linkage**

▲ The repeat unit for a polyamide. The blocks represent the carbon chains of the monomers. Can you see how to deduce it from the structure on the left?

The CONH group, where the monomers have joined, is called the **amide linkage**. So the polymer is a **polyamide**. (Proteins have this linkage too.)

Nylon, a polyamide
Nylon is a polyamide. So the drawings above represent its structure. The repeat unit for a polyamide is shown on the right.

226

POLYMERS

2 Making a polyester

Again, two different monomers are used to make a **polyester**. This time we will call them **C** and **D**. We show their carbon chains as blocks:

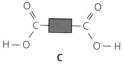

▲ Pumping a melted polymer through a spinneret (like a shower head), to make fibres. The fibres can then be woven.

Monomer **C** is a **dicarboxylic acid**, with a carboxyl functional group at each end (like A in example **1**).

Monomer **D** has an alcohol functional group (OH) at each end. So it is called a **diol**.

The reaction between them is shown below. Again some bonds are drawn vertically, to help you see how the polymer forms:

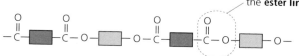

So the carbon atom of the carboxyl group bonds to the oxygen atom of the alcohol group, and a water molecule is eliminated. The reaction continues at the other ends of **C** and **D**. Thousands of monomer molecules join in this way. Here is part of the polymer molecule that forms:

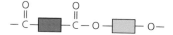

The key reaction is similar to the reaction between the acid and alcohol on page **219**, where water is eliminated to give an **ester**.

So the group where the monomers have joined is called an **ester linkage**. The polymer is a **polyester**. Its repeating unit is shown on the right.

▲ The repeat unit for a polyester. Is it correct? Check it against the drawing on the left, showing the structure of part of the molecule.

PET, a polyester

The polymer **PET** is a polyester. It is widely used to make water bottles. PET fibre is also used for clothing and bed linen, often woven with cotton. The drawing above shows its structure.

the starting monomers for PET
↓ polymerisation ↑ hydrolysis
the polymer PET + water

PET can broken down to its monomers again, by reaction with steam at high pressure and temperature. It is the reverse of polymerisation – the water that was eliminated is put back in. Chemical breakdown by adding water is called **hydrolysis**. The monomers can then be re-polymerised.

▲ PET can be broken down to its monomers, and then re-polymerised.

Q

1. List the differences between condensation polymerisation and addition polymerisation.
2. What do the blocks represent, in the diagrams in this unit?
3. a Draw and label the monomers used to make a polyamide.
   b Explain why each monomer needs *two* functional groups.
4. Draw a diagram to show the reaction that gives a polyamide.
5. Nylon is a polyamide. Draw the repeat unit for its structure, and circle the *amide linkage* in your drawing.
6. a Name the two types of monomers used for polyesters.
   b Which of these monomers is also used for polyamides?
7. PET is a polyester. Draw the repeat unit for its structure, and circle the *ester linkage* in your drawing.

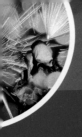

17.4 Plastics

Objectives: recall that plastics are polymers; give their general properties; know that properties can change with reaction conditions; give uses of named plastics

Plastics: synthetic polymers

In everyday life, synthetic polymers are usually just called **plastics**. (But when they made into thread, and woven into fabrics, we still call them synthetic polymers.)

Most are made from chemicals obtained from the naphtha fraction of petroleum (page 206), which is used as chemical feedstock.

The properties of plastics

Plastics have amazing general properties:

1. They do not usually conduct electricity or heat.
2. They are unreactive. Most are not affected by air, water, acids, or other chemicals. This means it is usually safe to store things in them.
3. They are usually light to carry – much lighter than wood, or stone, or glass, or most metals.
4. They don't break when you drop them. You have to hammer rigid plastics quite hard, to get them to break.
5. They are strong. This is because their long molecules are attracted to each other. Most plastics are hard to tear or pull apart.
6. They do not catch fire easily. But when you heat them, some soften and melt, and some char (go black as if burned).

They are everywhere …

Today there are *thousands* of plastics. They can be divided into a number of groups, of which polyamides and polyesters are just two.

Their properties make them useful in almost every area of life. Many are designed for specific purposes, by choosing suitable monomers and conditions.

But their unreactivity also means that plastics are very difficult to dispose of. This is a big problem. Find out more in the next unit.

▲ A synthetic polymer for sewing. (It is a polyester.)

a pair of polymer molecules

attraction increases with chain length

▲ As polymer chains get longer, the force of attraction between them increases. It gets harder to pull the polymer apart.

Different conditions, different properties

By changing the reaction conditions for polymerisation, you can change the properties of a plastic, to suit your needs. Poly(ethene) is a good example …

50 °C, 4 atm pressure, and a catalyst: the result is long chains, packed closely. So this poly(ethene) is called **high density**.

200 °C, 2000 atm pressure, and oxygen: you get branched chains that can't pack closely. So this poly(ethene) is **low density**.

We use high density Poly(ethene) for things like toys and bowls, and low density poly(ethene) for plastic bags.

POLYMERS

Examples of their uses

Here are examples of uses for some common plastics:

Polymer	Examples of uses
high density poly(ethene) (HDPE)	bowls, chairs, dustbins – and in artificial joints for the human body (used with metal)
low density poly(ethene) (LDPE)	plastic bags, disposable gloves, food wrap
polychloroethene (PVC)	water pipes, wellingtons, hoses, cover for electrical cables
polypropene	crates, ropes
polystyrene	used as **expanded polystyrene** for packaging, and in fast-food cartons, roofs, floors, and walls as a thermal insulator
PTFE (brand name Teflon)	coated on frying pans and irons to make them non-stick, used in making windscreen wipers, flooring, nail polish and eyeshadow
nylon, a polyamide	ropes, fishing nets and lines, tents, kites, curtains, umbrella fabric
PET, a polyester	bottles and other containers, clothing (often mixed with cotton), thread

▲ Nylon is also used for parasails like this one, and parachutes.

▲ Expanded polystyrene being used in house building. It is rigid but very light, and stops heat escaping. It is made from beads of polystyrene that contain gas. On heating, the gas expands.

▲ Teflon is smooth and slippery. It is used to coat irons to help them glide, and on frying pans to stop food sticking. It is also shiny, so is used in nail polish!

1. Plastics are *tetyncsih moreyslp*. Unjumble the jumbled words!
2. Look at the properties of plastics, on page 228. Which *three* do you think are the most important for:
   a fishing nets?   b plastic gloves?
   c hair dryers?   d kitchen bowls?
3. Choose one plastic item that you use. What properties does this plastic have? List them. (For example, is it rigid?)
4. The polymer called *Teflon* is used to coat frying pans, to make them non-stick. From this information, what can you deduce about the properties of Teflon? List them.
5. The plastic poly(chloroethene) is usually referred to as PVC.
   a Draw the monomer used for PVC. (Table on page 225.)
   b Which type of polymerisation produces PVC? addition condensation
   c Suggest reasons why PVC is suitable for water pipes.

17.5 The plastics problem

Objectives: explain why there is so much plastic waste; describe the environmental problems it causes in landfill sites, in the ocean, and when it is burned

Where has all the plastic gone?

The first truly synthetic plastic was invented in 1907. But production of plastics did not really take off until after World War II. Then it exploded.

Scientists estimated that the world had produced 8.3 billion tonnes of plastic by 2015. This pie chart shows what happened to it:

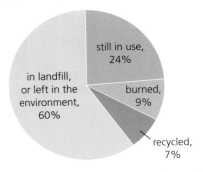

Plastic produced up to 2015: 8.3 billion tonnes
- still in use, 24%
- burned, 9%
- recycled, 7%
- in landfill, or left in the environment, 60%

▲ Bakelite was the first truly synthetic plastic. Invented in 1907, it is still made today. This is an old Bakelite telephone.

▲ Dumped illegally – mainly plastic items. The plastic will remain in the environment for a very long time.

As you can see, there was far more plastic waste than plastic still in use. Why? Because plastic is so stable that its waste breaks down only very slowly. It will decompose over time – but it could take hundreds of years.

Plastic in landfill sites

Much of the world's rubbish is dumped in pits or piled in heaps at **landfill sites**. Much of it is plastic. So these sites act as stores for plastic waste!

- A plastic bottle could last for 450 years in a landfill site.
- The more plastic waste there is, the more land we need to store it.
- Birds and other animals mistake plastic for food, at landfill sites. But they cannot digest it. It fills their stomachs, and they starve to death.
- When it does begin to break down, some plastic will release toxic chemicals that can drain into the soil and reach rivers.

▼ Plastic everywhere. The woman is collecting some to sell, in the landfill site. The puffer fish is eating a plastic bottle.

POLYMERS

Plastic in the ocean

Plastic reaches the ocean from many sources. Most is brought by rivers.

① Plastic particles from homes reach rivers via drains and sewage works. They include tiny **microbeads** that are added to toothpaste and face scrubs (to scrub us), and synthetic fibres from laundry and wet wipes.

② Plastic waste is dumped into rivers by careless people. Wind and floods help too.

③ The plastic litter that people leave on beaches is carried off by the waves.

④ Bits of plastic can leach from landfill sites, and reach rivers.

⑤ Plastic waste is dumped into the ocean from ships and fishing boats.

⑥ Every year, thousands of containers fall into the ocean from container ships, during storms. Some contain plastic goods.

So what happens to all this plastic in the ocean?

- Fish and other marine animals mistake plastic for food. The plastic fills their guts. They die of starvation.
- Some plastics **photodegrade** quite rapidly. They break into bits on exposure to sunlight. Eventually they end up as tiny **microparticles**.
- Marine animals can ingest microbeads and other microparticles. These accumulate in their bodies, and damage their health and fertility. And then we eat fish containing plastic.
- Ocean currents carry plastic waste along. It gets trapped where they change direction. So there are huge islands of plastic waste in the world's oceans, floating in a sea of microparticles.

▲ Currents will carry it around the ocean.

Why not burn it all?

As you saw from the pie chart, 9% of the plastic was burned. Burning gets rid of plastic fast … *and* releases toxic chemicals. These are linked to severe health problems, including damage to the heart, lungs, and nervous system, fertility issues, and harm to unborn babies.

Plastic waste can be burned safely in furnaces called **incinerators**. Harmful chemicals are trapped. But it is dangerous to burn plastic in the open air. The toxic chemicals end up in the air, soil, crops, and water. So people are exposed to them in many different ways.

What to do?

Plastics have brought huge benefits. But now plastic waste is a major problem. What can we do about it? Find out in the next unit.

▲ Burning plastic outdoors: harmful.

Q

1. It is more difficult to get rid of waste plastic than waste food. Explain why.
2. What is *a landfill site*?
3. Give three reasons why putting plastic waste in landfill sites is not an ideal way to dispose of it.
4. Define *microparticles*
5. Explain why plastic in the ocean harms marine life.
6. Look at the different ways plastic reaches the ocean. Choose any three, and suggest how to put an end to them.
7. Most plastic will burn if you heat it enough. Burning destroys it quickly. But in many countries it is illegal to burn plastic in the open air. Explain why.

17.6 Tackling the plastics problem

Objectives: give examples of single use plastic items; outline five ways to reduce plastic waste; list ways we can all help to protect the ocean from plastic waste

As you saw on page 230, over half of the plastic ever produced is still with us, as waste. It is expected that landfill sites and the environment will hold 12 billion tonnes of plastic waste by 2050, at this rate.

So what can we do to tackle the problem? Here are some approaches.

1 Cut down on single use plastic

Many plastic items are **single use**. Examples are plastic drinking straws, drinks bottles, and take-away coffee cups. **Wet wipes** for your skin have plastic fibres. The plastic bags from shops are often used just once.

It takes minutes to use these items – and then they hang around as waste for decades, or even centuries. This makes no sense.

So countries are moving to ban single-use plastic items. Many have already banned plastic bags. Are there bans in place where you live?

▲ Some single use plastic items.

2 Burn ... with care

Plastics contain a great deal of chemical energy. So they can provide a great deal of heat when they burn.

As you saw on page 231, burning plastics outdoors is harmful. But they can be burned in an incinerator, where toxic gases can be trapped. The heat produced can be used to heat homes, or to make steam to generate electricity. The waste gases must be monitored continually.

The number of waste incinerators around the world is growing. They trap toxic gases – but also produce **carbon dioxide**, linked to climate change.

Another solution is **pyrolysis,** where waste plastic is melted in almost no oxygen. The polymers break down, and can be turned into new products.

▲ Which way to go? In this modern incinerator, plastic is burned with other household waste, giving heat, ash, and gases ...

▲ ... while here, these plastic bottles made from PET will be melted down, and the PET will be reused.

POLYMERS

3 Make plastic easier to degrade
Plastics are being designed to degrade more easily, for disposable items. Their polymer chains contain links that are more easily attacked by microbes, water, oxygen, and sunlight.

Plastics that can be broken down by microbes are called **biodegradable**. The microbes produce carbon dioxide and water – or methane, if they are buried deep in landfill, without oxygen.

But if plastics break down only to microparticles, this is still a problem.

▲ A biodegradable plastic bag: it will break down along with the vegetable peelings and scrap paper inside.

4 Recycle
Some plastics are easy to **recycle**. The polyester PET is an example.

Bottles and other items made from PET are cleaned and shredded. The fragments are melted and filtered. The material is cooled into pellets, and dried. The pellets are then turned into other PET items. For example, back into bottles, or to make fibres for polyester carpets and clothing.

5 And finally ... reuse!
We can refill plastic water bottles, rather than buy new ones. More shops are allowing customers to bring in empty containers to refill – for example, empty shampoo bottles for a refill of shampoo.

▲ These labels are used on plastic items to identify the plastic.
1 = PET; 2 = high density poly(ethene); 3 = PVC; 4 = low density poly(ethene); 5 = polypropene; 6 = polystyrene; 7 = all other plastics.
PET is often labelled PETE.

What about the ocean?
We can all help to take care of the ocean, even if it is very far away. How?

Do not drop plastic litter (or any litter). Avoid single-use plastic items. Reuse plastic containers. If there is a recycling scheme in your area, use it. If not, dispose of waste with care.

All these will help to keep plastic out of the ocean. But what about the billions of plastic items in the ocean already? Some experts say we should collect them up. Some teams are working on this mammoth task already.

Bioplastics – the future?
- The plastics you met in this chapter are based on petroleum.
- But plastics are also being made from plant material. They are called **bioplastics**.
- An example is **PLA** (polylactic acid) which is made from maize, sugar cane, and oth bacteria. It is used for things like food containers, and plastic plates and cutlery.
- Not all bioplastics biodegrade easily. Some take a long time to break down.
- Scientists are working on bioplastics that will be better for the environment.

1. Give two examples of a single use plastic item.
2. Explain how banning single use plastic items helps the environment.
3. Waste gases from incinerators must be continually monitored. Why?
4. A plastic may be *biodegradable* and *photodegradable*. Explain the difference. (Glossary?)
5. Name the plastic, if the symbol on the item has this number: a 1   b 2   c 4
6. Explain how recyling PET bottles helps the environment.
7. Suggest reasons why *reusing* a bottle made of PET is better for the environment than *recycling* it.
8. Of those five ways to reduce the volume of plastic waste, which *two* do you think will have most impact? Explain.

17.7 Proteins

Objectives: define *amino acid*; describe how proteins are formed by polymerisation; draw the structure of a protein; explain why proteins are called *natural polyamides*

Proteins

Proteins are natural polymers, built up from molecules of **amino acids**.

All amino acids contain carbon, hydrogen, oxygen, and nitrogen. (Some contain other elements too.) Twenty-two different amino acids are found in the proteins in living things. We can represent them like this:

They are called amino acids because they have a **carboxyl functional group**, **COOH**, and an **amine functional group, NH$_2$**. Note that both are attached to the same carbon atom. Now look at **R**. It represents the **side chain** of the amino acid, which is different in each amino acid.

How amino acids join to make proteins

This reaction occurs in nature, including in your body. Only the functional groups take part. So we can draw the rest of the monomer as a block:

This shows four different amino acids joining:

... and another monomer can join here ... and so on

another monomer can join here ...

water molecules eliminated

water forms

the **amide linkage** as in nylon

So the carbon atom in the carboxyl group of one amino acid bonds to the nitrogen atom in the amine group of another. The reaction continues at the other ends of the monomers. Water molecules are eliminated.

From 60 to 6000 monomers can join like this, in any combination of amino acids, and in any order, to form a protein molecule. The result is an enormous number of proteins!

Note how similar this condensation polymerisation is to the one that gives synthetic polyamides (page 226). Proteins and polyamides both contain the **amide linkage**. So proteins are called **natural polyamides**.

▲ Three of the 22 amino acids found in proteins. Note how their side chains differ.

▲ Protein chains can be from a few hundred to over 500 000 atoms long. They fold in a specific way, to enable the protein to do a specific job.

▲ Imagine a very large number of beads, in 22 colours. Each colour represents a different amino acid. You can string the beads together in any order, in chains of any length, to make different proteins.

POLYMERS

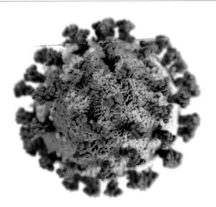

▲ A model of the coronavirus that caused the COVID-19 global pandemic, which began in 2019. (Hence the name.) It consists of 29 proteins. The spikes on its surface let it hack into body cells.

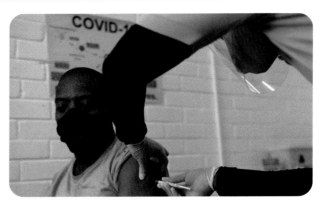

▲ This vaccine instructs body cells to produce the proteins in the corona spikes. The body then develops immunity to them. So if the COVID-19 virus comes along, it will be rejected.

The proteins in your body

There are more than 100 000 different proteins in your body – and countless trillions of protein molecules. So you contain a staggering number of polyamide linkages!

Apart from water and fat, your body is almost all protein. Your bones are mainly the soft protein collagen, with calcium phosphate for hardness. Your hair is mainly keratin. Your enzymes and hormones are proteins too.

In fact every living cell makes proteins. A polymer called DNA in each cell holds the instructions for making proteins. Another polymer, RNA, makes sure that the right amino acids join in the right order.

▲ Eggs are rich in protein. When you boil an egg the proteins are **denatured**: the chains lose their folded shape and clump together. So the egg goes hard.

How your body makes proteins from food
- First, you obtain proteins by eating foods that contain them. Eggs, for example.
- Next, your body breaks these proteins down to their amino acids. This takes place in your digestive system. Enzymes called **proteases** act as catalysts.
- Then the amino acids move to the liver. Here, some are polymerised to form proteins your body needs. The rest are carried by your blood from the liver to your body cells, where more proteins are made.
- Your body can in fact make many of the amino acid monomers itself. But there are nine **essential amino acids** it cannot make, so you must get these from food. A meal of rice and beans provides all nine!

1 Amino acids are represented by this general formula: $NH_2CH(R)COOH$.
  a Show their general structure as a displayed formula.
  b What does R represent?
  c In the amino acid molecule, which part is:
    i the amine group?   ii the carboxyl group?
  d Name the products that form when amino acids undergo condensation polymerisation.

2 a Draw the structure of part of a protein molecule, and draw a loop around one amide linkage in it.
  b If five amino acids join, how many water molecules will be eliminated?
  c It is not possible to draw a repeat unit for a protein molecule. Explain why.

3 a What is *hydrolysis*? (See page 227.)
  b Suggest what will form when a protein is hydrolysed.

Checkup on Chapter 17

Revision checklist

Core syllabus content
Make sure you can …
- [ ] explain these terms:
  *monomer   polymer   polymerisation
  natural polymer   synthetic polymer*
- [ ] draw a labelled diagram to show the addition polymerisation of ethene to make polyethene
- [ ] explain that plastics are synthetic polymers
- [ ] give at least five general properties of plastics
- [ ] give names and uses for at least two plastics
- [ ] explain why plastics are difficult to dispose of
- [ ] describe the environmental problems caused by plastic waste: *in landfill sites
  in the oceans   when they are burned*
- [ ] outline four ways in which we can reduce the amount of plastic waste in the environment

Extended syllabus content
- [ ] describe *addition polymerisation*, and
  – say what the key feature of the monomer is
  – draw the structure of a polymer molecule, when you are given the monomer
  – identify the monomer from the structure of the polymer
- [ ] describe *condensation polymerisation*, and
  – say what the key features of the monomers are
  – state the differences between condensation and addition polymerisation
- [ ] draw simple diagrams to show the monomers, and part of the polymer molecule, for:
  a polyamide   a polyester
  using blocks to represent carbon chains
- [ ] identify the *amide* and *ester* linkages
- [ ] state that nylon is a polyamide, and draw its structure and repeat unit
- [ ] state that PET is a polyester, and draw its structure and repeat unit
- [ ] state that PET can be converted back to its monomers, and repolymerised
- [ ] draw the structure of an amino acid, using *R* to represent the side chain
- [ ] show how a protein is formed from amino acids
- [ ] describe proteins as natural polyamides
- [ ] describe and draw the structure of a protein, and identify the amide linkage

Questions

Core syllabus content

1 In organic chemistry, many small units may join to form a large molecule. This represents a small unit:

a What are the small units called?
b What general name is given to a large molecule obtained when the small units join?
c This type of reaction is called p.......... ?
d Draw a diagram to represent the large molecule.

2 There are many different polymers, and they offer a wide range of properties to suit different uses.
a What is a *polymer*?
b Which of these are polymers?
   esters   plastics   petroleum   carboxylic acids
c Which of these could be used as monomers for making polymers?
   i   ethene, $CH_2=CH_2$
   ii ethanol, C_2H_5OH
   iii propane, C_3H_8
   iv styrene, $C_6H_5CH=CH_2$
   v   chloropropene, $CH_3CH=CHCl$
d What do all the compounds you chose in **c** have in common?
e Suggest a name for the polymer that forms, for each compound you chose in **c**.
f Name the type of chemical reaction in which polymers are formed.

3 This equation shows how two units join, in a typical *addition polymerisation*:

$$\begin{array}{c}H\ H\\|\ \ |\\C=C\\|\ \ |\\H\ H\end{array} + \begin{array}{c}H\ H\\|\ \ |\\C=C\\|\ \ |\\H\ H\end{array} \longrightarrow \begin{array}{c}H\ H\ H\ H\\|\ \ |\ \ |\ \ |\\-C-C-C-C-\\|\ \ |\ \ |\ \ |\\H\ H\ H\ H\end{array}$$

a i   Name the monomer in this reaction.
   ii Name the polymer that forms.
b Explain why this term is used for the reaction:
   i   addition   ii polymerisation
c The polymer that forms is used to make a wide range of items, including plastic bottles and shopping bags. Describe the environmental problems that arise when these items:
   i   are sent to land fill sites   ii are burned

236

Extended syllabus content

4 This diagram represents two units of an addition polymer called **poly(acrylamide)**:

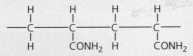

a Copy and complete this diagram to show the polymer in its short form:

$$\left(\right)_n$$

b Write the structural formula of the monomer.
c Suggest a name for the monomer.
d Is the monomer *saturated*, or *unsaturated*?

5 The polymer *Teflon* is made from this monomer, tetrafluoroethene:

$$\begin{array}{c} F \\ \diagdown \\ C=C \\ \diagup\diagdown \\ FF \end{array}$$

a Which feature of the monomer makes polymerisation possible?
b Which type of polymerisation occurs?
c Draw three units in the structure of the polymer molecule that forms.
d Give the chemical name for Teflon.

6 These diagrams represent the two groups of condensation polymers. We have labelled them X and Y.

group X
group Y

a In which ways is a condensation polymer:
   i similar to an addition polymer?
   ii different from an addition polymer?
b Which of the two groups of polymers are:
   i polyamides?   ii polyesters?
c Which group does it belong to, X or Y?
   i PET   ii nylon
d One condensation polymer can be converted back to its monomers. Then it can be re-polymerised.
   i Name this polymer.
   ii Suggest one advantage of being able to convert a polymer to its monomers, then re-polymerise.

7 One very strong polymer has this structure:

A hexagon with a circle in the middle represents a ring of six carbon atoms, with three C=C bonds.

a i Which type of polymerisation produced this polymer?
  ii Name the other product that formed.
b i Draw the formulae for the two monomers that were used to make this polymer.
  ii Add these labels to your drawings for **i**: *a diol*   *a carboxylic acid*
c i Name the linkage that joins the monomers.
  ii Show the atoms and bonds in this linkage.
d Which of these has a similar structure to the polymer shown above?
   PET   nylon   poly(phenylethere)
e The polymer above can be re-polymerised. Suggest how this could be carried out.

8 Here are three polymers in short form:

   A   B   C

a Which polymer is: i poly(ethene)?
  ii poly(ethanol)?   iii poly(propene)?
b Which type of polymerisation formed them?
c Show the displayed formulae for the monomers for A, B, and C.
d Draw two units for each polymer A, B, and C.

9 a Which important natural polymers contain the amide linkage?
b Their monomers can be shown as:

   i What is this type of compound called?
   ii What does R represent?
   iii What name is given to the COOH functional group in the monomer?
c The monomers join in a condensation reaction. Copy the diagram in **b**. On your diagram draw rings around the atoms or groups of atoms that are removed when the polymer forms.

18.1 Making a substance in the lab

Objectives: explain why separation methods are needed, when making substances in the lab; state that they increase the purity of the products; identify apparatus

Practical work involves ...

Practical work involves **materials**, **apparatus**, and **methods**. (Or **mam**!) This is true whether you are a carpenter, a brain surgeon, or an astronaut. It also applies to your work in the lab!

So let's outline a typical task, and look at all three.

The task: make crystals of copper(II) sulfate

The materials

Suitable **reactants**: copper(II) oxide and dilute sulfuric acid.

The **product**: hydrated crystals of copper(II) sulfate.

▲ Repairing the International Space Station. It needs materials, apparatus and methods – just like your work in the lab.

The apparatus and methods

How do you get from those reactants to that product? Here are the steps.

1 Add copper(II) oxide to **warm** sulfuric acid, for a faster reaction.

2 Filter the excess copper(II) oxide from the solution.
3 Remove most of the

water from the solution by **evaporation**.

4 Cool the solution and allow **crystallisation** to take place.

5 Filter the crystals from the solution.

6 Pat with filter paper to **dry** them. Success! Your task is complete.

238

SEPARATION AND PURIFICATION

Apparatus
This unit shows some items of apparatus. You will meet others too.
- Each item is designed for a particular job.
- Most of the items you will use are made of **glass**. Glass is not generally attacked by chemicals. It is easy to clean. It is relatively cheap. You can see through it too, which helps a lot!
- Most of the glass items come in different sizes. For example, **beakers**, **flasks**, **test-tubes**, and **measuring cylinders**. You choose a size to suit the amount of material you will use.
- Some items allow more accuracy than others. For example, to measure out 25 cm^3 of a liquid accurately, you would use a **volumetric pipette** rather than a **measuring cylinder**.

You need to know what each item of apparatus is for, and be able to identify it on diagrams. The diagrams of apparatus in this book will help.

Methods
Look at step **1** in the task. This is where countless trillions of invisible particles react to make your product. How exciting!

The other steps have one overall aim: to separate the product from unwanted substances, and present it as a solid, as pure as you can make it.

So let's look at steps **2 – 5** more closely.

- In step **2** you separate the unreacted copper(II) oxide by filtering. It is an **impurity**, and would ruin your product. (A pure product contains only the particles shown in its formula.)
- In step **3** you carry out an evaporation in order to make a **hot saturated solution**. You will learn why in Unit 18.2.
- Step **4** is where you see your crystals form. Amazing! Solid copper(II) sulfate is separating from the solution all by itself.
- In Step **5** you separate the crystals from the remaining copper(II) sulfate solution.
- In Step **6** you remove solution from the surface of the crystals.

So ... separation methods are an essential part of obtaining a product. Unwanted chemicals are removed, and the purity of the product increases. You will learn more about separation methods, and purity, in the rest of this chapter.

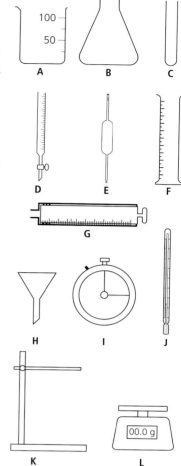

▲ Apparatus you will meet in the lab.

Q

1 Look at the steps in preparing copper(II) sulfate. The reaction takes place in step **1**. It is a *neutralisation*.
  a Write the word equation for the reaction.
  b Explain why *neither* of the products is an impurity.

2 In which of steps **2 – 6** is something removed? For each step you choose, state what is removed.

3 For step **4**, the copper(II) sulfate solution must be *concentrated*. How is this achieved?

4 How are the crystals dried?

5 Look at the items of apparatus drawn above. Name these:
  a A b B c C d H e K f G

6 25 cm^3 of sodium hydroxide solution is needed for a titration. Which will you choose to measure it out? Why?
  A, beaker   E, volumetric pipette
  F, measuring cylinder

7 Which item lets you add a liquid drop by drop, and measure the volume you used? Give its letter and name.

8 Suggest an experiment where you would use this item:
  a L   b K   c I   d J   e G

18.2 Solutions and solubility

Objectives: define *solution* and *saturated solution*; state that solubility is different for different substances, and generally increases with temperature

What is a solution?
When you mix sugar with water, the sugar seems to disappear.
This is because its molecules spread out among the water molecules:

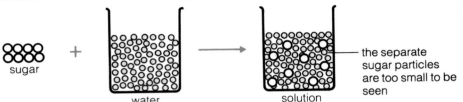

▲ A mixture of sugar and water forms a solution. No particles can be seen. Sugar is the solute. The solvent is water, so this is an aqueous solution.

The sugar has **dissolved** in the water, giving a mixture called a **solution**.
Sugar is the **solute**, and water is the **solvent**.

A solvent is a substance that dissolves a solute.
A solute is a substance that is dissolved in a solvent.
A solution is a mixture of one or more solutes dissolved in a solvent.
When water is the solvent, the solution is called an *aqueous* solution.

The only way to get the sugar back again is to evaporate the water away.

Some substances are insoluble in water
When you mix powdered chalk with water, it does not dissolve. You can separate it from the water again by filtering.

Why is solubility so different for sugar and chalk? Because their particles are different! Solubility is different for every substance. Look:

Compound	Mass (g) dissolving in 100 g of water at 25 °C
silver nitrate	241.3
calcium nitrate	102.1
sugar (glucose)	91.0
potassium nitrate	23.1
potassium sulfate	12.0
calcium hydroxide	0.113
calcium carbonate (chalk)	0.0013
copper(II) oxide	less than 0.00000001

decreasing solubility

▲ A mixture of chalk powder and water. This is *not* a solution. The chalk particles stay in clusters big enough to be seen. Over time, they sink to the bottom. (Find chalk in the table on the left.)

Silver nitrate is much more soluble than sugar in water – but potassium nitrate is a lot less soluble. Look at calcium hydroxide. It is only slightly or **sparingly** soluble. Its solution is called **limewater**.

The last two compounds in the table are described as **insoluble** because so little dissolves. Look at the value for copper(II) oxide.

The solubility of a substance in water is the amount that dissolves in 100 g of water at that temperature. It varies greatly.

Is there a pattern?
- The solubility of every substance in water is different.
- But there are some overall patterns. For example, *all* sodium compounds are soluble in water.
- Find out more on page 140.

SEPARATION AND PURIFICATION

Making a solution

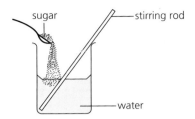
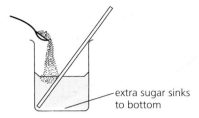
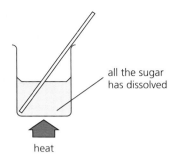

Sugar dissolves quite slowly in water at room temperature. If you stir the liquid, that helps. But if you keep on adding sugar …

… eventually no more of it will dissolve, no matter how hard you stir. The extra sinks to the bottom. The solution is now **saturated**.

What if you heat the solution? The extra sugar dissolves. Add more and it will dissolve too, until the solution becomes saturated again.

So sugar is **more soluble** in hot water than in cold water. **Solubility usually increases with temperature.**

A solution is called *saturated* when it can dissolve no more solute at that temperature.

Solubility on a graph

Look at the graph on the right. It shows how the solubility of copper(II) sulfate changes with temperature.

105 g of the salt will dissolve in 100 g of water at 90 °C, giving a saturated solution. (See **A**.) But only 34 g will dissolve at 25 °C. (See **B**.)

So if you cool the saturated solution from 90 °C to 25 °C, 71 g of copper(II) sulfate will come out of solution. (105 g – 34 g.) It will appear as **crystals**.

This explains why **crystallisation** works.

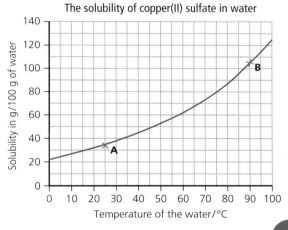

Other solvents

Solubility depends on the solvent too. Lots of substances are insoluble in water, but soluble in the organic solvent **ethanol**, for example. Ethanol evaporates easily, so it is used as the solvent in perfumes and aftershaves. Nail polish is a polymer, insoluble in water. But it does dissolve in the organic solvent **propanone** (acetone). So propanone is used in nail polish remover.

1 Explain each term:
   a soluble   b insoluble
   c aqueous solution d saturated solution
2 The table on page 240 shows solubility in water at 25 °C.
   a Which substance in the table is the most soluble?
   b About how many times more soluble is this substance than potassium sulfate, at 25 °C?
3 Calcium carbonate is called *insoluble*. Why?
4 Why is temperature given, for the table on page 240?
5 Look at the graph above. About how much copper(II) sulfate is contained in 100 grams of a saturated solution:
   a at 80 °C?   b at 40 °C?
6 You have a saturated solution of copper(II) sulfate at 80 °C. What will you see, if you cool it to room temperature?

18.3 Separating a solid from a liquid

Objectives: describe *filtration*, *crystallisation* and *evaporation* as ways to separate a solid from a liquid; choose a suitable method, given the solid and liquid

How to separate a solid from a liquid
Which method to use? It depends on whether the solid is in solution – and if so, how its solubility changes with temperature.

1 Filtering, to remove a solid
In the diagram below, chalk is being **filtered** from a mixture of chalk and water. The trapped chalk is called the **residue**. The water is the **filtrate**.

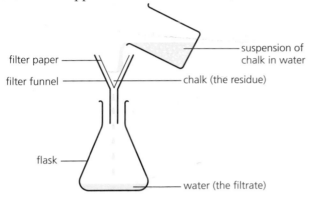

▲ Filtering in the kitchen.

> **The pores do it!**
> - Chalk particles cannot pass through the pores in filter paper.
> - Scientists have now developed filters with very tiny pores.
> - For example **nanofilters**, which can trap some ions – and even viruses.

2 Crystallisation, to obtain a solid from its solution
In Unit 18.1 you saw how **crystallisation** was used to separate solid copper(II) sulfate from its solution. It works because copper(II) sulfate is *less soluble at lower temperatures*. See the graph on page 241.

Let's look at the method in more detail.

 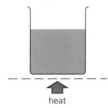 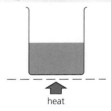

1 This is a solution of copper(II) sulfate in water. You want to obtain solid copper(II) sulfate from it.

2 So you heat the solution to evaporate water. It becomes more and more concentrated.

3 Eventually the solution becomes **saturated**. If you cool it now, crystals will start to form.

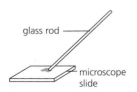 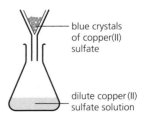

4 Check that it is ready by placing a drop on a microscope slide. Crystals will form quickly on the cool glass.

5 Leave the solution to cool. Crystals start growing as the temperature falls.

6 Remove the crystals from the solution by filtering. Then dry them with filter paper.

SEPARATION AND PURIFICATION

◀ Making a living from crystallisation. Seawater is led into shallow ponds. The water evaporates in the sun. He collects the sea salt, and sells it. (It may have sand in it. How could you remove the sand? There is a hint below!)

3 Evaporation, to remove *all* the solvent

The solubility of some substances changes very little as the temperature falls. So crystallisation will not work for them. Salt is an example.

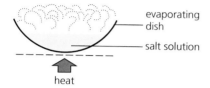

To obtain salt from an aqueous solution, you need to keep heating the solution, to evaporate the water.

When there is only a little water left, the salt will start to appear. Heat carefully until it is dry.

▲ Evaporating the water from a solution of salt in water.

Taking it further: separating two solids

You can use the ideas you met already to separate **two solids**. The key is to find a solvent that will dissolve *only one of them*.

For example, sugar is soluble in ethanol, but salt is not. So you can separate a mixture of sugar and salt like this:

1. Add ethanol to the mixture, and stir. The sugar dissolves. (Use only as much ethanol as you need.)
2. Filter the mixture. The salt is trapped in the filter paper, but the sugar solution passes through.
3. Rinse the salt with water, and dry it in an oven.
4. Separate the sugar from the solution by evaporating the ethanol. But **take care**: ethanol is flammable. When heating, use a water bath!

▲ Evaporating ethanol from a solution over a water bath (in this case a beaker of hot water) to protect it from the flame.

Q

1. Copper(II) oxide is insoluble in water. You are asked to separate it from water by filtering. Name:
   a. two items you need for filtering (one is disposable!)
   b. the residue you obtain   c. the filtrate
2. You are asked to separate sugar from its solution in water.
   a. Explain why filtering with filter paper will not work.
   b. Name a suitable separation method to use in the lab.
3. 23 g of potassium nitrate dissolves in 100 g of water at 25 °C. 169 g will dissolve at 80 °C.
   Outline the steps to obtain crystals of potassium nitrate from its aqueous solution. Your answer can be brief bullet points.
4. 36 g of sodium chloride dissolves in 100 g of water at 20 °C. 38 g dissolves at 80 °C. Crystallisation is *not* used to separate it from water. Explain why, and suggest another method.

18.4 Separating by distillation

Objectives: describe and explain how *simple distillation* and *fractional distillation* work; give examples of fractional distillation in industry

Distillation
Distillation is a way to separate substances from a liquid mixture, by boiling and then condensing. Here we look at the two different types.

Simple distillation
Simple distillation is used to separate the solvent from a solution, leaving the solute behind. You could use it to obtain purified water from salt water, for example. This is the apparatus:

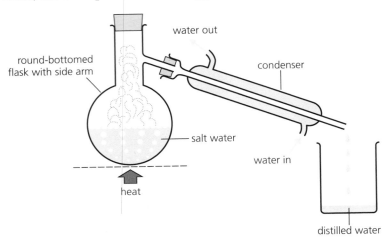

▲ Before chemists there were **alchemists**. Alchemy developed in Egypt, India, and China, before reaching Europe in the 8th century. Alchemists gave us many lab techniques, including distillation.

These are the steps:
1 Heat the solution in the flask. As it boils, water vapour rises into the condenser, leaving salt behind.
2 The condenser is cold, so the water vapour condenses to water in it.
3 The water drips into the beaker. It is called **distilled water**.

   A **residue** of sodium chloride and other salts will remain in the flask.

Is it pure?
- Distilled water is not 100% pure.
- Salts have been left behind.
- But it may contain other solvents from the original water, and impurities from the apparatus, as well as gases from the air.

▲ Simple distillation has been used for centuries in making perfume. These rose petals will be heated. Their scented oils will turn into gases, and condense in a pipe connected to the vat.

▲ Some **desalination plants** use distillation to turn seawater into drinkable water. (There are other ways to remove salt too.) The sea provides the water supply in many dry coastal places.

Fractional distillation

This is used to separate liquids from each other. **It makes use of their different boiling points.** You could use it to separate ethanol and water, for example.

The apparatus is shown on the right. It is like the apparatus for simple distillation, but with a glass column added, filled with glass beads. Note the thermometer too.

These are the steps:

1 Heat the mixture in the flask. At about 78 °C, the ethanol begins to boil. Some water evaporates too. So a mixture of ethanol and water vapours rises up the column.
2 The vapours condense on the glass beads. The beads heat up.
3 When the beads reach about 78 °C, ethanol vapour will no longer condense on them. Only the water vapour does. The water drips back down into the flask.
4 The ethanol vapour passes into the condenser, and condenses. Pure liquid ethanol drips into the beaker.
5 Eventually, the thermometer reading rises above 78 °C. This is a sign that all the ethanol has gone. So you can stop heating.

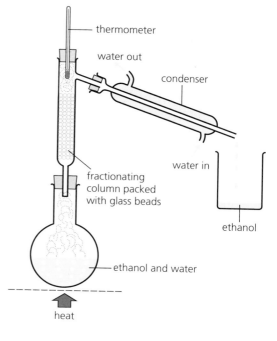

Fractional distillation on a large scale

Fractional distillation is carried out on a large scale in industry:

- to **refine** petroleum into different groups of compounds, or **fractions**. The petroleum is heated. The vapours rise up a tall steel fractionating column, and condense on trays at different heights. See page 206.
- in producing **ethanol**. Some ethanol is made by fermenting sugar cane and other plant material. At the end, the solids are filtered off. Pure ethanol is then obtained by fractional distillation of the filtrate.
- to separate nitrogen, oxygen and the noble gases from air. Air is cooled to −200 °C. By then it is liquid, except for neon and helium. These are removed. Then the liquid is warmed up slowly. The other gases boil off one by one, and are collected.

▲ The fractionating tower at a factory which makes ethanol by fermentation of cassava.

1 How would you obtain pure water from seawater? Describe the apparatus, and explain how the method works.
2 Why are *condensers* called that? What is the cold water for?
3 The apparatus you described in **1** would not be used to separate ethanol and water. Give a reason.
4 Look at the apparatus for fractional distillation, above.
  a Explain the purpose of the glass beads.
  b Explain why a thermometer is essential.
5 a Where does ethanol separate, in the photo above?
  b Give one other use for fractional distillation in industry.

18.5 Paper chromatography (part I)

Objectives: explain *why* paper chromatography works; state that it can be used to separate, purify, and identify substances; describe how to carry it out

Paper chromatography
Paper chromatography is a simple and very useful technique. It is used:
- to separate substances from a mixture
- to check whether a substance is pure
- to identify substances in a mixture.

Let's look at each of these in turn.

1 Separating substances in a mixture
We use **chromatography paper**, which is like filter paper.

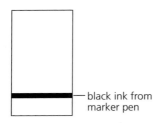
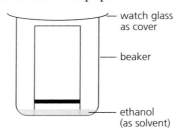
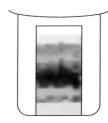

A thick line is drawn on a piece of chromatography paper, using a black marker pen.

The paper is stood in a beaker with a little ethanol. The ink line must be *above* the level of the ethanol.

The ethanol rises up the paper, dissolving the ink on the way. Bands of different colours appear.

So the coloured substances that make up the 'black' ink have separated.

How does it work?
To separate substances using paper chromatography, you need a solvent which can dissolve them all. Ethanol is a good choice here.

- The ethanol moves up the paper by capillary action. (A liquid will creep up a paper tissue or a cloth in the same way.)
- When it reaches the ink line, the ethanol dissolves the substances in the ink, and they travel with it.
- Even though they are all dissolved, *each substance travels at a different speed*. It depends on their attraction to the ethanol *and* the paper. The substance that is the most soluble in ethanol will travel fastest.
- The result is that the substances separate. The paper showing the separated substances is called a **chromatogram**.

Collecting the separate substances
You could cut a strip of each colour from the dry chromatogram. Dip it in a little solvent to dissolve the substance. Then evaporate the solvent. So now the mixture is fully separated.

You will end up with very tiny samples! You could then analyse each one.

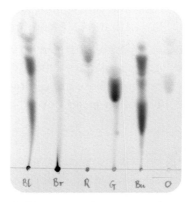

▲ Here paper chromatography was carried out on ink from six marker pens. The ink colours were: Bl, black; Br, brown; R, red; G, green; Bu, blue; O, orange. Was any ink just a single colour?

SEPARATION AND PURIFICATION

2 Checking whether a substance is pure

To see if substance **X** is pure, you first prepare a concentrated solution of **X**. Then carry out paper chromatography as below.

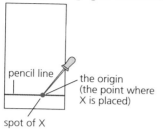 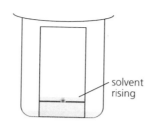 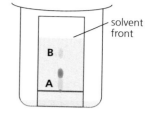

Draw a pencil line on the paper. Put a drop of the solution of **X** on the line, and let it dry. Its position on the line is called the **origin**.

Stand the paper in a little solvent as before, with the solvent level below the pencil line. The solvent rises, and **X** begins to dissolve.

After a time, **X** has separated into two blobs, A and B. So **X** is *not* pure. The smaller one, B, shows the presence of an impurity.

If there was only one spot, it would show that **X** was pure.

Note that the line is drawn on the paper in pencil, not pen, because the ink in the pen might dissolve.

3 Identifying substances in a mixture

You have an unknown mixture of substances. Let's call it **Y**. You think it contains substances **A**, **B**, **C**, and **D**, but you are not certain.

You can use paper chromatography to find out. *You will compare Y with the known substances.* Let's assume they are all soluble in propanone.

> **Neat ...**
> - Paper chromatography usually gives trails of colour.
> - But in diagrams, the colours are sometimes shown as neat spots, as below!.

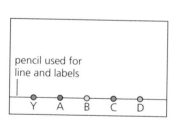 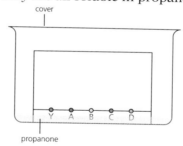 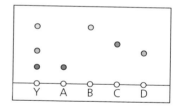

Make concentrated solutions of **Y**, **A**, **B**, **C**, and **D**, in propanone. Put a spot of each along a pencil line on chromatography paper. Label them.

Stand the paper in a little propanone in a covered container. The solvent rises up the paper. When it is near the top, remove the paper.

Y has separated into three spots. Two are at the same height as **A** and **B**, so **Y** must contain **A** and **B**. Does it also contain **C** and **D**?

This method could be used to identify the coloured substances in a leaf, or in a coloured sweet such as a fruit gum or jelly baby.

Q

1 Look at the experiment at the top of page 246.
   a Why must the ink line be *above* the ethanol level?
   b Why do the colours separate?
   c Which is more soluble in ethanol, the blue substance or the yellow substance? Explain your choice.
   d Suggest a reason why ethanol is used instead of water.

2 What is the *origin*, in paper chomatography?

3 Look at the chromatogram at the bottom of page 246. Which of the six inks share the substance that is:
   a yellow?   b red?   c purple?

4 In the last chromatogram above, does **Y** (the mixture) contain **C** and **D**? Give your evidence.

18.6 Paper chromatography (part II)

Objectives: describe how to use paper chromatography to separate and identify colourless substances; define R_f; explain why R_f values make identification possible

Identifying colourless substances
In the last unit, paper chromatography was used for *coloured* substances. Now you will learn how to use it for *colourless* substances! You will see how to make the substances visible, and how to identify them.

First, the theory ...
As you saw, paper chromatography depends on each substance in a mixture interacting differently with the paper and solvent.

1 These dots represent two different substances dissolved in a solvent. Imagine that both are colourless. (The solvent is not shown in the drawing.)

2 The substances travel along the paper at different speeds, because of their different solubilities in the solvent, and attraction to the paper. So they start to separate.

3 They are now separated. They are sprayed with a chemical to make them visible. The distance they have travelled, compared to the solvent, is used to identify them.

And now the procedure ...
Test-tubes **A–E** on the right contain five colourless solutions of **amino acids** in water.

The solution in **A** contains several amino acids. They have been obtained by the hydrolysis of a protein.

The other solutions contain just one amino acid each.

Your task is to identify *all* the amino acids in **A–E**.

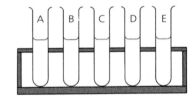

▲ The five mystery solutions.

1 Place a spot of each solution along a pencil line (the **base line**) on slotted chromatography paper. See below. (The slots keep the samples separate.) Label each spot in pencil at the *top* of the paper:

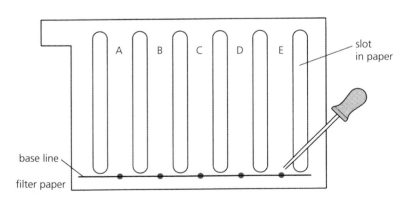

248

SEPARATION AND PURIFICATION

2 Place a suitable solvent in the bottom of a beaker. (For amino acids, a mixture of butanol, ethanoic acid and water is suitable.)
3 Roll the chromatography paper into a cylinder and place it in the beaker. Cover the beaker.
4 The solvent rises up the paper. When it has almost reached the top, remove the paper.
5 Mark a pencil line on it to show where the solvent has reached: the **solvent front**. (You cannot tell yet where the amino acids are, because they are colourless.)
6 Put the paper in an oven to dry out.
7 Next spray it with a **locating agent** to make the amino acids show up. **Ninhydrin** is a good choice here. (Use it in a fume cupboard!) After spraying, warm the paper in an oven for 10 minutes. The spots turn purple. You now have a proper chromatogram.
8 Mark a pencil dot at the centre of each spot. Measure the distance from the origin to the dot, and from the origin to the solvent front.

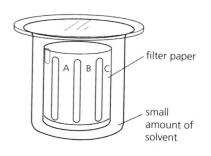

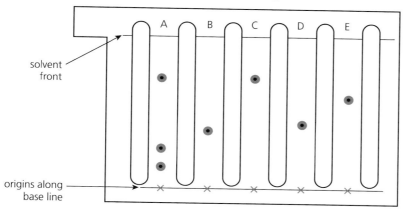

9 Now work out the R_f value for each amino acid. Like this:

$$R_f = \frac{\text{distance from origin to dot}}{\text{distance from origin to solvent front}}$$

(R_f stands for **retention factor.** *Retention* means *holding back*.)

10 Finally, look up R_f tables to **identify** the amino acids. Part of an R_f table for the solvent you used is shown on the right. The method works because **the R_f value of a compound is always the same for a given solvent, under the same conditions**.

R_f values for amino acids (for a mixture of butanol / ethanoic acid / water as solvent)

amino acid	R_f value
cysteine	0.08
lysine	0.14
glycine	0.26
serine	0.27
alanine	0.38
proline	0.43
valine	0.60
leucine	0.73

1 a What is the *solvent front*, in paper chromatography?
   b The chromatography paper must be removed before the solvent reaches the top of it. Why?
2 a Define *locating agent*.
   b Why is a locating agent needed, in an experiment to separate amino acids by chromatography?
3 a Define R_f.   b What makes R_f values so useful?
4 For the chromatogram above:
   a Which of the amino acids in **B–E** were also present in **A**?
   b Using a ruler, work out the R_f values for the amino acids in **A–E**.
   c Now use the R_f table above to name them.

249

18.7 Checking purity

Objectives: explain how melting and boiling points are used to assess purity, and to identify pure substances; recall that paper chromatography is used to check purity

What is a pure substance?

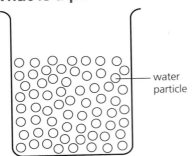

This is a beaker of water. The water contains only water particles, nothing else. So it is 100% **pure**.

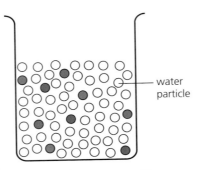

This water is not pure. It contains particles of other substances. These other substances are **impurities**.

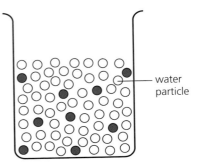

This water is not pure either. The impurities in it are harmful. They could make you ill.

How can you tell if something is pure?

There are two ways to check the **purity** of a substance:

- find the melting and boiling points of the substance
- carry out paper chromatography.

Using melting and boiling points to check purity

Melting and boiling points will tell you whether a substance is pure.

- A pure substance has a definite, sharp, melting point and boiling point. They are different for each substance. You can look them up in tables.
- When a substance contains an impurity:
  - its melting point falls, and its boiling point rises
  - it melts and boils over a range of temperatures, not sharply.
- The more impurity there is in a substance:
  - the bigger the change in melting and boiling points
  - the wider the temperature range over which it changes state.

▲ Purity might not matter too much for substances you make in the school lab. But it is critical for products like vaccines.

For example:

substance	melting point /°C	boiling point /°C
sulfur	119	0
water	445	100

These are the melting and boiling points for two pure substances, sulfur and water. You could check them in data tables.

The sulfur in this sample melts sharply at 119 °C and boils at 445 °C. That tells you it is pure.

The water in this beaker freezes around −0.5 °C and boils around 101 °C. So it is not pure.

SEPARATION AND PURIFICATION

Why do melting and boiling points change?
As you know, the particles in a solid are held in a lattice. To melt the solid, heat is needed to break up the lattice.

When there are particles of an impurity in the lattice, it is easier to break up. So the solid melts at a lower temperature – and less sharply. The more impurity, the bigger the effect.

In the liquid, the impurity makes it *harder* for particles to leave the liquid and form a gas. So the boiling point is higher – and less sharp.

Compare the two heating curves in this graph.

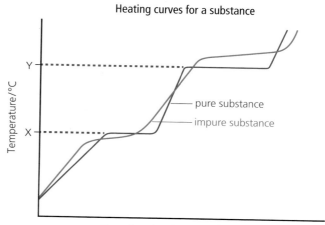

Heating curves for a substance

Using paper chromatography to check purity
Paper chromatography will separate different substances in a mixture. So it follows that it can be used to check purity, as page 247 showed. Chromatograms A and B on the right are for two samples of the same substance. Which sample contains an impurity?

How can purity be increased?
The product of a chemical reaction will have particles of other things mixed with it, even if only in tiny amounts. For example, particles of a reactant that have not reacted. This is an impurity.

Increasing the purity means separating more unwanted substances.

- Rinsing a solid product with distilled water will wash away some impurities.
- Repeating a separation technique will help to remove impurities. You could dissolve crystals in a pure solvent, and then recrystallise.

Purity is critical for many products, such as vaccines and food flavourings. So they go through several stages of purification in factories. Methods may include filtering through ultrafine membranes, repeat crystallisation, and chromatography (but using materials other than paper). Each method helps to keep us safe.

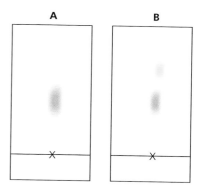

▲ Chromatograms for two samples of the same substance. Which one contains an impurity?

▲ Synthetic additives are used in sweets and other foods, to add colour and flavour. They require a high level of safety and purity. The WHO has set standards.

Melting and boiling points as ID
Each pure substance has a unique pair of melting and boiling points. So these can be used to identify it (like your fingerprints can identify you). Measure the melting and boiling points, and then search for that pair of values in a table of data. This will tell you what the substance is.

1 Define: **a** a pure substance **b** an impurity
2 Pure potassium nitrate melts sharply at 334 °C.
   You have two samples of potassium nitrate, **X** and **Y**.
   X melts in the range 332.5 – 333.5 °C.
   Y melts in the range 331 – 333.5 °C.
   Which sample contains more impurity?
3 Pure ethanol boils at 78 °C. A sample of ethanol has been contaminated. What can you say about its boiling point?
4 Look at the two chromatograms above. Which one shows the presence of an impurity? Explain your choice.
5 You have an unknown compound. It melts at 0 °C and boils at 100 °C, at normal pressure. Identify it!

The chromatography detectives

▲ After a crime, the forensic detectives move in. They collect fingerprints, hairs, fibres of clothing, dried blood spots, and anything else that can be used in evidence.

▲ Back in the forensic lab, the evidence is examined using a range of techniques. Chromatography is one of them.

The key ideas in chromatography.
Much of chromatography is detective work.
You already met paper chromatography. There are many other kinds. But the key ideas are always the same.

- There are two phases:
  - a non-moving or **stationary** phase, such as paper.
  - a moving or **mobile** phase. This carries the mixture of substances you are examining.
- The substances in the mixture separate because each has a different level of attraction to each phase.
- You can then identify the separated substances. Depending on the type of chromatography, you can collect them too.

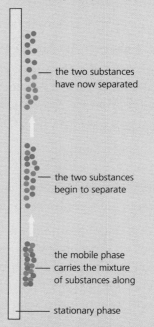

▲ The two phases in chromatography.

Different types of chromatography
The different types of chromatography differ in their phases, and how analysis is carried out. Compare these:

The stationary phase could be …	The mobile phase could be …	To identify the substances, you could …
• paper • a thin coat of an adsorbent substance on a glass plate, or inside a tube • plastic beads packed into a tube	• a liquid solvent or mixture of solvents that can dissolve the substances • an inert (unreactive) gas to carry gaseous substances along; this is called **gas chromatography**	• include known substances for comparison, or work out R_f values, as in paper chromatography • pass the substances through a machine that will analyse them for you

Paper chromatography is the main type used in the school lab. It is also used, along with other types of chromatography, to catch criminals.

SEPARATION AND PURIFICATION

Chromatography in crime detection

Chromatography is widely used in crime detection. For example, it is used to analyse samples of fibre from crime scenes, check people's blood for traces of illegal drugs, and examine clothing for traces of explosives.

This shows how a blood sample could be analysed for traces of illegal drugs, or a poison, using gas chromatography. Follow the numbers:

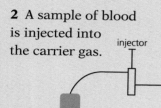
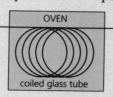

2 A sample of blood is injected into the carrier gas.

3 The mixture goes into a hot oven, where the blood sample forms a vapour.

1 An inert carrier gas, such as helium or nitrogen, is fed into the system.

4 The vapour moves over the stationary phase: an adsorbent substance lining a coiled glass tube.

5 The separated substances pass into a mass spectrometer, where they are analysed.

6 The data is fed into a computer. The police study it. They might make an arrest ...

Other uses

Chromatography is used *on a small scale* in labs, to:

- identify substances, such as the amino acids in a protein
- check the purity of substances
- help in crime detection (as above)
- identify pollutants in air, or in samples of river water.

It is used *on a large scale* in industry, for example to:

- separate pure substances for use as medical drugs, food flavourings, and similar, from tanks of reaction mixtures
- separate single compounds from a group of similar compounds (a fraction) in the petroleum industry.

So chromatography, in all its forms, is a powerful and versatile tool.

▲ Injecting a sample into the carrier gas, at the start of gas chromatography. Will it lead to an arrest?

▲ Whose biro was used for the forged signature? Not looking good for suspect 4!

◀ Collecting water samples, to analyse for pollutants. The factories that dumped them could then be identified – and fined.

Checkup on Chapter 18

Revision checklist

Core syllabus content
Make sure you can …
- [ ] define and use these terms:
  solute   solvent   solution
  aqueous solution   saturated solution
  solubility
- [ ] give at least two examples of solvents
- [ ] state that most solids become more soluble as the temperature of the solvent rises
- [ ] describe these methods for separating mixtures, and sketch and label the apparatus:
  filtration
  crystallisation
  evaporation
  simple distillation
  fractional distillation
  paper chromatography
- [ ] explain why each of the above methods works
- [ ] choose a suitable separation method for a given mixture, and explain your choice
- [ ] identify coloured substances present in a mixture, by comparing spots on a chromatogram
- [ ] define: pure substance   impurity
- [ ] state that removing unwanted substances increases the purity of a product
- [ ] state that these can be used to check purity:
  – melting and boiling points
  – paper chromatography
- [ ] state that melting and boiling points can also be used to identify pure substances
- [ ] identify and name all the items of apparatus shown in the photos and diagrams in this chapter

Extended syllabus content
Make sure you can also …
- [ ] explain what a *locating agent* is
- [ ] describe how to carry out chromatography, to identify colourless substances
- [ ] define R_f
- [ ] identify the substances in a mixture, given a chromatogram and a table of R_f values

Questions

Core syllabus content

1 This question is about ways to separate and purify substances. Match each term on the left with the correct description on the right.

A evaporation	i a solid appears as the solution cools
B condensing	ii used to separate a mixture of two liquids
C filtering	iii the solvent is removed as a gas
D crystallising	iv this method allows you to recycle a solvent
E distillation	v a gas changes to a liquid, on cooling
F fractional distillation	vi separates an insoluble substance from a liquid

2 Seawater can be purified using this apparatus:

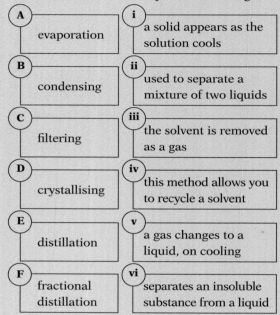

a i What is the maximum temperature recorded on the thermometer, during the distillation?
   ii How does this compare to the boiling point of the seawater?
b In which piece of apparatus does evaporation take place? Name it.
c i Which is the condenser, A, B, or C?
   ii Where does the supply of cold water enter?

3 Gypsum is insoluble in water. You are asked to purify a sample of powdered gypsum that is contaminated with a soluble salt.
  a Which of these pieces of apparatus will you use?
  Bunsen burner, filter funnel, tripod, distillation flask, conical flask, pipette, thermometer, condenser, gauze, stirring rod, filter paper, beaker
  b Write step-by-step instructions for the procedure.

4 Argon, oxygen, and nitrogen are obtained from air by fractional distillation. Liquid air, at −250 °C, is warmed up, and the gases are collected one by one.
  a Is liquid air a mixture, or a pure substance?
  b Explain why fractional distillation is used, rather than simple distillation.
  c During the distillation, nitrogen gas is obtained first, then argon and oxygen. What can you say about the boiling points of these three gases?

5 You have to separate a mixture of salt and sugar. Sugar is soluble in ethanol. Salt is not.
  a What will be the first step in this separation?
  b Draw a diagram to show how you will separate the salt.
  c How could you obtain sugar from the sugar solution, *without* losing the ethanol?
  d Draw a diagram of the apparatus for **c**.
  e Suggest how to separate a mixture of sand and salt.

6 A student investigated a pigment from flower petals, using paper chromatography. The pigment is soluble in ethanol, but not in water. This shows the chromatography paper before and after the process:

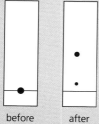

before   after

  a Describe the steps the student will have taken to carry out the chromatography, starting with a *solution* of the pigment.
  b What can you deduce about the pigment, from the chromatogram?
  c What would the chromatogram look like, if the pigment had been a pure substance?

7 In a chromatography experiment, eight coloured substances were spotted onto a piece of filter paper. Three were the basic colours red, blue, and yellow. The others were unknown substances, labelled A–E. This shows the resulting chromatogram:

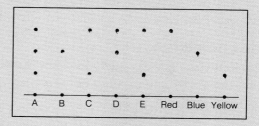

  a Which one of substances A–E contains only one basic colour?
  b Which contains all three basic colours?
  c The solvent was propanone. Which of the three basic colours is the most soluble in propanone?

Extended syllabus content

8 The diagram below shows a chromatogram for a mixture of amino acids.

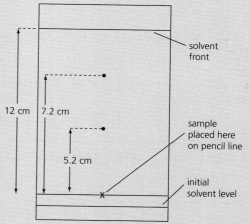

The solvent was a mixture of water, butanol, and ethanoic acid.
  a Using the table of R_f values on page 249, identify the two amino acids.
  b Which of them is *less* soluble in the solvent?
  c How will the R_f values change if the solvent travels only 6 cm?

9 You have three colourless solutions. Each contains an amino acid you must identify. Explain how to do this using chromatography. Use the terms R_f and *locating agent* in your answer, and show that you understand what they mean.

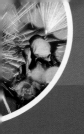

19.1 The scientific method, and you

Objectives: describe the scientific method; identify *independent* and *dependent* variables in experiments; list the skills needed for successful lab work

The lab: the home of chemistry
All the chemistry in this book has one thing in common. It is all based on real experiments, carried out in labs around the world, over the years – and even over the centuries. The lab is the home of chemistry!

How do chemists work?
Like all scientists, chemists follow the **scientific method**. This flowchart shows the steps. The handwritten notes are from a student.

1. You observe something that makes you ask yourself a question.
   Kitchen cleaner X is better at removing grease than kitchen cleaner Y. Why?

2. Come up with a **hypothesis** – a reasonable statement that you can test.
   Sodium hydroxide is used in kitchen cleaners to help remove grease.
   Cleaners X and Y contain sodium hydroxide – and X is better at removing grease.
   My hypothesis: X contains more sodium hydroxide than Y does.

3. Plan an experiment to test the hypothesis.
   I plan to do titrations to test my hypothesis. See the details in the next unit.

4. Carry out the experiment, and record the results.
   See the results in the next unit.

5. Analyse the results.
   You can help me do this, in the next unit.

6. Did the results support your hypothesis?
   See the next unit.

7. Share your conclusions with other people.
   The teacher wants to see them!

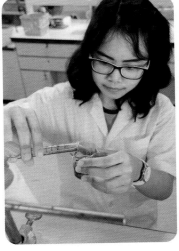

▲ Step into the lab, and try the scientific method ….

Peer review
- You share your work with your teacher.
- Around the world, chemists share their work with other chemists, in chemical journals.
- Other chemists will repeat the work to see if they get the same results. The results are **validated**, or else **rejected**.

Planning an experiment: the variables
Suppose you want to investigate the effect of temperature on the rate of a reaction.
- *You* change the temperature. So temperature is the **independent variable**. It is the *only* thing you change as you do the experiment.
- If the rate changes as you change the temperature, then the rate *depends on* the temperature. It is a **dependent variable**.

In many experiments you do, there will be an independent variable. You control it – and you keep everything else unchanged.

The golden rule …
When you investigate something in the lab, **change only one thing at a time**, and see what effect it has.

EXPERIMENTS AND TESTS IN THE LAB

The skills you use
When you plan and carry out an experiment, you use many different skills:

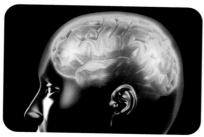

Thinking Use your brain before, during, and after the experiment. That is what brains are for. (They really like being used.)

Observing This is a very important skill. Chemists have made some amazing discoveries just by watching very carefully.

Using apparatus and techniques Weigh things, measure out volumes of liquids, measure temperature, do titrations, prepare crystals ….

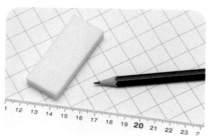

Working accurately Sloppy work will ruin an experiment. Follow the instructions. Measure things carefully. Think about safety too.

Doing some maths You may have to do some calculations using your results. And drawing a graph can help you see what is going on.

Writing up You may have to write a report on your experiment, and give conclusions. And say how the experiment could be improved!

The experiments you do
Often, you will not get a chance to plan an experiment for yourself. Instead, the teacher will tell you what to do. So you might miss out steps **1–3** in the flowchart on page 256.

But even if you pick up at step **4**, you are still using the scientific method, and gaining practice in it. You are following in the footsteps of many famous scientists, who have changed our lives by their careful work in the lab.

One day, you may become a scientist yourself.

Even a famous one!

▲ Famous? The German chemist Robert Bunsen (1811–1899) invented the Bunsen burner – and flame tests for metals!

1. Define *hypothesis*.
2. You can test a hypothesis by doing an experiment. Do you think this counts as a hypothesis?
    a. I am late for class again.
    b. The solubility of sugar is different in different solvents.
    c. December follows November.
    d. Adding *more* catalyst makes a reaction go even faster.
3. Explain in your own words the term *independent variable*.
4. Which would be the independent variable, in an experiment to test the statement in **2b**?
5. Outline an experiment to test the statement in **2d**, for the decomposition of hydrogen peroxide using manganese(IV) oxide as catalyst (page 112). Hint: use a gas syringe!
6. Skills for lab work are given above. List them in what you think is their order of importance, most important first.

19.2 Writing up an experiment

Objectives: be able to write up an experiment, and evaluate it for reliability of results; describe how to carry out a titration; explain the role of the indicator

Comparing those kitchen cleaners

In step **2** of the scientific method in the last unit, a student put forward a hypothesis. Here you can read how the student tested the hypothesis. But the report is not quite finished. That is *your* task.

An experiment to compare the amount of sodium hydroxide in two kitchen cleaners

Introduction

I noticed that kitchen cleaner X is better than kitchen cleaner Y at removing grease. The labels show that both kitchen cleaners contain sodium hydroxide. This chemical is used in many cleaners because it reacts with grease to form soluble sodium salts. These go into solution in the washing-up water.

My hypothesis

Kitchen cleaner X contains more sodium hydroxide than kitchen cleaner Y does.

Planning my experiment

I plan to titrate a sample of each cleaner against dilute hydrochloric acid, using methyl orange as indicator. This is a suitable method because the sodium hydroxide in the cleaner will neutralise the acid. The indicator will change colour when neutralisation is complete.

To make sure the results are valid, I will use exactly the same volume of cleaner, and the same concentration of acid, and the same number of drops of indicator each time. I will add the acid in the same way. I will do the titration at room temperature. The only thing I will change is the type of cleaner.

I will wear safety goggles, since sodium hydroxide and hydrochloric acid are corrosive.

The experiment

25 cm³ of cleaner X were measured into a conical flask, using a volumetric pipette. 5 drops of methyl orange were added, and the solution turned yellow.

A burette was filled to the upper mark with hydrochloric acid of concentration 1 mol/dm³. The initial level of the acid was noted.

The acid was allowed to run into the conical flask. The flask was continually and carefully swirled. As the acid dripped in, the solution showed flashes of red. When the end-point was near, the acid was added drop by drop. When the solution changed from yellow to red, the titration was stopped. The final level of the acid was recorded.

The experiment was repeated with cleaner Y.

Indicators !

- For acid/alkali titrations, the indicator must show a clear colour change, when the final drop that completes the neutralisation is added.
- Litmus will not work. The colour change is not sharp enough.
- But methyl orange works, and so would thymolphthalein. (See page 129.)

▲ A single drop of acid has just caused a clear colour change in the methyl orange indicator. It is the **end-point**. All the alkali in the cleaner has been neutralised.

EXPERIMENTS AND TESTS IN THE LAB

The results

For X:
Initial level of acid in the burette   0.0 cm³
Final level   22.2 cm³
Volume of acid used   22.2 cm³

For Y:
Initial level of acid in the burette   22.2 cm³
Final level   37.5 cm³
Volume of acid used   15.3 cm³

Analysis of the results

The same volume of each cleaner was used. The sodium hydroxide in X neutralised 22.2 cm³ of acid. The sodium hydroxide in Y neutralised 15.3 cm³ of acid. This means that solution X ...

My conclusion

These results ...

To improve the reliability of the results

I would ...

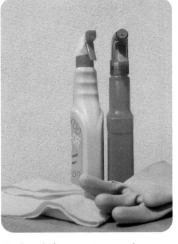

▲ One is better at removing grease. Might it have a higher concentration of sodium hydroxide?

In the question section below, you will have the chance to complete the student's analysis and conclusions, and come up with suggestions for ensuring that the results were reliable.

For example, would *you* have repeated the titrations? How many times?

1 In this experiment, was there:
   a an independent variable? If so, what was it?
   b a dependent variable? If so, what was it?
2 a Look at the apparatus below. Which pieces did the student use in the experiment? Give their letters and names.
   b When measuring out solutions for titration, a volumetric pipette is used instead of a measuring cylinder. Why?
   c Why is a conical flask used rather than a beaker, for the titration?
   d Why are burettes used for titrations?
   e Which is *more* accurate for measuring liquids?
      i a burette   ii a volumetric pipette
      Explain clearly why you think so.
3 What is the purpose of an indicator, in titrations?

4 a Suggest a reason why universal indicator is not used in titrations (page 131).
   b Suggest a different indicator the student could have used, and describe the colour change in the flask.
5 Now complete the student's *Analysis of the results*.
6 Complete the *Conclusion*, by saying whether or not the results supported the hypothesis.
7 How would you improve the reliability of the results?
8 The student did not give the equation for the reaction, in the report. Write the missing equation.
9 How would you modify the experiment, to compare liquid scale-removers for kettles? (They contain acid.)
10 Next week the student will do an experiment to see whether neutralisation is *exothermic*, or *endothermic*. Which item below will the student *definitely* use?

A 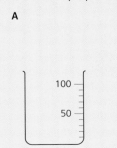   B    C    D    E    F

19.3 Preparing, and testing, gases

Objectives: outline how to produce gases in the lab; describe the ways to collect a gas, and when to use them; recall the tests, and results, for the named gases

Preparing gases in the lab

The usual way to make a gas in the lab is to displace it from a solid or solution, using apparatus like this. The table gives some examples.

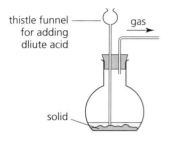
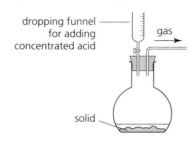

To make ...	Place in flask	Add	Reaction
carbon dioxide	calcium carbonate (marble chips)	dilute hydrochloric acid	$CaCO_3\ (s) + 2HCl\ (aq) \longrightarrow CaCl_2\ (aq) + H_2O\ (l) + CO_2\ (g)$
hydrogen	pieces of zinc	dilute hydrochloric acid	$Zn\ (s) + 2HCl\ (aq) \longrightarrow ZnCl_2\ (aq) + H_2\ (g)$
oxygen	manganese(IV) oxide (as a catalyst)	hydrogen peroxide	$2H_2O_2\ (aq) \longrightarrow 2H_2O\ (l) + O_2\ (g)$

But to make ammonia, you can heat any ammonium compound with a base such as sodium hydroxide or calcium hydroxide – using both reactants in solid form.

Collecting the gases you have prepared

The table below shows four ways to collect a gas.

The method to choose will depend on how much of the gas you need, whether it is heavier or lighter than air, whether it is soluble in water, and whether you need to measure its volume.

Measuring gas volume
- You can collect a gas over water, in a gas jar.
- But to measure the volume of the gas *roughly*, use a measuring cylinder instead.
- To measure its volume *accurately*, use a gas syringe.

method	upward displacement of air	downward displacement of air	over water	gas syringe
use when ...	the gas is heavier than air	the gas is lighter than air	the gas is sparingly soluble in water	to measure the volume accurately
apparatus				
examples	carbon dioxide, CO_2 sulfur dioxide, SO_2 hydrogen chloride, HCl	ammonia, NH_3 hydrogen, H_2	carbon dioxide, CO_2 hydrogen, H_2 oxygen, O_2	any gas

EXPERIMENTS AND TESTS IN THE LAB

Tests for gases

You have a sample of gas. You think you know what it is, but you're not sure. So you need to do a test. Below are some tests for common gases. Each is based on particular properties of the gas, including its appearance, and sometimes its smell.

Reminder
Aqueous solution means a solution in water.

Gas	Description and test details
ammonia, NH_3	
Properties	Ammonia is a colourless alkaline gas with a strong sharp smell.
Test	Hold damp red litmus paper in it.
Result	The indicator paper turns blue. (You may also notice the sharp smell.)
carbon dioxide, CO_2	
Properties	Carbon dioxide is a colourless, weakly acidic gas. It reacts with **limewater** (an aqueous solution of calcium hydroxide) to give a white precipitate of calcium carbonate: $CO_2(g) + Ca(OH)_2(aq) \longrightarrow CaCO_3(s) + H_2O(l)$
Test	Bubble the gas through limewater.
Result	The limewater turns cloudy or milky.
chlorine, Cl_2	
Properties	Chlorine is a green poisonous gas which bleaches dyes.
Test	Hold damp indicator paper in the gas, in a fume cupboard.
Result	The indicator paper bleaches (turns white).
hydrogen, H_2	
Properties	Hydrogen is a colourless gas which combines violently with oxygen when lit.
Test	Collect the gas in a tube and hold a lighted splint to it.
Result	The gas burns with a squeaky pop.
oxygen, O_2	
Properties	Oxygen is a colourless gas. Fuels burn much more readily in it than in air.
Test	Collect the gas in a test-tube and hold a glowing splint to it.
Result	The glowing splint immediately bursts into flame.
sulfur dioxide, SO_2	
Properties	Sulfur dioxide is a colourless, poisonous, acidic gas, with a choking smell. It will reduce the purple manganate(VII) ion to the colourless manganese(II) ion.
Test	Soak a piece of filter paper in acidified aqueous potassium manganate(VII). Place it in the gas. (Acidified means that a little dilute acid – usually hydrochloric acid – has been added.)
Result	The colour on the paper changes from purple to colourless.

1 **a** Sketch the *complete* apparatus you will use to prepare and collect carbon dioxide gas. Label all the parts.
  b How will you then test the gas to confirm that it is carbon dioxide?
  c Write the equation for a positive test reaction.
2 **a** Hydrogen cannot be collected by upward displacement of air. Why not?
  b Hydrogen burns with a squeaky pop. Write a balanced equation for the reaction that takes place.
3 **a** Name two substances you could use to make ammonia.
  b Ammonia cannot be collected over water. Why not?
  c The test for ammonia is …… ?
4 It is not a good idea to rely on smell, to identify a gas. Suggest at least two reasons why.
5 To measure the rate of the reaction between magnesium and hydrochloric acid, you will collect the hydrogen that forms. Which is better to use for this: a measuring cylinder over water, or a gas syringe? Give more than one reason.

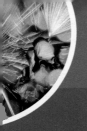

19.4 Testing for cations

Objectives: describe flame tests, chemical tests, and results, for the named cations; understand that chemical tests for metal cations depend on forming a precipitate

Testing a salt
You are given a tiny amount of a salt, and asked to find out what it is. Don't panic! Put on your detective hat, and do some tests!

Remember, a salt is made up of positive ions, or **cations**, and negative ions, or **anions**. There are different tests for each.

Tests for cations (+)
In this unit we look at tests for cations: metal ions, and the ammonium ion. Let's start with flame tests.

Flame tests
A **flame test** works well for some metal cations – and especially Group I. You start with the solid salt. These are the steps:

- First, clean a platinum or nichrome wire. To do this, dip the wire into concentrated hydrochloric acid, then hold it in a hot Bunsen flame.
- Next, moisten the clean wire by dipping it into the acid again. Then dip it into the salt, so that some salt sticks to it.
- Hold it in the clear part of a blue Bunsen flame, and observe the colour.

Here are the results for six cations:

> **Remember CAP!**
> **C**ations **A**re **P**ositive.
> They would go to the **cat**hode (−).

▲ Carrying out a flame test. The lilac colour indicates the potassium cation (K^+).

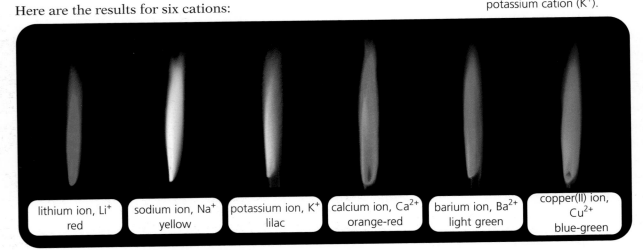

| lithium ion, Li^+ red | sodium ion, Na^+ yellow | potassium ion, K^+ lilac | calcium ion, Ca^{2+} orange-red | barium ion, Ba^{2+} light green | copper(II) ion, Cu^{2+} blue-green |

Chemical tests
The table on the next page shows chemical tests for some other cations.

- **To test for the ammonium ion**, you add aqueous sodium hydroxide to the solid salt, or its solution, to drive off ammonia gas. Then test the gas.
- **To test for a metal cation**, you precipitate its hydroxide. (Most hydroxides are insoluble, or only slightly soluble.) You do this by adding dilute aqueous sodium hydroxide, or ammonia, to a solution of the salt. Then observe the colour and behaviour of the precipitate.
- Note that the first four cations below require only *one* test.
- For the second four you divide the salt solution into two portions and do *two* tests. Test one portion with dilute aqueous sodium hydroxide. Then test the other with aqueous ammonia, to confirm the result.

> **More about flame tests**
> - In the flame, electrons in the cations take in heat energy, and jump to higher energy levels.
> - Then they fall back again, giving out energy as light of a specific colour.

EXPERIMENTS AND TESTS IN THE LAB

Cation	Test	If the cation is present	Ionic equation for the reaction
ammonium NH_4^+	Add dilute aqueous sodium hydroxide. Heat gently.	Ammonia gas is given off. (It turns litmus blue.)	$NH_4^+ (aq) + OH^- (aq) \longrightarrow NH_3 (g) + H_2O (l)$
iron(II) Fe^{2+}	Add dilute aqueous sodium hydroxide, *or* aqueous ammonia.	A pale green precipitate forms.	$Fe^{2+} (aq) + 2OH^- (aq) \longrightarrow Fe(OH)_2 (s)$
iron(III) Fe^{3+}	Add dilute aqueous sodium hydroxide, *or* aqueous ammonia.	A red-brown precipitate forms.	$Fe^{3+} (aq) + 3OH^- (aq) \longrightarrow Fe(OH)_3 (s)$
zinc Zn^{2+}	Add dilute aqueous sodium hydroxide, *or* aqueous ammonia.	A white precipitate forms in each case. It dissolves again if you add **excess** (more) of the sodium hydroxide or aqueous ammonia, giving a colourless solution.	$Zn^{2+} (aq) + 2OH^- (aq) \longrightarrow Zn(OH)_2 (s)$ The precipitate dissolves again in excess sodium hydroxide because zinc hydroxide is **amphoteric**. (It reacts with acids *and* alkalis.) It dissolves in excess ammonia because a soluble **complex ion** forms: $Zn(OH)_4^{2-}$. (A complex ion has a number of ions or molecules around a metal ion.)
aluminium Al^{3+}	Add dilute aqueous sodium hydroxide to one portion.	A white precipitate forms. It dissolves again if you add excess sodium hydroxide, giving a colourless solution.	$Al^{3+} (aq) + 3OH^- (aq) \longrightarrow Al(OH)_3 (s)$ The precipitate dissolves in excess sodium hydroxide because aluminium hydroxide is amphoteric.
	Add aqueous ammonia to the other portion.	A white precipitate forms. Adding excess ammonia has no effect.	
calcium Ca^{2+}	Add dilute aqueous sodium hydroxide to one portion.	A white precipitate forms. Excess sodium hydroxide has no effect.	$Ca^{2+} (aq) + 2OH^- (aq) \longrightarrow Ca(OH)_2 (s)$
	Add aqueous ammonia to the other portion.	A very slight white precipitate, or none.	
copper(II) Cu^{2+}	Add dilute aqueous sodium hydroxide to one portion.	A pale blue precipitate forms. Excess sodium hydroxide has no effect.	$Cu^{2+} (aq) + 2OH^- (aq) \longrightarrow Cu(OH)_2 (s)$
	Add aqueous ammonia to the other portion.	A pale blue precipitate forms. It dissolves in excess ammonia, giving a deep blue solution.	The precipitate dissolves again in excess ammonia because a soluble **complex ion** forms.
chromium(III) Cr^{3+}	Add dilute aqueous sodium hydroxide to one portion.	A green precipitate forms. It dissolves in excess sodium hydroxide.	$Cr^{3+} (aq) + 3OH^- (aq) \longrightarrow Cr(OH)_3 (s)$
	Add aqueous ammonia to the other portion.	A grey-green precipitate forms. Excess ammonia has no effect.	

1. The other name for a positive ion is … ?
2. How would you carry out a *flame test*?
3. Which metal cation will give this flame colour?
   a yellow   b blue-green   c red
4. In the table above, all the precipitates that form are h……?
5. How can you tell whether a cation is Fe^{2+}, or Fe^{3+}?
6. a Using only sodium hydroxide, you cannot tell whether a cation is Al^{3+} or Zn^{2+}. Why not?
   b Which further test would you do? Explain why.
7. a Neither sodium hydroxide nor aqueous ammonia can be used to identify potassium ions. Explain this.
   b Suggest a different test to use, to identify potassium ions.

19.5 Testing for anions

Objectives: understand that the chemical tests for anions rely on releasing a gas, or forming a precipitate; recall the tests and results for the named anions

Now for the anions

In the last unit you saw how to test an unknown salt, to see if it contains certain positive ions, or **cations** – for example NH_4^+, K^+, Al^{3+}.

In this unit we look at tests to identify the negative ion, or **anion**, that the unknown salt contains. For example Cl^-, NO_3^-.

- Some of the tests work by producing a gas. So you start with the solid salt, *or* its aqueous solution. Then test the gas that forms.
- The others work by producing a precipitate. For those, you start with the aqueous solution (just as you did for cations).

The tests are listed below, starting with the halide ions.

> **Remember AN!**
> **A**nions are **N**egative.
> They'd go to the **an**ode (+).

Tests for anions (−)

Halide ions (Cl^-, Br^-, I^-)

- To a small amount of the solution, add an equal volume of dilute nitric acid. Then add aqueous silver nitrate.
- Silver halides are insoluble. So if halide ions are present, a precipitate will form. The colour tells you which one. Look at this table:

Precipitate	Indicates presence of …	Ionic equation for the reaction
white	chloride ions, Cl^-	$Ag^+ (aq) + Cl^- (aq) \longrightarrow AgCl (s)$
cream	bromide ions, Br^-	$Ag^+ (aq) + Br^- (aq) \longrightarrow AgBr (s)$
yellow	iodide ions, I^-	$Ag^+ (aq) + I^- (aq) \longrightarrow AgI (s)$

The colours are shown in the photo below.

◀ The precipitates in the test for halides.

> **Precipitates**
> A *precipitate* is an insoluble product of a reaction. It's easy to see – which is useful in testing.
> - You mix a known solution with the unknown one.
> - If an insoluble product appears, that gives you useful clues about the unknown substance.
>
> Look back at the table about the solubility of salts on page 140.

EXPERIMENTS AND TESTS IN THE LAB

Sulfate ions (SO_4^{2-})
- To a small amount of the solution add an equal volume of dilute nitric acid. Then add barium nitrate solution.
- Barium sulfate is insoluble. So if sulfate ions are present a white precipitate will form.
- The ionic equation for the reaction is:
  $$Ba^{2+}(aq) + SO_4^{2-}(aq) \longrightarrow BaSO_4(s)$$

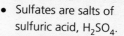

Sulfates and sulfites
- Sulfates are salts of sulfuric acid, H_2SO_4.
- Sulfites are salts of sulfurous acid, H_2SO_3.

Sulfite ions (SO_3^{2-})
Sulfites are salts that contain the sulfite ion, SO_3^{2-}. It is similar to the sulfate ion, but with just three oxgyen atoms.
- To a small amount of the solution add a small amount of acidified aqueous potassium manganate(VII). This is purple.
- The added solution turns colourless.
- The ionic equation for the reaction is:
  $$5SO_3^{2-}(aq) + 2MnO_4^-(aq) + 6H^+(aq) \longrightarrow$$
  purple
  $$5SO_4^{2-}(aq) + 2Mn^{2+}(aq)$$
  colourless

The potassium manganate(VII) has acted as an oxidising agent.

▲ Sulfites such as sodium sulfite and calcium sulfite are added to dried fruit and other foodstuffs, to preserve them. (They prevent the growth of bacteria.)

Nitrate ions (NO_3^-)
- To a small amount of the solid or solution, add a little dilute aqueous sodium hydroxide, and then some small pieces of aluminium foil. Heat gently.
- If ammonia gas is given off, the substance contained nitrate ions.
- The ionic equation for the reaction is:
  $$8Al(s) + 3NO_3^-(aq) + 5OH^-(aq) + 2H_2O(l) \longrightarrow$$
  $$3NH_3(g) + 8AlO_2^-(aq)$$

Carbonate ions (CO_3^{2-})
- To a small amount of the solid or solution, add a little dilute hydrochloric acid.
- If the mixture bubbles and gives off a gas that turns limewater milky, the substance contained carbonate ions. The gas is carbon dioxide.
- The ionic equation for the reaction is:
  $$2H^+(aq) + CO_3^{2-}(aq) \longrightarrow CO_2(g) + H_2O(l)$$

▲ Testing for carbonates. The test-tube on the right contains limewater. The carbon dioxide is turning it milky.

1. What are *anions*? Give two examples from this unit.
2. A test will produce a *pctpiteeiar* if the product is *ilbsoenul*. Unjumble the words in italics!
3. Silver nitrate is used in the test for halides. Why?
4. How would you test for the anion in calcium sulfite?
5. Nitrates are not tested by forming a precipitate. Why not?
6. Where do the OH^- ions come from, in the test for nitrate ions?
7. a Why is acid used, in testing for carbonates?
   b Limewater is also used in the test.
   i What is limewater?   ii Why is it a good choice?

Checkup on Chapter 19

Revision checklist

Core syllabus content
Make sure you can …
- [ ] identify common pieces of laboratory apparatus, and say what they are used for, including:
  beaker   test-tube   conical flask
  volumetric pipette   burette   measuring cylinder
  gas jar   gas syringe   condenser
  thermometer   filter funnel   water trough
- [ ] arrange these pieces of apparatus in order of accuracy (as here) for measuring out a volume of liquid:
  measuring cylinder   burette   volumetric pipette
- [ ] describe how to carry out these procedures:
  filtration   crystallisation
  simple distillation   fractional distillation
  titration   paper chromatography
  (See also Chapter 18 and page 139.)
- [ ] explain what these are, in experiments:
  an independent variable   a dependent variable
- [ ] explain why measurements are often repeated, in experimental work
- [ ] describe how to prepare these gases in the lab:
  hydrogen   oxygen   carbon dioxide   ammonia
  and name suitable reactants to use
- [ ] give the test for these gases:
  hydrogen   oxygen   carbon dioxide
  ammonia   chlorine   sulfur dioxide
- [ ] give another term for: cation   anion
- [ ] say how to carry out a flame test for metal cations, and state the flame colour for these cations:
  Li^+ Na^+ K^+ Ca^{2+} Ba^{2+} Cu^{2+}
- [ ] explain that in the chemical tests for anions and cations, a precipitate is formed *or* a gas is given off
- [ ] describe chemical tests to identify these cations:
  Cu^{2+} Fe^{2+} Fe^{3+} Al^{3+} Zn^{2+} Ca^{2+} Cr^{3+} NH_4^+
- [ ] describe chemical tests to identify these anions:
  halide ions (Cl^-, Br^-, I^-)   sulfate ion, SO_4^{2-}
  sulfite ion, SO_3^{2-} nitrate ion, NO_3^-
  carbonate ion, CO_3^{2-}
- [ ] describe a test for water (page 118)
- [ ] state that melting and boiling points can be used to test whether a substance is pure (page 250)

Extended syllabus content
Make sure you can also …
- [ ] give ionic equations for the tests for cations and anions

Questions

For all students

1 A sample of soil from a vegetable garden was thoroughly crushed, and water added as shown:

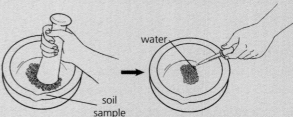

a Using a conical flask, filter funnel, filter paper, universal indicator, and dropping pipette, show how you would measure the pH of the soil.
b How would you check that the results for this sample were valid for the whole garden?

2 This apparatus is used to collect gases in the lab.
a Copy the drawing, and label the water, water trough, measuring cylinder, delivery tube, flask, and dropping funnel.

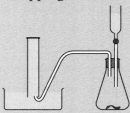

b The apparatus above can be used for preparing the gases hydrogen and carbon dioxide, but *not* sulfur dioxide. Explain why.

3 This apparatus is used to measure the rate of a reaction:

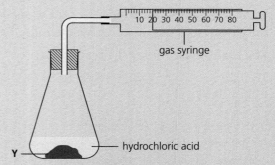

a Suggest a suitable substance to use as Y.
b Which other piece of apparatus is needed?
c Outline the procedure for this experiment.
d You must be careful not to use too much of Y and the acid. Why?

EXPERIMENTS AND TESTS IN THE LAB

4 Two solutions, W and Y, are tested with universal indicator paper.
Solution W: the indicator paper turns red.
Solution Y: the indicator paper turns orange.
  a i Which solution could have a pH of 1, and which could have a pH of 5?
  ii What type of solution is Y likely to be?

Further tests are carried out in test-tubes.

TEST A
A piece of magnesium is added to solution W.
  b i What will you observe in the test-tube?
  ii What is formed as a result of the reaction.
  iii How will solution Y compare, in this reaction?

TEST B
A solid, which is a sodium compound, is added to solution W. A gas is given off. It turns limewater milky.
  c i What colour will the solid be?
  ii Name the gas released.
  iii Suggest a name for the solid.

TEST C
A few drops of barium nitrate solution are added to a solution of W. A white precipitate forms.
  d i Name the white precipitate.
  ii Identify solution W.

5 In some countries, potassium sulfite is used as a food preservative. It is not permitted in others. An unknown white solid food preservative is to be tested, to see if it is potassium sulfite.
  a First, a flame test is carried out on the sample.
    i Describe how you would do the flame test.
    ii What colour will be seen in the flame, if the sample is potassium sulfite?
  b The sample is heated gently with dilute hydrochloric acid. Sulfur dioxide gas is released.
    i Write the formula for this gas.
    ii What test would you carry out, to confirm this gas?
    iii How would the results for **i** and **ii** differ, if the preservative had been sodium sulfite?

6 Ammonium nitrate (NH_4NO_3) is an important fertiliser. The ions in it can be identified by tests.
  a Name the cation present, and give its formula.
  b Describe a test that will confirm the presence of this cation.
  c Name the anion present and give its formula.
  d Describe a test that will confirm the presence of this anion.

7 A solution of a metal chloride was tested by adding aqueous sodium hydroxide to one portion, and aqueous ammonia to another. The results were:

Test	Observation
aqueous sodium hydroxide added	white precipitate forms; it dissolves in excess sodium hydroxide
ammonia solution added	white precipitate forms; it dissolves in excess ammonia

  a The above is a test for a particular cation.
    i What is a *cation*?
    ii Which cation has been identified in the test?
    iii Name a cation that would give a similar result for sodium hydroxide, but not for the ammonia solution.
    iv Name a cation that forms a white precipitate with aqueous sodium hydroxide, but does not dissolve in excess sodium hydroxide.
  b i Give the formula of the anion present in the compound tested above.
    ii Describe and give the result of the test you would carry out to confirm the presence of this anion.

8 A sample of mineral water contained these ions:

Name of ion	Concentration (milligrams/dm^3)
calcium	55
chloride	37
hydrogen carbonate	248
magnesium	19
nitrate	0.05
potassium	1
sodium	24
sulfate	13

  a Make two lists, one for the anions and the other for the cations present in this mineral water.
  b i Which metal ion is present in the highest concentration?
    ii What mass of that metal would be present in a small bottle of water, volume 50 cm^3?
  c Which of the ions will react with barium nitrate solution to give a white precipitate?
  d Of the metal ions, only calcium can be identified by a precipitation test. Why is this?
  e A sample of the water is heated with sodium hydroxide and aluminium foil. Ammonia gas could not be identified, even though the nitrate ion is present. Suggest a reason.

Answers for Chapters 1–19

All the questions that appear in this book were written by the authors. For answers to the IGCSE practice questions on pages 287–310, access online at www.oxfordsecondary.com/complete-igcse-science

Chapter 1

page 3 **1** The dust particles are colliding with the gas particles in the air. **2** The purple colour spreading through the water shows that particles left the solid. That means the solid must be made of particles. **3** It spreads upwards because the bromine particles are struck by the particles in air, and bounce away in all directions, including upwards. **4 a** In diffusion, particles mix and spread by colliding and bouncing off in all directions. **b** The smell is carried by particles from the perfume. These collide with gas particles in the air and bounce off in all directions. Overall, they move to where they are less concentrated. That is why the smell spreads. **5 a** atoms, molecules and ions **b** Atoms are single particles, while molecules contain two or more atoms joined together.

page 5 **1** any two from the properties for each, at the top of page 4 **2 a** ice **b** steam **3 a** condense **b** melt **4 a** oxygen **b** diamond **c** table salt **d** mercury, ethanol, water **e** because their particles are different

page 7 **1** You could copy drawings from page 6 of this book. (For the solid and liquid, use the simpler drawings in the bottom row.) Add notes like those in the table on page 6. **2 a** You cannot pour solids because the particles are held in a lattice, and cannot slide past each other. **b** You can pour liquids because the particles can slide past each other. **3** The drawing could be like this one:

As the liquid cools the particles lose energy and move more and more slowly …

… and eventually they settle into a regular pattern. A solid has formed.

4 The water evaporates. **5** Evaporation is a slower process / takes place over a range of temperatures below the boiling point / occurs only at the surface of the liquid. Boiling is a faster process / takes place at a specific temperature (the boiling point) / occurs throughout the liquid. **6 a i** a gas **ii** a liquid **b** Cool it to below its freezing (melting) point, −219 °C.

page 9 **1** a graph that shows how the temperature of a substance changes as it is heated **2** The horizontal lines show where melting and boiling is taking place. So extend the horizontal lines to the vertical axis. The melting and boiling points are where they meet the axis. **3** Heat energy is being taken in, to overcome the forces holding the particles together in the solid. **4** Boiling takes more energy. Look at the horizontal lines. It takes longer (and therefore takes more heat energy) to change from liquid to gas than from solid to liquid. **5 a** 5 minutes **b** Particles release heat energy as they move closer together, and this cancels the cooling. **6 a** Iron is boiling (turning from liquid to gas). **b** between B and C **c** melting point around 1500 °C, boiling point around 2800 °C (actual figures are 1538 °C and 2862 °C)

page 11 **1** particles hitting the walls of the container **2 a** It increases. **b** It decreases. **3 a** It decreases. **b** It increases. **4 a** The particles move faster and strike the piston with more force, so the piston moves upwards. **b** The particles slow down and strike the piston with less force, so the piston moves downwards. **5** There are now more particles in a smaller space, so they hit the walls more often. **6 a** Ammonia particles move along by colliding with the gas particles in the air; they bounce off in all directions, including forward. **b** They have a greater mass, so do not bounce as far after collisions. **7** Hydrogen is the lightest of all gases.

page 12 Core 1 a melting **b** solidifying (freezing) **c** condensing **d** boiling (or evaporating) **2 a i** The liquid at the bottom of the beaker would be purple. **ii** All the liquid would be purple. **b** Purple particles leave the crystal. Water particles collide with them, causing them to spread. **c** dissolving and diffusion **3 a** The particles are held in a fixed position. **b** The particles can slide around but are attracted to each other so stay close. **c** The particles collide and bounce away – out of the container. **d** There are no gaps between the particles so they cannot be pushed closer together. **e** As you blow more gas particles into the balloon, the sides of the balloon are struck more often. The pressure makes the balloon expand. (Eventually it will burst.) **4 a** 17 °C **b** remains steady **c** More energy is needed to separate the particles completely (in boiling) than to break down the lattice structure (in melting). **d** melting point is not 0 °C, boiling point is not 100 °C **e** like the one shown, but with the horizontal parts at exactly 0 °C and 100 °C (see also page 8)

page 13 Core 5 a liquid **b** It is ethanol. **c** like the one shown, but with the horizontal parts at exactly 0 °C and 100 °C (and see page 9) **6 a** The particles are held in fixed positions. **b** The particles will take in the heat energy, and travel faster. They will strike the walls more often, and with more energy, so pressure rises. **c** The poisonous particles diffuse through the air in all directions, as a result of collisions. (Wind will carry them too.) **d** The dust particles are being struck by the gas particles in air. **7 a** example v **b** It should be like the drawing for potassium manganate(VII) on page 3. **c** Blue particles leave the crystal. Water particles collide with them, causing them to spread.

Extended 8 a At 78 °C, the arrangement is changing from one where particles move freely and bounce apart to one where they are very closer together and only slide past each together. The reason for the change is that, as they cool, the particles lose energy and bounce less far after they collide. So the forces of attraction between them take over. **b i** The particles move more slowly as the gas

cools, and bounce less far apart. **ii** The particles move around more slowly as the liquid cools. (When it cools to the freezing point, a lattice will form.) **9 a** Hydrogen is leaving the tube faster than air particles are entering. **b** Hydrogen diffuses faster. **c** The mass of a hydrogen particle is less than the masses of the particles in air. **d** The water level will fall, because air particles enter the tube faster than carbon dioxide particles leave it.
10 a helium and methane: they have the lowest relative masses **b** helium **c** Yes. Its relative molecular mass is greater than those of nitrogen and oxygen, the two main gases in air. **d** It must be greater than 16, but less than 28.

Chapter 2

page 15 **1 a** An atom is a particle of matter that cannot be broken down further by chemical means. **b** An element contains only one kind of atom. **2** 94 **3 a** It is a table of the elements in order of their proton number, and arranged to show their patterns. **b** See the Periodic Table on page 15. **4 a** calcium **b** magnesium **c** nitrogen **5** metals **6 a** Select any three elements from *the same numbered group* that lie to the left of the ziz-zag line. (For example, lithium, sodium, and potassium.) **b** Select any three elements *from the same numbered group* that lie to the right of the ziz-zag line. (For example, chlorine, bromine, and iodine.) **7** It is made of hydrogen and oxygen atoms bonded together. **8 a** A compound is made up of different atoms bonded together. **b** A mixture is made up of different substances that are not bonded together.

page 17 **1** protons, electrons, neutrons **2 a** proton **b** neutron **c** electron **3** fluorine **4** They have equal numbers of protons and electrons so the charges cancel. **5 a** the number of protons in the nucleus of an atom **b** the total number of protons and neutrons in the nucleus of an atom **6 a** atomic number **b** mass number **7** proton number (atomic number) **8** In order: carbon atom with 6p, 6e, 6n; oxygen atom with 8p, 8e, 8n; magnesium atom with 12p, 12e, 12n; aluminium atom with 13p, 13e, 14n; copper atom with 29p, 29e, 35n

page 19 **1 a** The diagrams should show 3 electron shells, with 2 electrons in the first (inner) shell, 8 in the second, and 3 in the third. **b** 2,8,3 **c** aluminium **2 a** 2,1 **b** 2,8,2 **c** 1 **3** Group V **4** three shells **5** It makes them unreactive.**6 a** 2,8,18,8 **b** It is unreactive.

page 21 **1 a** atoms of the same element, with different numbers of neutrons **b** carbon-12, $^{12}_{6}C$; carbon-13, $^{13}_{6}C$; carbon-14, $^{14}_{6}C$ **2** oxygen-16, 8p 8n; oxygen-17, 8p 9n; oxygen-18, 8p 10n **3** $^{1}_{1}H$, $^{2}_{1}H$, $^{3}_{1}H$ **4 a** relative atomic mass **b** the carbon-12 atom **c** It stands for *relative* and tells you that the mass is relative to 1/12 of the mass of a carbon-12 atom. **5** See the definition on page 21. **6 a** 47.1 **b** They have the same number of electrons, and therefore the same number of outer shell electrons.

page 24 **Core 1 a i** Mg **ii** about 100 **b i** the Periodic Table **ii** element number 12, in Group II, Period 2 **iii** any one of: beryllium, calcium, strontium, barium, radium **c i** any element on the right of the zig-zag line in the Periodic Table **ii** Compounds are made up of atoms of different elements, bonded together. **iii** The different substances in mixtures are not bonded together.
2 A p **B** n **C** n **D** e **E** e **F** e **G** n **H** n **I** p **J** p **3 a i** X is the symbol of the element **ii** *y* is the nucleon number / mass number. **iii** *z* is the proton number / atomic number. **b i** 60 **ii** 34 **iii** 0 **iv** 10 **v** 146 **c** $^{80}_{35}Br$

page 25 Core 4 a This table gives the answers; m stands for metal and n for non-metal:

	Al	B	N	O	P	S
i	m	n	n	n	n	n
ii	3	2	2	2	3	3
iii	III	III	V	VI	V	VI
iv	13	5	7	8	15	16
v	13	5	7	8	15	16
vi	2,8,3	2,3	2,5	2,6	2,8,5	2,8,6
vii	3	3	5	6	5	6

b The elements in each pair below will have similar properties, because they have the same number of outer shell electrons: aluminium and boron; nitrogen and phosphorus; oxygen and sulfur **5 a** B has one more neutron than A **b** isotopes **c** $^{10}_{5}B$ and $^{11}_{5}B$ **d** 10 **e** heavier **f i** Both have the same elecronic configuration: 2,3. **ii** Isotopes of an element have the same number of electrons and the same electronic configuration. **6 a i** same **ii** different **b** sodium, since its relative atomic mass is a whole number (and note that the value for magnesium is usually rounded off to 24, for school chemistry) **7 a** 38 **b** 5 **c** 2 **8 a i** 2,8 **ii** It is a stable arrangement (a full outer shell of electrons) so the element will be unreactive. **b** Group VIII **c** neon **d** argon / krypton / xenon / radon

Extended 9 a electron, proton, neutron **b** 1, 2 and 3 respectively **c** … contain the same number of protons and electrons, but different numbers of neutrons. **d** hydrogen; the average mass is very close to 1
10 a i 38 **ii** 40 **b** 70 **c** Using these approximate percentages, and the formula from page 21:
$$A_r = \frac{(60 \times 69)}{100} + \frac{(40 \times 70)}{100} = 69.4.$$ This is close to 69.723, so consistent. **11 a** 19.9% **b** $A_r = \frac{(19.9 \times 10)}{100} + \frac{(80.1 \times 11)}{100}$ = 10.801 (or 10.8)

Chapter 3

page 27 1 An element contains only one type of atom. A compound is made up of atoms of different elements bonded together. **2** H_2O, CO_2 **3 a** HCl **b** NH_3 **4** Iron and sulfur are not chemically combined in the mixture, but are in the compound. **5** There is a colour change, and two liquids turn to (soft) solids. **6** Energy is given out in form of light and heat, and the silvery metal turns to a white powder. **7** + means *reacts with*, and → means *to give* **8** magnesium + oxygen → magnesium oxide **9 a / c** physical (no new substance formed) **b** chemical (a new substance formed)

page 29 **1** They have full outer shells of electrons. They do not need to gain or lose electrons, so there is no need to react. **2** They react with other atoms in order to gain, lose, or share outer electrons, and reach a stable arrangement, like the noble gases. **3** The drawings should be as on page 29. **4** an atom or group of atoms with a charge **5** It has 17 protons and 18 electrons so the overall charge is 1−. **6** They have full outer shells of electrons, so there is no need to gain or lose any. **7 a** cation **b** anion **8 a** Br^-, O^{2-} **b** K^+, Mg^{2+} **9** bromide ion

page 31 **1** The drawing should be as on page 30. **2** the strong electrostatic attraction between ions of opposite charge **3** It is a giant structure containing millions of ions, in which sodium ions alternate with chloride ions in a regular lattice. Its formula is NaCl because there is one Na^+ ion for each Cl^- ion **4 a** The atom loses two electrons to form an ion. So the ion has two more protons than electrons, giving a charge of 2+. **b** Opposite charges attract. **c** There are two Cl^- ions for each Mg^{2+} ion. **5** $AlCl_3$

page 33 **1** cations **2** It has a positive charge. **3** It woud take too much energy to gain or lose 4 electrons. **4 a** K^+, Cl^- **b** Ca^{2+}, S^{2-} **c** Li^+, S^{2-} **d** Mg^{2+}, F^- **5 a** KCl **b** CaS **c** Li_2S **d** MgF_2 **6** transition / more / than **7 a** $CuCl_2$ **b** Fe_2O_3 **8** copper(I) chloride, iron(II) sulfide, magnesium nitrate, ammonium nitrate, calcium sulfite **7 a** Na_2SO_4 **b** KOH **c** $AgNO_3$

page 35 **1 a** to obtain full outer shells of electrons **b** a covalent bond **2 a** two or more atoms held together by covalent bonds **b** diatomic **3** five from: hydrogen H_2 / oxygen O_2 / nitrogen N_2 / fluorine F_2 / chlorine Cl_2 / bromine Br_2 / iodine, I_2 **4** The drawings should be like those on pages 34 and 35. **5** A nitrogen atom needs three more electrons to reach a full outer shell. So in N_2, the two nitrogen atoms each *share* three electrons to form the triple bond. **6** It would take too much energy for the atoms of one of the non-metals to give up electrons to achieve full outer shells. It takes less energy for the non-metal atoms to share electrons with each other.

page 37 **1 a** a compound formed by atoms of different elements sharing electrons **2** The drawings should be like those on page 36. **3** one (it needs a share of one other electron to achieve a full outer shell) **4** The drawing should be like the one on page 37. **5** methane, ethene **6** four (two from each carbon atom)

page 39 **1** The particles are held in a fixed pattern extending in all directions. **2 a** ions **b** molecules **3** To conduct electricity, ions must be free to move. They cannot move in solid sodium chloride. But the lattice breaks up in water and the ions are released, so the solution can conduct. **4 a** molecular **b** no **c** any ionic compound, for example sodium chloride or magnesium oxide **d** A low melting point indicates weak forces between molecules in a lattice; and molecules have no charge, so the compound will not be able to conduct electricity. **5** The molecules are held in a regular lattice, with weak forces between them. As the ice is heated the molecules vibrate more vigorously about their positions. Eventually they gain enough energy to break away from the lattice, to form a liquid (water).

page 41 **1** It is molecular. The melting point of a giant structure would be much higher, since the atoms are held in the lattice by strong covalent bonds. **2** different forms of the same element **3 a** Each carbon atom in diamond is held in the lattice by four strong covalent bonds. **b** cutting tool **4 a** Its layers can slide over each other easily. **b** It has free electrons. **5** dark colour / soft / layers slide off easily to leave a mark on paper **6** as a lubricant (for **a**); as electrodes (for **b**) **7** Both are hard, with high melting points, because they are giant structures with strong covalent bonds between the atoms. Neither conduct electricity because there are no free electrons to carry charge.

page 43 **1** a regular lattice of metal ions in a sea of delocalised electrons **2** It is the electrostatic attraction between the positive metal ions in the lattice and the sea of delocalised electrons. **3** can be hammered into shape **4** The layers of ions can slide over each other, and from new bonds in their new position. **5 a** The delocalised electrons can move through the lattice, carrying the current. **b** Yes; the electrons are still free to move. **6** Some examples are: **a** car and aircraft bodies / metal signs / metal roofs **b** wiring in computers / metal cables / metal guitar strings **c** terminals in torch batteries / wiring in computers and phones / outdoor electricity cables **7** The diagram should be like the one for copper on page 42.

page 44 Core 1 a 3; the drawing should show two shells, with 2 electrons in the inner shell and 1 in the outer shell **b** by losing its partly-filled outer shell **c** The drawing should show just one shell with 2 electrons; the symbol is Li^+. **d** 9; the drawing should show two shells, with 2 electrons in the inner shell and 7 in the outer shell **e** by gaining electrons, to obtain a full outer shell **f** The drawing should be as for **d** but with 8 electrons in the outer shell; the symbol is F^-. **g** The diagram should show the outer electron of the lithium atom moving to the fluorine atom. **h** lithium + fluorine → lithium fluoride **2 a** ammonia, NH_3 **b** electrons **c** covalent **d** Choose from hydrogen chloride, water, methane, or any other compound formed between non-metals. **3 a** a gas **b** It is made of molecules; the evidence is its low melting and boiling points. **c i** covalent **ii** The drawing should be like the first one on page 36, but with bromine in place of chlorine. **d** HBr **e i** two from hydrogen fluoride / hydrogen chloride / hydrogen iodide **ii** HF / HCl / HI

page 45 Core 4 a D and G: good conductors in both solid and liquid states **b i** B and E: soluble, conduct electricity when liquid but not when solid **ii** electrostatic attraction between ions of opposite charge **c** Ionic compounds have high melting points, because of the strong bonds between ions. **d** A and C **e** F **f i** ionic **ii** covalent **iii** ionic **iv** covalent

Extended 5 a i loses electrons **ii** 3 **iii** positive ion **iv** 3+ **b i** gains electrons **ii** 3 **iii** negative ion **iv** 3− **c i** 2,8 for both **ii** They are the same. After forming ions ,both have an outer shell of 8 electrons, and the same number of shells. **d** phosphorus / arsenic **6 a** Both are solid. **b** giant covalent **c** also giant covalent, given its high melting and boiling points

d weak **e** No. Its very high melting point indicates a giant structure. Carbon dioxide is a gas at room temperature, which indicates a molecular compound. **7 a** giant structure **b** ionic **c i** four **ii** one to one **d i** ZnS **ii** Yes. The charges on the ions are 2+ and 2−. They balance. **e** any combination of a Group II metal (or transition element with ions with a charge of 2+) with a group VI non-metal, **OR** any combination of a Group I metal (or transition element with ions with a charge of 1+) with a Group VII non-metal, **OR** any combination of a Group III metal (or transition element with ions with a charge of 3+) with a group V non-metal **8 a** The metal consists of a lattice of metal ions in a sea of delocalised electrons. The ions are held together by their attraction to the sea of delocalised electrons. **b i** The delocalised electrons can move through the lattice of metal ions, forming a current of electricity. **ii** Layers of metal ions can slide over each other; new bonds then form in the new positions.

Chapter 4

page 47 **1 a** sodium fluoride **b** hydrogen fluoride **c** hydrogen sulfide **d** beryllium bromide **2 a** HBr **b** CF_4 **3 a** HBr **b** NaF **c** MgO **d** correct **e** BeO **4 a** BaI_2 **b** H_2O **5** phosphorus chloride, PCl_3 (also called phosphorus trichloride) **6** It contains two oxygen atoms for every silicon atom.

page 49 **1** + means *and*, and → means *to give* **2 a** $2Na\,(s) + Cl_2\,(g) \rightarrow 2NaCl\,(s)$ **b** $H_2\,(g) + I_2\,(g) \rightarrow 2HI\,(g)$ **c** $2Na\,(s) + 2H_2O\,(l) \rightarrow 2NaOH\,(aq) + H_2\,(g)$ **d** $2NH_3\,(g) \rightarrow N_2\,(g) + 3H_2\,(g)$ **e** $4Al\,(s) + 3O_2\,(g) \rightarrow 2Al_2O_3\,(s)$ **3 a** sodium + chlorine → sodium chloride **b** hydrogen + iodine → hydrogen iodide **c** sodium + water → sodium hydroxide + hydrogen **d** ammonia → nitrogen + hydrogen **e** aluminium + oxygen → aluminium oxide **4 a** reactions **b** and **d** **b** reaction **c** **5** $2Al\,(s) + 3Cl_2\,(g) \rightarrow 2AlCl_3\,(s)$

page 51 **1 a** The relative atomic mass for an element is the average mass of its naturally occurring isotopes, relative to one-twelfth of the mass of a carbon-12 atom. **b** because *relative* means *compared with* something else **c** A_r **2** It is the standard atom against which all other atoms are compared. **3** They are the small numbers below the symbols. See the Periodic Table and key on page 312. **4** because the electrons lost or gained in forming an ion have negligible mass **5 a** 71 **b** 64 **c** 58 **6 a** 102 **b** 132

page 53 **1 a** 95 g **b** 35.5 g **c** 47.5 g **2 a** 64 g **b** 32 g **3** 75% carbon, 25% hydrogen **4** Ca 40%, C 12%, O 48% **5 a** 90% pure **b** 1.5 g **6 a** 24 g **b** 15 g

page 54 Core **1 a** H_2O **b** CO **c** CO_2 **d** SO_2 **e** SO_3 **f** NaCl **g** $MgCl_2$ **h** HCl **i** CH_4 **j** NH_3 **2 a** PbO_2 **b** Pb_3O_4 **c** KNO_3 **d** N_2O **e** N_2O_4 **f** $NaHCO_3$ **g** Na_2SO_4 **h** $Na_2S_2O_3$ **3 a** 1 copper, 1 oxygen **b** 2 copper, 1 oxygen **c** 1 aluminium, 3 chlorine **d** 1 hydrogen, 1 nitrogen, 3 oxygen **e** 1 calcium, 2 oxygen, 2 hydrogen **f** 2 carbon, 2 oxygen, 4 hydrogen **g** 2 nitrogen, 4 hydrogen, 3 oxygen **h** 2 nitrogen, 8 hydrogen, 1 sulfur, 4 oxygen **i** 3 sodium, 2 phosphorus, 8 oxygen **j** 1 iron, 1 sulfur, 11 oxygen, 14 hydrogen **k** 1 cobalt, 2 chlorine, 12 hydrogen, 6 oxygen

page 55 Core **4 a** HBr **b** PCl_3 **c** H_2O_2 **d** C_2H_2 **e** N_2H_4 **f** XeF_6 **g** H_2SO_4 **h** H_2SO_3 **5 a** phosphorus oxide **b** P_4O_6 **6 a** zinc + hydrogen chloride (hydrochloric acid) → zinc chloride + hydrogen **b** sodium carbonate + sulfuric acid → sodium sulfate + carbon dioxide + water **c** magnesium + carbon dioxide → magnesium oxide + carbon **d** zinc oxide + carbon → zinc + carbon monoxide **e** chlorine + sodium bromide → sodium chloride + bromine **f** calcium oxide + nitric acid → calcium nitrate + water **7 a** $N_2 + 2O_2 \rightarrow 2NO_2$ **b** $K_2CO_3 + 2HCl \rightarrow 2KCl + CO_2 + H_2O$ **c** $C_3H_8 + 5O_2 \rightarrow 3CO_2 + 4H_2O$ **d** $Fe_2O_3 + 3CO \rightarrow 2Fe + 3CO_2$ **e** $Ca(OH)_2 + 2HCl \rightarrow CaCl_2 + 2H_2O$ **f** $2Al + 6HCl \rightarrow 2AlCl_3 + 3H_2$ **8 a i** Zn 65, C 12 **ii** ZnO 81, CO_2 44 **b i** 32.5 g **ii** 40.5 g **9 a** 46 **b** 80 **c** 98 **d** 36.5 **e** 142 **10 a** 40 **b** 78 **c** 132 **d** 138 **e** 278

Extended **11 a** calcium chloride **b** $CaCl_2$ **12 a i** 27.2 g **ii** 2.72 g **b** 50 g **c** 80% **13 a** 17.5% **b i** 2185 kg (2.185 tonnes) **ii** 375 kg **c** 91.7% **14 a i** 60% **ii** 21.2% **iii** 20.1% **b** increases

Chapter 5

page 57 **1** 6.02×10^{23} atoms **2** 12.04×10^{23} molecules **3** Avogadro constant **4 a** 1 g **b** 127 g **c** 35.5 g **d** 71 g **5 a** 32 g **b** 64 g **6** 68 g **7** 138 g **8 a** 9 moles **b** 3 moles **9** 40 **10 a** 6.02×10^{23} **b** 6.02×10^{23} **c** 12.04×10^{23}

page 59 **1 a** 2 moles **b i** 32 g **ii** 8 g **2 a i** iron **ii** 8.8 g **b** 6.4 g

page 61 **1** room temperature and pressure (20 °C, 1 atmosphere) **2** the volume occupied by 1 mole of the gas molecules **3** 24 dm³ **4 a** 168 dm³ **b** 12 dm³ **c** 0.024 dm³ (or 24 cm³) **5 a** 12 dm³ **b** 2.4 dm³ **6 a** 3000 cm³ **b** 100 cm³ **7 a** 2 dm³ **b** 0.5 dm³ **8 a** 12 dm³ **b** 12 dm³ **9 a** 12 dm³ **b** 6 dm³

page 63 **1 a** copper(II) sulfate **b** water **2** the amount of solute that is dissolved in 1 dm³ of solution **3 a** 500 cm³ **b** 2000 cm³ **4 a** 1 mole **b** 1 mole **5 a** 2 mol/dm³ **b** 1.5 mol/dm³ **6** 500 cm³ (or 0.5 dm³) **7** 5 cm³ (or 0.005 dm³) **8 a** 20 g **b** 0.5 g **9 a** 0.5 mol/dm³ **b** 0.25 mol/dm³

page 65 **1 a** 4 **b** 4 g **2** found by experiment **3** FeS **4** SO_3

page 67 **1** $MgCl_2$ **2** AlF_3 **3** The empirical formula gives *the simplest whole number ratio* of the atoms or ions in a compound. The molecular formula gives *the exact numbers of atoms* in the molecule. The two formulae may be the same, for example for CH_4. **4** CH **5** C_2H_4 **6 a** C_7H_{16} **b** C_7H_{16} **7** P_4O_6

page 69 **1 a** It is the actual amount obtained, given as a % of the theoretical amount (which is calculated from the equation). **b** It is the mass of pure substance in the product, given as a % of the total mass of the product. **2** 76.7% **3** 63% **4** 172 g **5** 88%

page 70 Extended **1 a** iron(III) oxide + carbon monoxide → iron + carbon dioxide **b** 160 **c** 2000 moles **d** 2 moles **e** 4000 moles **f** 224 kg **2 a** calcium carbonate → calcium oxide + carbon dioxide

b 0.5 moles **c i** 11.2 g **ii** 8.8 g **iii** 4.8 dm³ or 4800 cm³
3 a i 4 moles **ii** 19 moles **b** 4.75 moles **c** 114 dm³
d 227 g **e** 502.2 dm³ **f** A small volume of liquid will produce a very large volume of gas. This creates a massive, dangerous, pressure wave.

page 71 Extended 4 a 0.5 moles **b** 25 cm³
c i 75 cm³ **ii** 50 cm³ **5 a** sulfuric acid **b** 0.025 moles **c** 0.025 moles **d** Fe + H_2SO_4 → $FeSO_4$ + H_2; Fe^{2+} **e** 0.6 dm³ **6 a** 106.5 g **b** 3 moles **c** 1 mole **d** $AlCl_3$ **e** 0.1 mol/dm³ **7 a** 45.5 cm³ **b** 41.7 cm³ **c** 62.5 cm³ **8 a** P_2O_3 **b** 41.3 g **c** P_4O_{10} (empirical formula P_2O_5) **d** The molecular formula of the other oxide is in fact P_4O_6. **9 a** Zn_3P_2 **b** 24.1% **10 a** 64 g **b** 4 moles **c** 2 moles **d** MnO_2 **e** 632.2g **11** The missing molecular formulae are: **a** N_2H_4 **b** C_2N_2 **c** N_2O_4 **d** $C_6H_{12}O_6$ **12 a** carbon and hydrogen **b** CH_2 **c** A is C_3H_6 and B is C_6H_{12}. **13 a** 217 g **b** 20.1 g of mercury, 1.6 g of oxygen **c** 94.5% **14 a** 0.05 moles **b** 4.2 g **c** 84%

Chapter 6

page 73 1 a ... oxygen is gained **b** ... oxygen is lost **c** ... take place together **2** Your arrows should show that magnesium is oxidised to magnesium oxide, and sulfur dioxide is reduced to sulfur. **3** from **red**uction + **ox**idation **4** Methane gains oxygen – it is oxidised. Since oxidation and reduction always take place together, this means that oxygen is reduced. **5** Your arrows should show that magnesium is oxidised, and oxygen is reduced.

page 75 1 a Oxidation is gain of oxygen, or loss of electrons **b** Reduction is loss of oxygen, or gain of electrons. **2** It shows how a reactant gains or loses electrons, in a reaction. **3 a** 2Ca (s) + O_2 (g) → 2CaO (s) **b** Calcium loses electrons to oxygen. **c** 2Ca → 2Ca²⁺ + 4e⁻, O_2 + 4e⁻ → 2O²⁻ **4 a** 2K (s) + Cl_2 (g) → 2KCl (s) **b** 2K → 2K⁺ + 2e⁻, Cl_2 + 2e⁻ → 2Cl⁻
5 a Br_2 + 2e⁻ → 2Br⁻, 2I⁻ → I_2 + 2e⁻
b The ionic equation is: Br_2 + 2I⁻ → 2Br⁻ + I_2

page 77 1 It is the number of electrons that each atom of an element has lost, gained, or shared, in forming a compound. **2 a** Fe 0 **b** H +I, Cl –I **c** Ca +II, Cl –I **d** Na +I, O –II, H +I **e** Ca +II, C +IV, O –II **3 a** hydrogen + oxygen → water **b** You should have these oxidation numbers: in H_2 and O_2, both 0; in H_2O, +I for H, –II for O **c** Oxidation numbers have changed, so it is a redox reaction. **d** Hydrogen is oxidised, and oxygen reduced. **4 i a** potassium bromide → potassium + bromine **b** oxidation numbers: in KBr, +I for K and –I for Br; in the products K and Br_2, both 0 **c** Oxidation numbers have changed, so it is a redox reaction.
d Potassium is reduced and bromine oxidised.
4 ii a potassium iodide + chlorine → potassium chloride + iodine **b** oxidation numbers: in KI, +I for K and –I for I; in Cl_2, 0; in KCl, +I for K and –I for Cl; in I_2, 0
c Oxidation numbers have changed. **d** Iodine is oxidised and chlorine reduced. **5 a** C (s) + O_2 (g) → CO_2 (g)
b The oxidation numbers for the reactants are both 0. The oxidation numbers in carbon dioxide are +IV for carbon and –II for oxygen. Oxidation numbers have changed, so it is a redox reaction. **6** Yes. Oxidation numbers are *always* zero in elements, but *never* zero in compounds. So there is always a change in oxidation numbers during the reaction, which means it is a redox reaction.

page 79 1 a An oxidising agent oxidises another substance during a reaction, and is itself reduced.
b A reducing agent reduces another substance during a reaction, and is itself oxidised. **2 a** oxidising agent, oxygen; reducing agent, magnesium **b** oxidising agent, iron(III) oxide; reducing agent, carbon monoxide
3 a oxidising agent, chlorine; reducing agent, iron
b oxidising agent, copper(II) sulfate; reducing agent, iron
4 a In $KMnO_4$, the oxidation number of manganese is +VII. It has a strong drive to become the more stable ion with oxidation number +II. To achieve this reduction, it acts as a strong oxidising agent. **b** When aqueous potassium iodide is oxidised by an oxidising agent, the solution goes from colourless to red-brown. So this is a good visual test for the presence of an oxidising agent.

page 80 Core 1 a A <u>calcium</u> + oxygen → calcium oxide
B <u>carbon monoxide</u> + oxygen → carbon dioxide
C <u>methane</u> + oxygen → carbon dioxide + water
D copper(II) oxide + <u>carbon</u> → copper + carbon dioxide
E <u>iron</u> + oxygen → iron(III) oxide F iron(III) oxide + <u>carbon monoxide</u> → iron + carbon dioxide **b** In each equation above, the underlined reactant is oxidised, and the other reactant is reduced. **2 a** A yes (oxygen gained and lost) B yes (oxygen gained and lost) C no (it is a neutralisation) D yes (oxygen gained and lost)
E yes (oxygen gained and lost) **b i** oxidised: A Mg, B C, D Fe, E C. **ii** reduced: A CO_2, B SiO_2, D CuO, E PbO.

Extended 3 a magnesium chloride **b** Mg (s) + Cl_2 (g) → $MgCl_2$ (s) **c i** magnesium **ii** Mg → Mg^{2+} + 2e⁻
d i chlorine **ii** Cl_2 + 2e⁻ → 2Cl⁻

page 81 Extended 4 a lithium fluoride
b 2Li (s) + F_2 (g) → 2LiF (s)
c

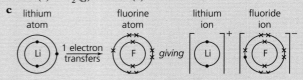

d i lithium **ii** Li → Li^+ + e⁻ **e** F_2 + 2e⁻ → 2F⁻
5 a displacement **b** solution turns orange **c i** Cl_2 + 2e⁻ → 2Cl⁻ **ii** reduced (it gains electrons) **d** 2Br⁻ → Br_2 + 2e⁻ **e** chlorine; it took the electrons from the bromide ions **f i** the iodide ion **ii** Br_2 (aq) + 2I⁻ (aq) → 2Br⁻ (aq) + I_2 (aq) **6 a** –I **b** The reactants are colourless but iodine is coloured. **c i** oxidised **ii** 2I⁻ → I_2 + 2e⁻ **d i** –I **ii** It goes down to –II in the compound water. **iii** H_2O_2 (aq) + 2H⁺ (aq) + 2e⁻ → $2H_2O$ (l) **7 a i** +III **ii** –III **iii** +IV **iv** +III **v** +I **vi** +II **b** Copper is a transition element, and transition elements can have more than one oxidation number in their compounds. **8 a** An acid must be present. **b** from purple to colourless **c** the Fe^{2+} ion
d If you add another drop of potassium manganate(VII) solution, the purple colour remains. **e** Fe^{2+} → Fe^{3+} + e⁻
9 a i +VI **ii** +VI **b** There is no change in oxidation number. **10 a i** silver chloride **ii** insoluble
b Oxidation numbers do not change so it is not a redox

reaction. **c** $2AgCl\ (s) \rightarrow 2Ag\ (s) + Cl_2\ (g)$. There are changes in oxidation numbers: Ag +I to 0, Cl −I to 0. So this is a redox reaction.

Chapter 7

page 83 **1** a substance that allows electricity to pass through it **2** Base your circuit on the middle photo at the top of page 83, but with a little mercury in the beaker. The carbon rods conduct. The bulb will light, showing that mercury conducts too. **3** The delocalised electrons can flow through the metal as a current. **4** Again use a circuit as in the middle photo at the top of page 83, but with dry salt in the beaker. The bulb will not light because the ions in the salt are not free to move. **5** It means breaking down a substance into two or more products. **6** It will conduct, and at the same time it will break down to the metal sodium, and chlorine gas. **7** It is a molecular compound (or mixture of molecular compounds).

page 85 **1 a** the process of breaking down a compound by passing a current through it **b** the liquid through which the current is passed **2 a** ionic compounds; the ions carry the current **b** molten or in solution (ions cannot move in solids) **3 a** to carry current into and out of the electrolyte **b** cathode **4** unreactive **5 a** sodium and chlorine **b** lead and sulfur **6 a** Bubbles of a colourless gas (hydrogen) will form at the cathode. Bubbles of a yellow-green gas (chlorine) will form at the anode. **b** Bubbles of hydrogen will form at the cathode, and bubbles of a colourless gas (oxygen) will form at the anode. **7 a** the same observations as for concentrated aqueous sodium chloride in **6a** **b** Bubbles of hydrogen will form at the cathode, and bubbles of oxygen at the anode. **c** A red-brown solid (copper) will form at the cathode, and bubbles of oxygen at the anode.

page 87 **1** At the cathode. The negative cathode attracts the positive ions, and supplies the electrons for reduction to take place. **2 a** $2Cl^- \rightarrow Cl_2 + 2e^-$ **b** $2O^{2-} \rightarrow O_2 + 4e^-$ **3 a** Sodium is more reactive, so has a stronger drive to remain in solution as ions. **b** from water molecules which dissociate **4** *Dilute*: the solution of sodium chloride remains. *Concentrated*: a solution of sodium hydroxide forms. **5** copper metal and chlorine (copper is less reactive than hydrogen) **6** In the answers for **b** and **c** below, the two half-equations have been balanced to give the same number of electrons in each. But you can also answer without balancing the equations.
a $2H^+ + 2e^- \rightarrow H_2$ $2Cl^- \rightarrow Cl_2 + 2e^-$
b $4H^+ + 4e^- \rightarrow 2H_2$ $4OH^- \rightarrow O_2 + 2H_2O + 4e^-$
c $2Cu^{2+} + 4e^- \rightarrow 2Cu$ $4OH^- \rightarrow O_2 + 2H_2O + 4e^-$

page 89 **1 a** The solution loses its colour because the blue copper ions are reduced to copper atoms at the cathode. **b** The solution keeps its colour because the concentration of ions in the solution does not change. While copper ions are discharged at the cathode, the anode dissolves to give more copper ions. **2** One electrode dissolves and the other gets larger. So they are taking part in the electrolysis, not just conducting current. **3** using electricity to coat one metal with another **4** to improve appearance and to prevent corrosion **5** to prevent steel from rusting / reacting with the soup **6 a** nickel **b** the cutlery **c** a solution of a soluble nickel salt such as nickel nitrate (all nitrates are soluble)

page 90 **Core** **1 a** The ions in the solid are not free to move. **b** bubbles of gas **c** bromine **d** lead **2 a** E **b** C **c i** F **ii** A **iii** C **iv** E **d i** B and D **ii** hydrochloric acid / sodium chloride (or the chloride of any other metal above hydrogen in the order of reactivity – except for lead chloride, which is insoluble)

page 91 **Core** **3 a** It contains ions which are free to move. **b i** They do not react. **ii** carbon and platinum **c** anode **d** bubbles of gas **e** A, chlorine, pale yellow-green gas; B, hydrogen, a colourless gas (so you would not see it) **f** Sodium metal would form at the cathode rather that hydrogen. The gas formed at the anode would be the same (chlorine).

Extended **4 a i** Na^+, Cl^-, H^+, OH^- **ii** Cu^{2+}, Cl^-, H^+, OH^- **b** The ions are attracted to the electrode of opposite charge. So negative ions (anions) move to the anode, and positive ions (cations) to the cathode.
c *For the sodium chloride solution*: **i** $2Cl^- \rightarrow Cl_2 + 2e^-$ **ii** $2H^+ + 2e^- \rightarrow H_2$. *For the copper(II) chloride solution*: **i** $2Cl^- \rightarrow Cl_2 + 2e^-$ **ii** $Cu^{2+} + 2e^- \rightarrow Cu$.
d Both contain chloride ions in a concentrated solution. **e i** Oxygen will then be given off in preference to chlorine. **ii** $4OH^- \rightarrow O_2 + 2H_2O + 4e^-$ **f** Each time, the ions of the *less reactive* substance are discharged. So H^+ ions are discharged in preference to Na^+ ions, and Cu^{2+} ions in preference to H^+ ions. **g** Concentrated hydrochloric acid, or a concentrated solution of any reactive metal chloride (such as potassium chloride) will give same result.
h copper(II) chloride **5 a i** Positive ions move through the solution to the cathode (−), where they accept electrons; negative ions move to the anode (+), where they give up electrons. **ii** In the wires, the electrons flow from the anode, through the battery, to the cathode.
b i / ii $Li^+ + e^- \rightarrow Li$ *reduction* $2Cl^- \rightarrow Cl_2 + 2e^-$ *oxidation* $2LiCl\ (l) \rightarrow 2Li\ (l) + Cl_2\ (g)$ *redox*
6 a lithium ion Li^+, chloride ion Cl^-, hydrogen ion H^+, hydroxide ion OH^- **b i** Oxygen is formed at the anode: $4OH^- \rightarrow O_2 + 2H_2O + 4e^-$ **ii** Hydrogen is formed at the cathode: $2H^+ + 2e^- \rightarrow H_2$ **c** dilute hydrochloric acid / dilute sulfuric acid / sodium hydroxide solution / any compound of a reactive metal, in a dilute solution
d Chlorine will form at the anode, instead of oxygen.
7 a chromium sulfate / nitrate / chloride **b i** Your diagram should be like the one on page 89, but with chromium in place of silver, and iron as cathode, in a solution of the chromium salt. **ii** They will travel through the wires, from the chromium anode to the iron cathode.
c The anode dissolves: $Cr \rightarrow Cr^{3+} + 3e^-$. Nickel deposits on the cathode: $Cr^{3+} + 3e^- \rightarrow Cr$ **d** at the anode
e It does not change because for each chromium ion deposited on the cathode, another forms at the anode.
8 a anode: $Ni \rightarrow Ni^{2+} + 2e^-$ cathode: $Ni^{2+} + 2e^- \rightarrow Ni$
b at the cathode, where the nickel ions gain electrons
c The anode gets smaller as it dissolves. **d** because for each nickel ion deposited at the cathode, another forms at the anode. **e** electroplating metal objects, for example steel cutlery, with nickel

Chapter 8

page 93 **1 a** 1 heat **2 a** In an exothermic reaction, the products have less energy than the reactants do. **b** In an endothermic reaction, the products have more energy than the reactants do. **3** Heat (thermal energy) is given out in the reaction between the silver and chloride ions. This heat goes to the surroundings, which include the solution in the beaker. **4** The reaction between the particles in the beaker takes in thermal energy from the surroundings, including the water under the beaker. This freezes because its temperature falls below 0 °C. **5 a** See diagram 2 on page 93. **b** See diagram 1 on page 92. **6** in their bonds **7** Yes; the thermal energy that causes the temperature rise can only have come from the chemical energy released when silver chloride forms.

page 95 **1** It shows how the energy changes during a reaction. **2** step 1, bond breaking; step 2, bond making **3 a** Energy must be taken in first, to break bonds. So the energy level rises. **b** There is a change in energy between the start and end of the reaction. / The energy level falls again after it rises, because energy is given out when new bonds form. **4** The energy taken in to break bonds is more than the energy given out when new bonds are made. **5 a** It is the minimum energy that colliding particles must have, in order to react; E_a. **b** It is the overall energy change during a reaction; ΔH. **6** because a certain amount of energy is needed to break bonds before a reaction can take place **7** It will be low: it will not take much energy to break weak bonds. **8** Choose from diagrams 3 and 4 on page 95.

page 97 **1** Enthalpy change (ΔH) = energy in to break bonds – energy out in forming bonds. **2** the kilojoule (a unit of energy) **3** $2H_2 + O_2 \rightarrow 2H_2O$; energy needed to break bonds = $(2 \times 436) + 498 = 1370$ kJ **4 a** the bonds that break: C=C (612) C–H (413) H–H (436); the bonds that form: C–C (346) C–H (413) **b** energy in [612 + 4(413) + 436] – energy out [346 + 6(413)] = – 124 kJ **5** A lot of energy is needed to break a triple bond: 946 kJ / mol.

page 99 **1 a** Both produce water, both give out energy, and both are redox reactions. **b** The energy is given out in different forms; the reaction in the fuel cell takes place slowly, while the combustion of hydrogen in air can be explosive. **2** No: it is water. **3** providing current for electric cars / electricity for buildings **4** No polluting gases are released into the atmosphere. **5** Water is a renewable energy source (rain keeps on falling) while the fossil fuels are non-renewable. Hydrogen is made by reacting the fossil fuel natural gas (methane) with steam (see page 122). The second product of that reaction is carbon dioxide, which causes global warming.
6 electrolysis

page 100 **Core** **1** *Exothermic* means the reaction gives out thermal energy to the surroundings. **b** It will rise. (The water around the reacting particles is part of the surroundings.) **c** the reactants **2 a** beaker, spatula, thermometer **b** a fall of 4 °C for ammonium nitrate, and a rise of 20 °C for calcium chloride **c i** calcium chloride **ii** There was a temperature rise in the solution. **iii** The energy level of the ions in solution is lower than their energy level in the solid.

d The answers are:

solution	Estimated temperature of solution / °C	
	ammonium nitrate	calcium chloride
i	17	65
ii	23	35
iii	21	45

3 a Add the label *energy*. **ii** X, reactants Y, products **b** endothermic **c** The temperature will fall. (The reacting particles take in thermal energy from their surroundings.)

page 101 **Extended** **4 a** The energy given out when new bonds form is *greater than* the energy needed to break bonds.
b i

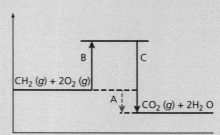

ii Some energy must be put in, to start the bonds breaking. A spark or flame can provide this energy. **c i** Energy is given out, overall.
ii exothermic **d** 55.6 kJ of energy is given out, overall.
5 a

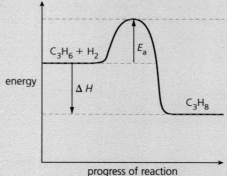

b i the enthalpy change **ii** Thermal energy is transferred to the surroundings. **c i** It is the minimum energy the colliding particles must have in order to react. **ii** E_a
iii See the diagram above. **6 a** bonds broken: 1 N–N 4 N–H 1 O=O **b** new bonds formed: 1 N≡N 4 O–H
c i 2220 kJ / mol **ii** 2801 kJ / mol **d** –581 kJ / mol
e exothermic **f** from the reacting particles to the surroundings (for example to the surrounding air)
7 a i A **ii** a negative sign **b** B **c** D **d** The energy needed to break the bonds in hydrogen peroxide is less that the energy released in making new bonds to form water and oxygen molecules. **e** The energy needed to break bonds is greater than the energy released in making new bonds. So the reaction takes in heat from its surroundings. The enthalpy change has a positive sign.
8a iii and **iv** are valid reasons. **b ii** A scarcity of refill stations is a temporary drawback. As the demand for fuel-cell cars rises, more refill stations will appear. (The cars will also fall in price – but by the time they become a lot cheaper, petrol-fuelled cars may have been banned.)

Chapter 9

page 103 **1** c a d b e **2** 60 km/hour **3** Only **b** would be suitable. **a** is unsuitable because zinc is a solid. **c** is unsuitable because the centimetre is a unit of length, and the zinc is in irregular pieces. (But it is simpler to focus on hydrogen produced.) **4 a** iron + sulfuric acid → iron(II) sulfate + hydrogen **b** Measure the rate at which iron or sulfuric acid is used up, or iron(II) sulfate or hydrogen is formed.

page 105 **1 a i** to remove the layer of magnesium oxide or any other impurity **ii** because the reaction starts the instant the reactants are mixed, and you need to measure the reaction time accurately **iii** to prevent hydrogen escaping into the air **b** to collect the hydrogen and let you measure its volume **2** It is over at the point where the curve goes flat. **3 a i** 29 cm^3 **ii** 39 cm^3 **b** 1.5 minutes **c i** 5 cm^3 of hydrogen per minute **ii** 0 cm^3 of hydrogen per minute (reaction over) **4** gas leaks in apparatus / poorly cleaned magnesium / not inserting stopper immediately / timer started too late or too early

page 107 **1 a i** 60 cm^3 **ii** 60 cm^3 **b** Faster reaction (B) gives steeper curve. **2** because the same amount of magnesium was used each time (with the acid in excess) **3** the time taken for the cross to disappear **4** A reaction goes *faster* when the concentration of a *reactant* is increased. It also goes *faster* when the *temperature* is raised. **5** For example: turning up the heat when something is cooking too slowly / applying paint in warm weather, when the drying reactions go faster / keeping unripe fruit in a warm place to help it ripen. **6** The rate slows. Ways we make use of this: storing food and vaccines in fridges, to stop them going off / deep-freezing some foods (such as pizza) / transporting food in refrigerated trucks / storing samples of body tissue in liquid nitrogen (very cold)

page 109 **1** For experiment 1 (large chips): **a** 0.55 g **b** 0.33 g /minute. For experiment 2 (small chips): **a** 0.95 g **b** 0.5 g / minute. **2** Increasing the surface area increases the rate of reaction. **3** mass of marble / temperature / concentration of acid / volume of acid **4** They have a greater surface area, giving faster reactions. **5** Hydrogen is very light so it would be very difficult to measure mass loss. **6** The particles are forced into a smaller volume, so the number of particles per unit volume increases. **7** The reaction will be slower. **8** Pressure has no effect on the volume of solids. The number of particles per unit volume of the solid will not change.

page 111 **1** Two particles can react together only if they *collide* and the *collision* has enough *energy* to be *successful*. **2** It is one where the colliding particles have sufficient energy (the activation energy) for a reaction to occur. **3** There are more gas particles per unit volume. So the frequency of collisions increases, which means the frequency of successful collisions increases. **4 a** There are twice as many acid particles, so the frequency of collisions between acid particles and magnesium atoms increases. So the frequency of successful collisions increases too. **b** The particles take in heat energy and move faster, so they collide more often, and the collisions have more energy. So the frequency of successful collisions increases. **c** Stirring helps to remove bubbles of hydrogen from the metal surface, and brings more acid particles to it – so the frequency of successful collisions increases. **d** Many more magnesium atoms are exposed when the metal is powdered, so the frequency of successful collisions increases.

page 113 **1** It is a substance that increases the rate of a reaction and is unchanged at the end of the reaction. **2** A catalyst does not change **b** or **c**. **3 a** It is not used up in the reaction. **b** There was no change in mass because catalysts are not used up. **4** a protein made in living cells, which acts as a catalyst for biological reactions; for example catalase / amylase **5** Without enzymes, most reactions in our bodies would be far too slow at body temperature. **6** Enzymes are destroyed (denatured) at high temperatures. **7 a** the minimum energy that colliding particles must have, for reaction to occur **b** Catalysts lower the activation energy so colliding particles need less energy to react. **8** Products are obtained faster. So a factory will be able to produce much more of a product per week, for example. With a catalyst, a reaction may even go fast enough at a lower temperature, so the factory saves on fuel costs.

page 116 **Core** **1 a** It holds the magnesium ribbon until you are ready to start the reaction. **ii** to measure gas volume **b** Shake the flask to get the magnesium into the acid. **2 a** more than is needed to react with that amount of magnesium **b** The graph on page 106 will help you. **c i** 14 cm^3 of hydrogen per minute in the first minute **ii** 9 cm^3 of hydrogen per minute in the second minute **iii** 8 cm^3 of hydrogen per minute in the third minute **d** Magnesium is used up as the reaction proceeds, so there is less for the acid to react with. **e** 40 cm^3 **f** 5 minutes **g** 8 cm^3 of hydrogen per minute **h** lower the temperature **3 a** Catalase acts as a catalyst. **b** The concentration of the acid is increased, so the reaction rate increases too. **c** The powder has a much larger total surface area than the ribbon, so burns much faster.

page 117 **Core** **4 a i** a gas syringe **ii** a stopwatch / timer **b** A: steeper curve **c** Here curves X and Y are shown on the graph:

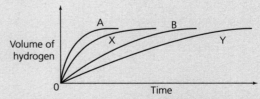

5 a a substance that increases the rate of a reaction and is unchanged at the end of the reaction **b** See the diagram on page 106. **c** oxygen **d** $2H_2O_2\,(aq) \rightarrow 2H_2O\,(l) + O_2\,(g)$ **e** Your graph should be similar in shape to those in this chapter. **f** The rate is greatest at the start, and decreases as the reaction proceeds. **g** It decreases. **h** water and copper(II) oxide **i** 0.5 g **j** The sketch should show a steeper curve, but still ending at 58 cm^3 of oxygen. **k** manganese(IV) oxide, or the enzyme catalase

Extended **6 a** carbon dioxide **b** You could measure the loss of mass of the flask over time, or collect the gas and measure its volume over time. **c i** The rate will increase. **ii** The rate will decrease. **d i** As the temperature rises the particles gain energy, move faster, and collide more frequently, with more energy. So the frequency of successful collisions increases, and the rate rises. **ii** Adding water reduces the concentration of the acid particles. So successful collisions between acid particles and calcium carbonate are less frequent, and the rate decreases. **e** Yes, it will increase. More particles are exposed, so successful collisions will be more frequent. **7 a** It will decrease. **b** zinc: all the iodine was used up **c** As the reaction proceeds the number of iodine particles falls, so the frequency of successful collisions decreases. **d**

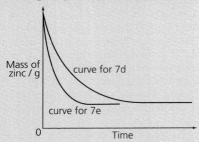

e Particles have more energy, so they move faster. So collisions are more frequent, and a higher percentage of them have the activation energy, E_a, needed for reaction to occur. **f i** It lowers the activation energy. **ii** A collision now needs less energy to be successful. So a higher percentage of collisions are successful, so the reaction rate rises. **8 a** Increasing the pressure means there are more gas particles per unit volume. So the frequency of successful collisions increases. So the rate rises. **b** There are no gases involved in the reaction. An increase in pressure has very little effect on the volume of a liquid.

Chapter 10

page 119 **1** one that can go backwards as well as forwards **2 a** $CuSO_4$ (no water molecules in the solid) **b** add water **3 a** The blue compound in the paper has reacted with water, which has bonded into the crystal structure, turning the compound pink. **b** The colour will turn blue again. **4 a** You can tell by the $\rightleftharpoons$ sign. **b** ammonia + hydrogen chloride → ammonium chloride **5 a** $N_2(g) + 3H_2(g) \rightarrow 2NH_3(g)$ $2NH_3(g) \rightarrow N_2(g) + 3H_2(g)$ **b** No. At equilibrium the rate of the forward and reverse reactions will be equal. **c** Both the forward and reverse reactions continue to take place.

page 121 **1** It is a reversible reaction. Reversible reactions reach equilibrium, and do not complete. **2** as near to 100% product as possible (that is, a yield of 100%) **3** It increases the yield because there are fewer molecules on the right side of the equation. (Pushing more molecules into a smaller space favours the forward reaction.) **4** A catalyst does not affect yield. **5** The position of equilibrium will shift to the left for both **a** and **b**. Both favour the reverse reaction, so the yield will fall. But **c** favours the forward reaction, so equilibrium will shift to the right and the yield will rise.

page 123 **1** the Haber process **2 a** air **b** natural gas **3 a** in the converter **b** iron **c** to give several chances for the gases to react **d** A liquid takes up a much smaller volume and is easier to transport. **4 a** A pressure of 400 atmospheres would be expensive to achieve, and there are safety concerns; the reaction would be too slow at 250 °C. **b** 200 atm and 450 °C **c** 28% **5** removing ammonia as it forms, and recycling the unreacted gases

page 125 **1 a** sulfur dioxide obtained by burning sulfur or roasting sulfide ores; oxygen from air **b** vanadium(V) oxide **2 a** $2SO_2(g) + O_2(g) \rightarrow 2SO_3(g)$ **b** $2SO_3(g) \rightarrow 2SO_2(g) + O_2(g)$ **3** to give an acceptable reaction rate **4 a** This gives several chances for the gases to react. **b** This prevents equilibrium being reached, and encourages the forward reaction to continue. **5 a** Cold water is piped to the catalyst beds to keep the temperature under control. / Sulfur dioxide is dissolved in concentrated sulfuric acid rather than in water (which would give a violent reaction). **b** using a low pressure (2 atm) / recycling unused gases

page 126 **Core** **1 a** to condense water vapour **b** The blue solid turns white. **c** It can go backwards as well as forwards. **d** Add water to the white solid left in the test-tube after heating. It will turn blue. **e** $CuSO_4.5H_2O(s) \rightleftharpoons CuSO_4(s) + 5H_2O(l)$ **2** Statements **b**, **c** and **d** are not correct. **3 a i** Heating releases water from the hydrated crystals. The water condenses at the top of the tube. **ii** pink → blue **iii** anhydrous cobalt(II) chloride, $CoCl_2$ **b** $CoCl_2.6H_2O(s) \rightleftharpoons CoCl_2(s) + 6H_2O(l)$ **c** It is used as a test for water, which turns blue cobalt(II) chloride pink.

page 127 **Extended** **4 a i** Nothing can enter or leave the reaction container. **ii** The forward and reverse reactions take place at the same rate, so the quantities of reactants and products present do not change. **iii** In industry, the aim is to increase the yield rather than let the reaction reach equilibrium. **b** Ammonia is removed (so that more will form) and unused gases are recycled. **c** A catalyst has no effect on these. (It does not affect yield.) **5 a** The gas mixture will turn purple. **b** There will be no further change in the intensity of the purple colour. **c** Diagram A is correct. It shows an endothermic reaction, where an increase in temperature favours the forward reaction. **d** no effect, because the number of gas molecules is the same on both sides of the equation **6 a** Equilibrium will shift left. **b** The forward reaction must be endothermic. **7 a i** the Contact process **ii** $2SO_2(g) + O_2(g) \rightleftharpoons 2SO_3(g)$ **b** burning sulfur, and roasting sulfide ores **c i** The forward reaction is exothermic, so the higher temperature favours the reverse (endothermic) reaction, giving a lower % conversion (a lower yield). **ii** It would be too slow, so not economic. **d i** The higher pressure favours the side of the equation with fewer molecules, giving a higher yield of sulfur trioxide. **ii** The costs would be too high. **e i** It increases the reaction rate of both the forward and reverse reactions. **ii** no change in % converted **8 a** actions **i** and **iii** **b** The number of molecules is the same on both sides of the equation, so equilibrium will not shift in response to a change in pressure. **c** Increasing the pressure will speed up both the forward and reverse reactions, so equilibrium is reached faster. (The molecules are pushed closer together so the frequency of successful collisions increases.) This is a benefit for industry. **9 a** natural gas (methane) **b** $N_2(g) + 3H_2(g) \rightleftharpoons 2NH_3(g)$ **c** exothermic; ΔH has a negative sign **d i** 4 molecules of reactants give 2 molecules of product. So an increase in pressure favours the forward reaction, and the yield of ammonia rises. **ii** Raising the temperature favours the endothermic (reverse) reaction, so the yield of ammonia falls. **e i** Increasing the pressure increases the rate of both reactions because molecules are pushed closer together, so the frequency of successful collisions rises. **ii** Increasing the temperature increases the rate of both reactions, because collisions have more energy, so the frequency of successful collisions rises. **f** to give an acceptable rate **g i** iron **ii** no change in yield

Chapter 11

page 129 **1** reacts with substances it comes into contact with **2** See the list on page 128. **3** an oxide or hydroxide of a metal **4 a** They are soluble bases. **b** any two from the examples named on page 129 **5** It indicates, by its colour, whether a solution is acidic or alkaline.
6 a acidic **b** alkaline **c** alkaline **7** from colourless in acids to blue in alkalis (and vice versa)

page 131 **1** Their solutions contain H^+ ions. **2** Their solutions contain OH^- ions. **3** pH 5 **4 a** weakly alkaline **b** weakly acidic **c** neutral **d** strongly acidic **e** strongly alkaline **f** more acidic than **b** but not strongly acidic **5** the solution in **b** **6** B **7** universal indicator **8** Y **9** The pH will rise, because adding water lowers the concentration of H^+ ions.

page 133 **1 a** zinc + sulfuric acid → zinc sulfate + hydrogen **b** sodium carbonate + sulfuric acid → sodium sulfate + water + carbon dioxide **2** the one in **a**
3 calcium chloride, Ca^{2+}, Cl^- **4** Use this reaction: zinc oxide + nitric acid → zinc nitrate + water **5 a** Both reactions produce calcium chloride and water. **b** Only the reaction with the carbonate produces carbon dioxide.
6 $H^+ (aq) + OH^- (aq) → H_2O (l)$ **7 a** It is a base, so it will neutralise the acidity. **b** water (since neutralisation takes place)

page 135 **1 a** It is an equation showing only the ions involved in the reaction. **b** $H^+ (aq) + OH^- (aq) → H_2O (l)$ **2** ions that are present in the solution but not involved in the reaction **3** The hydrogen atom contains 1 proton and 1 electron (no neutron). When it becomes an ion it loses the electron, leaving just a proton. **4 a** They transfer a hydrogen ion (a proton) to a base.
b They accept a hydrogen ion (a proton) from an acid.
5 Oxidation numbers remain unchanged: H +I, O –II, Cl –I, Mg +II. **6 a i** $2H^+ (aq) + 2Cl^- (aq) + Mg^{2+} (s) + O^{2-} (s) → Mg^{2+} (aq) + 2Cl^- (aq) + H_2O (l)$ **ii** Cross out the spectator ions, which are $Mg^{2+} (aq)$ and $2Cl^- (aq)$.
iii $2H^+ (aq) + O^{2-} (s) → H_2O (l)$ **7** $2H^+ (aq) + CO_3^{2-} (aq) → H_2O (l) + CO_2 (g)$ **8** $Mg(OH)_2 (s) + 2HCl (aq) → MgCl_2 (aq) + 2H_2O (l)$

page 137 **1** Show that it reacts with an acid to form a salt. **2** iron – it reacts more vigorously than copper does **3** missing words: *basic, acidic* **4** carbon, sulfur, phosphorus (least reactive first) **5** They are acidic oxides which dissolve in water to form acids. **6** It will react with both acids and bases; examples are zinc oxide and aluminium oxide.

page 139 **1** zinc / zinc oxide / zinc carbonate and hydrochloric acid **2 a** reaction too slow, since lead is not very reactive **b** Use lead oxide / carbonate and nitric acid.

3 There was more zinc than was needed to react with the acid (so some will remain unreacted).
4 a to find the exact volumes of acid and alkali that react together, with no excess of either **b** evaporate the solution to remove water **c** It would be an impurity.

page 141 **1** See the table at the top of page 140.
2 carbonates **3 a** It contains water molecules chemically bonded into the crystals. **b** $CuSO_4$ **4** any soluble salt (see the table at the top of page 140) **5** Choose suitable soluble salts using the table at the top of page 140. Such as:
a calcium nitrate + sodium sulfate **b** magnesium sulfate + potassium carbonate **6 a** $Ca^{2+} (aq) + SO_4^{2-} (aq) → CaSO_4 (s)$ **b** $Mg^{2+} (aq) + CO_3^{2-} (aq) → MgCO_3 (s)$ **7 a** They have water molecules chemically bonded into their crystal structure. There are 10 water molecules for each unit of sodium sulfate. If the crystals are heated, the water will be driven off. **b** the water bonded within the crystals in hydrated salts

page 143 **1** The end-point in a titration is where a single drop of added liquid changes the colour of the indicator fully (not just flashes of colour). **2** 50 cm³ **3 a** methyl orange / thymolphthalein **b** $2NaOH (aq) + H_2SO_4 (aq) → Na_2SO_4 (aq) + H_2O (l)$ **c** 1.6 mol/dm³

page 144 Core **1** The correct choices are, in order: bases, sulfates, metals, hydrogen, carbonates, carbon dioxide, good, red, colourless, pH number, lower.
2 a B **b** carbon dioxide **c** A and B react together to form a soluble salt. **3 a i** copper(II) oxide **ii** calcium oxide **iii** phosphorus(V) oxide **b** The paper turns blue. **c** ammonia **d i** copper(II) oxide + sulfuric acid → copper(II) sulfate + water **ii** neutralisation
e sulfur dioxide / sulfur trioxide / nitrogen dioxide

page 145 Core **4 i** The missing chemicals are, in order: **b** zinc chloride, hydrogen **c** sulfuric acid
d hydrochloric acid, sodium carbonate, carbon dioxide **e** sulfuric acid, iron, hydrogen **f** alkali, sodium nitrate, water **g** sulfuric acid, water **h** carbonate, sulfuric acid, copper(II) carbonate, water
4 ii a $Ca(OH)_2 (aq) + 2HNO_3 (aq) → Ca(NO_3)_2 (aq) + 2H_2O (l)$ **b** $Zn (s) + 2HCl (aq) → ZnCl_2 (aq) + H_2 (g)$
c $2KOH (aq) + H_2SO_4 (aq) → K_2SO_4 (aq) + 2H_2O (l)$
d $Na_2CO_3 (s) + 2HCl (aq) → 2NaCl (aq) + H_2O (l) + CO_2 (g)$
e $Fe (s) + H_2SO_4 (aq) → FeSO_4 (aq) + H_2 (g)$
f $NaOH (aq) + HNO_3 (aq) → NaNO_3 (aq) + H_2O (l)$
g $CuO (s) + H_2SO_4 (aq) → CuSO_4 (aq) + H_2O (l)$
h $CuCO_3 (s) + H_2SO_4 (aq) → CuSO_4 (aq) + H_2O (l) + CO_2 (g)$
5 a i carbon dioxide **ii** copper(II) carbonate + ethanoic acid → copper(II) ethanoate + water **iii** No more gas bubbles off, and some solid remains in the beaker. **b i** copper(II) carbonate; after reaction, some remains in the beaker **ii** ethanoic acid **c** to make sure the reaction is quick **d** copper(II) carbonate
e • put some ethanoic acid into a beaker
- add a spatula measure of copper(II) carbonate powder, and stir
- continue to add copper(II) carbonate until no further reaction
- filter the mixture into an evaporating basin, to remove unreacted copper(II) carbonate

- put the evaporating basin on a tripod and gauze, and evaporate some of the water, to obtain a saturated solution
- leave to cool and crystallise
- remove the crystals by filtering or decanting
- rinse and dry the crystals

f copper(II) oxide or copper(II) hydroxide

Extended 6 a sulfuric acid **b** $MgO\ (s) + H_2SO_4\ (aq) \rightarrow MgSO_4\ (aq) + H_2O\ (l)$ **c i** strong **ii** the hydrogen ion, H^+ **d i** a chemical that neutralises an acid to form a salt and water **ii** $O^{2-} + 2H^+ \rightarrow H_2O$ **7 i** for example, sodium chloride (or any other soluble chloride) and silver nitrate **ii** precipitation **b i** $Ag^+\ (aq) + Cl^-\ (aq) \rightarrow AgCl\ (s)$ **ii** sodium ion (or ion of other chosen metal), nitrate ion **8 a** contains water of crystallisation **b** $Na_2CO_3\ (aq) + 2HCl\ (aq) \rightarrow 2NaCl\ (aq) + H_2O\ (l) + CO_2\ (g)$ **c** 0.014 moles **d i** 0.007 moles **ii** 0.742 g **e i** 1.258 g **ii** 0.07 moles **f** 10 moles **g** $Na_2CO_3.10H_2O$

Chapter 12

page 147 1 a period **b** group **2 a** symbol N, non-metal, proton number 7, mass number 14, Group V, Period 2 **b** symbol Mg, metal, proton number 12, mass number 24, Group II, Period 3 **c** symbol Ar, non-metal, proton number 18, mass number 40, Group VIII, Period 3 **3** from Group III (with only one non-metal, boron) to Group VIII **4** The number of outer shell electrons is equal to the group number, in Groups I to VII. **5** Their outer shells are full, and this is a stable arrangement. **6 a** francium **b** americium **c** einsteinium

page 149 1 a soft **b** reactive **2 a** Melting points gradually change as you go down the group. **b** the density of potassium **3** Their oxides and hydroxides form alkaline solutions. **4** All contain the same number of electrons in their outer shells. **5** They need to lose only one electron to gain a full outer shell. **6** too soft, and too reactive **7 a** $2K\ (s) + Cl_2\ (g) \rightarrow 2KCl\ (s)$ **b** colourless **c** Yes. It contains potassium ions and chloride ions which are free to move, carrying the current. **8 a** There will be very vigorous reaction; the products will be hydrogen, which bursts into flame, and an alkaline solution of rubidium hydroxide. **b** It will burn very brightly; the product will be an ionic white solid, rubidium chloride.

page 151 1 See the table at the top of page 150. **2 a** Reactivity decreases down the group. **b** No, it is the opposite. **3** All three products are coloured solids with the formula FeX_3. **4 a** is correct; the reaction shows that it is more reactive than chlorine, and reactivity increases up Group VII. **5** Their atoms need just 1 more electron to give a full, stable, outer shell. They have a strong drive to obtain it. **6 a** bromine + potassium iodide → potassium bromide + iodine; the solution will turn red-brown **b** Bromine is more reactive than iodine – it has a stronger drive to obtain electrons. So it displaces iodide ions from the solution. **7 a** a solid, to fit with the trend down the group **b** coloured **c** harmful

page 153 1 The number increases by 1 each time. **2 a** They need to *lose* 3 electrons to obtain a full (stable) outer shell. **b** They need to *gain* 2 electrons to obtain a full outer shell. **c** They need to *gain* just 1 electron to obtain a full outer shell. **3** N^{3-} **4 a** helium, neon, and radon (which is radioactive) **b** The given data shows that boiling points increase down the group; actual values for the other noble gases are helium −269 °C, neon −246 °C, and radon −61.7 °C. **5** has both metallic and non-metallic properties **6 a** a change from metal to non-metal / reactivity falls and then rises / boiling points rise and then fall / change in oxides: basic → amphoteric → acidic **b** Yes, elements from the same group have similar properties, so the trend across the numbered groups in Period 4 should be the same (that is, excluding the transition elements)

page 155 1 iron, copper, nickel etc. **2 a** hard **b** high density **c** high melting point **d** unreactive with water **3 a** They are used for their colour. **b** It is strong and hard (and especially when made into alloys – see later). **4 a** a substance that increases the rate of a reaction and is unchanged at the end of the reaction **5** It is a liquid at room temperature. **6** one **7 a** CuCl **b** $FeCO_3$ **8 a** copper(II) chloride **b** nickel(II) oxide **c** iron(III) bromide

page 158 Core 1 a Period 2 **b** neon (Ne); its atoms already have a full (stable) outer electron shell so do not need to react to obtain one. **c** Group I (the alkali metals) **d** K (potassium) **2 a** Melting point and boiling points decrease, reactivity increases. **b** The actual values for rubidium are: melting point 39 °C, boiling point 688 °C. It is extremely reactive. **c i** 5 **ii** 37 **iii** 1 **3 a** unreactive **b** Choose from these: helium 2 neon 2,8 argon 2,8,8 krypton 2,8,18,8. (You will not be asked about electronic configurations beyond element 20 of the Periodic Table, in the exam.) **c i** exists as single atoms **ii** Their atoms already have full outer electron shells so do not need to share electrons. **d** Their atoms gain, lose, or share electrons, and in this way obtain full outer electron shells, like the noble gases.

page 159 Core 4 a the halogens **b i** like chlorine but more explosively **ii** hydrogen + fluorine → hydrogen fluoride **c i** like chlorine but less explosively **ii** $H_2\ (g) + Br_2\ (g) \rightarrow 2HBr\ (g)$ **5 a** the second row of the Periodic Table (begins with lithium) **b i** The modern Group I does not contain H, Cu, or Ag. **ii** It contains reactive metals with similar properties; by contrast, copper and silver are unreactive and hydrogen is not a metal. **iii** alkali metals **c i** halogens **ii** It is a metal. **iii** It is one of the transition elements. **d** to allow for elements that had not yet been discovered at that time **e** The noble gases were not discovered until later – largely because they are unreactive.

Extended 6 a A alkali metal, **B** halogen, **C** transition element, **D** Group II (alkaline earth metal), **E** halogen, **F** transition element **b i A** towards the bottom of the group, **B** towards the top of the group **ii** The outer shell contains 1 electron in **A**, and 7 in **B** **c** Atoms react in order to gain full outer shells of electrons. The easier it is to gain or lose electrons to achieve this, the more reactive an element will be. If its atoms already have a full outer shell, an element will be unreactive. **d A** rubidium or caesium, **B** chlorine, **C** any transition element, e.g. iron, **D** magnesium, **E** bromine, **F** iron **e** transition elements **C** and **F** **7 a i** Melting points increase down the group. **ii** Density increases down the group. **b** The actual values are melting point 450 °C, density 6.2 g/cm³. **c** Polonium is a metal.

Chapter 13

page 161 **1** See the physical properties on page 160. **2 a** tin **b** 2.64 times more dense **3** Mercury has a much lower melting point than iron – it is a liquid at room temperature. **4 a** Metals are good conductors of electricity, while non-metals (apart from carbon) do not conduct. **b** Metals are good conductors of heat, while non-metals are not. **c** Metals can be hammered into different shapes, but non-metals break into pieces if hammered. **d** Metals usually have high melting and boiling points, while most non-metals have low melting and boiling points. **5** phosphorus and sulfur **6** Most metal oxides are basic (some are amphoteric), but non-metal oxides are acidic. **7** Unlike other non-metals, carbon (graphite) conducts electricity.

page 163 **1 a** potassium **b** potassium **c** All three react to give hydrogen and the metal hydroxide. **2** magnesium, zinc and iron **3** Hydrogen is driven out from these compounds. **4 a** zinc; it displaces hydrogen from hydrochloric acid **b** hydrogen; copper does not displace hydrogen from hydrochloric acid **5** All form oxides. **6 a** yes **b** potassium most reactive, gold least reactive

page 165 **1** carbon **2 a i** no reaction **ii** The carbon will reduce the zinc oxide. **b** zinc oxide + carbon → zinc + carbon dioxide **3** aluminium **4** Iron takes copper's place as ions in the solution. **5** Copper is the more reactive metal, so it displaces silver. It forms blue copper ions, while the silver is deposited on the wire. **6 a** $Cu(s) \rightarrow Cu^{2+}(aq) + 2e^-$, $Ag^+(aq) + e^- \rightarrow Ag(s)$ **b** $Cu(s) + 2Ag^+(aq) \rightarrow Cu^{2+}(aq) + 2Ag(s)$

page 167 **1 a** See the list on page 166. **b** K and Na in Group I, Ca and Mg in Group II, Al in Group III, and the rest are transition elements **c** Group I **d** in the block of transition elements **2** It is too reactive to remain uncombined. **3** Heat the oxides with carbon *or* with a more reactive metal than iron (but this would not be economic). **4 a** Carbon is less reactive than aluminium. **b** Heat the oxide with a more reactive metal than aluminium (for example potassium) or carry out electrolysis (as in industry). **5** Aluminium instantly forms a thin protective coat of oxide, which prevents further corrosion. **6** $2Al + Fe_2O_3 \rightarrow Al_2O_3 + 2Fe$

page 169 **1 a** $Fe_2O_3.2H_2O$ **b** It contains water bonded into its crystals. **2** oxygen and water **3** Place iron nails or iron turnings in two test-tubes, one containing damp oxygen, and the other damp nitrogen. Observe the extent of rusting. **4** The coating of grease prevents oxygen and moisture from reaching the iron. **5 a** Another metal corrodes instead of iron, to protect the iron. **b** They are both more reactive than iron. **c** Its protective coat of oxide prevents further corrosion. **6 a** zinc **b** It creates a barrier between iron and the atmosphere, but will also provide sacrificial protection if the barrier gets damaged.

page 170 Core 1 a the mass of a substance per cm³ **b** sodium, calcium, magnesium, aluminium, iron, copper, lead, gold **c** iron (density is 158 ÷ 20 = 7.9 g/cm³) **d i** sodium, calcium, magnesium, aluminium, iron, copper, gold **ii** 0.97 g/cm³ **iii** 19.3 g/cm³ **iv** Yes. In general the more reactive the metal, the lower the density. **e** aluminium because it has low density and is unreactive (due to its oxide layer) **2 a** iron + oxygen + water → hydrated iron(III) oxide **b** to speed up the rusting process (so you are not relying only on water vapour) **c** because rusting uses up the oxygen from the air in the test-tube **d** 20.8% **e** the same result, since the amount of oxygen in the test-tube will not change **f** Grease acts as a barrier to keep oxygen away from the damp iron.

page 171 Core 3 a zinc (or iron or magnesium); zinc + copper(II) sulfate → zinc sulfate + copper **b** iron (or zinc); iron + hydrochloric acid → iron(II) chloride + hydrogen **c** sodium (or potassium or lithium); sodium + water → sodium hydroxide + hydrogen **d** magnesium; magnesium + steam → magnesium oxide + hydrogen **4 a** pairs **ii** and **iii** **b** carbon + <u>copper(II) oxide</u> → copper + carbon dioxide; magnesium + <u>carbon dioxide</u> → magnesium oxide + carbon

Extended 5 a magnesium + copper(II) sulfate → magnesium sulfate + copper **b** because magnesium is more reactive that copper **c i** $Mg \rightarrow Mg^{2+} + 2e^-$ **ii** oxidation **d i** $Cu^{2+} + 2e^- \rightarrow Cu$ **ii** reduction **iii** magnesium **e i** $Mg + Cu^{2+} \rightarrow Mg^{2+} + Cu$ **ii** redox **f i** yes **ii** no **iii** yes **g i** For iron + copper(II) sulfate: brown copper appears, the solution goes from blue to pale green. For zinc + silver nitrate: zinc disappears and a silvery metal forms, but the solution remains colourless. **ii** $Fe(s) + Cu^{2+}(aq) \rightarrow Fe^{2+}(aq) + Cu(s)$; $Zn(s) + 2Ag^+(aq) \rightarrow Zn^{2+}(aq) + 2Ag(s)$ **6 a** <u>reduction</u> and <u>oxidation</u> **b i** magnesium **ii** copper(II) oxide **c** 2 electrons are transferred from each magnesium atom to each copper ion **d** Magnesium is more reactive than copper so has a stronger drive to give up electrons and exist as ions. **e i** calcium, sodium, potassium **ii** It loses electrons more easily than magnesium does. **7 a i** The metal disappears while gas bubbles off; the colourless solution goes green, due to the Cr^{3+} ion. **ii** $2Cr(s) + 6HCl(aq) \rightarrow 2CrCl_3(aq) + 3H_2(g)$ **b i** The powder disappears, a brown solid appears (copper), and the blue solution turns green. **ii** $2Cr(s) + 3Cu^{2+}(aq) \rightarrow 2Cr^{3+}(aq) + 3Cu(s)$ **c i** zinc **ii** $Cr_2O_3 + 3Zn \rightarrow 3ZnO + 2Cr$ **d** $Cr^{3+} + 3e^- \rightarrow Cr$ **e** It is less reactive than sodium. **8** Reactions **a** and **d** occur. **9 a i** It must be more reactive than iron. **ii** zinc or magnesium **iii** $Zn \rightarrow Zn^{2+} + 2e^-$, $Mg \rightarrow Mg^{2+} + 2e^-$ **9 iv** By giving up electrons, the second metal prevents the steel from giving up electrons and becoming oxidised. **b** sacrificial protection

Chapter 14

page 173 **1** oxygen and silicon **2** aluminium and iron **3** Potassium, sodium, magnesium and calcium, which are among the most reactive metals, together form 12% of Earth's crust. This is much more than the total % of all the other metals, apart from aluminium and iron. So they are *relatively plentiful among the metals*. (And even 1% of Earth's crust is an enormous quantity.) **4** rock from which a metal is obtained **5** It is unreactive and has not formed compounds with other elements. **6 a** hematite, iron(III) oxide **b** bauxite, aluminium oxide **7 a** Carbon is less reactive than aluminium.   **b** Carbon / carbon monoxide can be used – and cost much less than electrolysis.   **c** The iron(III) oxide loses oxygen, giving iron.

page 175   **1** iron ore / hematite, limestone and coke   **2** It is hot air, needed to start the coke burning.   **3 a** carbon + carbon dioxide → carbon monoxide   **b** iron(III) oxide + carbon monoxide → iron + carbon dioxide   **c** calcium carbonate → calcium oxide + carbon dioxide   **4 a** calcium silicate   **b** It forms in the reaction between calcium oxide and the silicon(IV) oxide from sand.   **5** No, it still contains some carbon and sand.   **6 a** oxygen   **b** iron(III) oxide

page 177   **1 a** bauxite   **b** aluminium oxide   **c** electrolysis   **2 a** It allows electrolysis to be carried out at a lower temperature.   **b** It is used as a solvent for the aluminium oxide.   **3 a** They are negatively charged, so are attracted to the positive anode.   **b** $6O^{2-}(l) \rightarrow 3O_2(g) + 12e^-$ (or $2O^{2-} \rightarrow O_2 + 4e^-$)   **c** It is made of carbon, and the oxygen reacts with it.   **4 a** It lines the cell.   **b** $4Al(l) + 12e^- \rightarrow 4Al(l)$ **c** It is drained from the bottom of the cell.

page 179   **1** An anchor needs to be heavy and unmoveable on the sea floor; aluminium is too light.   **b** Copper's high density means it would need more support than aluminium, with shorter distances between pylons.   **2 a** Aluminium is non-toxic / light / can be rolled into thin sheets / resists corrosion.   **b** Copper is malleable (so can be made into pipes, and bent) / non-toxic / resists corrosion.   **3** Zinc protects steel from rusting (by acting as a barrier to air and moisture, and by sacrificial protection)   **4** Silver is scarce compared to copper / much more expensive.

page 181   **1** a mixture of a metal with at least one other element   **2** Pure iron is too soft, ductile, and corrodable for most uses. Turning it into an alloy makes it much more useful.   **3 a** an alloy of iron   **b** stronger / harder / resistant to corrosion / more attractive in appearance   **4 a** bronze   **b** The atoms of tin must enter the copper lattice.   **c** Bronze is harder and more sonorous than copper, and more resistant to corrosion.   **d** See the table on page 181.

page 184   Core   **1 a** iron(III) oxide, aluminium oxide, sodium chloride   **b i** iron   **ii** sodium   **iii** aluminium   **c** iron, aluminium, sodium (in order of increasing reactivity)   **d i** by electrolysis   **ii** too expensive   **e i** by heating with carbon   **ii** Oxygen is removed from the iron(III) oxide.   **f i** electrolysis **ii** heating with carbon   **2 a** low density (light)   **b** resistant to corrosion   **c** low density and good electrical conductivity   **3 a** See the diagram on page 174.   **b i** hematite, coke (coal), limestone   **ii** Hematite is the iron ore; coke is used to produce carbon monoxide, the gas which brings about the reduction; limestone is used to remove acidic impurities such as sand.   **c i** slag   **ii** road building   **d i** nitrogen / carbon dioxide   **ii** It is hot, so is used to heat the incoming air blast.   **e i** iron(III) oxide + carbon monoxide → iron + carbon dioxide **ii** The iron(III) oxide loses oxygen.   **iii** carbon monoxide

page 185   Core   **4 a** rock from which a metal is extracted **b** contains a low % of the metal   **c** 990 kg   **d i** 25%   **ii** Copper has many important uses, but it is not an abundant metal. The price will rise if the demand for it grows, making a low-grade ore economical to mine.   **e i** less reactive   **ii** carbon or carbon monoxide   **iii** reduction   **f** good conductor of electricity, ductile   **5 a** mixture where at least one other substance is added to a metal, to improve its properties   **b** A is stainless steel and B is brass.   **c i** alloy A (stainless steel)   **ii** hard, resists corrosion, shiny appearance, non-toxic   **d** Pure iron would not be hard or strong enough, and would rust, so the joint would wear away.   **e** Brass is: harder and stronger than copper / an attractive colour / shiny.   **f** good electrical conductivity / good thermal conductivity

Extended   **6 a i** $C + CO_2 \rightarrow 2CO$ **ii** $Fe_2O_3 + 3CO \rightarrow 2Fe + 3CO_2$ **iii** $Al^{3+} + 3e^- \rightarrow Al$ **b** In each case a high temperature is used / molten metal forms and is drawn off.   **c i** to remove acidic impurities   **ii** as the electrodes   **iii** to dissolve the aluminium oxide   **d** $CaCO_3 \rightarrow CaO + CO_2$ then $CaO + SiO_2 \rightarrow CaSiO_3$ **e** Impurities must be removed from aluminium ore (bauxite) before electrolysis, and that costs money. Electrolysis needs electricity, which is expensive. The carbon anodes must be replaced regularly. By contrast impurities can be removed from hematite in the blast furnace, and the fuel for the extraction is coal, which is cheap.   **f i** See the diagrams on page 181.   **ii** With different atoms in the metal lattice, the layers can no longer slide over each other.

Chapter 15

page 187   **1** See points on page 186.   **2** We are harmed by: waste material in air including particulates / the gases in air that cause global warming / microbes and toxic waste substances in water.   **3** There is no right or wrong answer. You should justify your answer in terms of how widespread and damaging the negative impact is, and what can be done to reduce it. Many would consider our impact on the climate to be the biggest issue.   **4 a** tiny particles (solid and liquid) suspended in air   **b** waste from toilets and sinks in homes   **c** substances added to soil to help plants grow   **d** microscopic organisms such as bacteria and viruses   **5** more drinking water needed, so more water pumped from rivers and aquifers / more sewage to be disposed of / more food needed, so more intensive farming with greater use of fertilisers (and water for irrigation) / more fuel needed, so more land needed for wind farms, solar farms and so on, competing with crop land and animal habitats

page 189   **1** through their gills   **2 a/b** plastics, dumped by humans / nitrates from fertilisers / phosphates from fertilisers, shampoos and detergents   **3** Rivers, and the rain that feeds rivers, dissolve calcium compounds from the limestone rock.   **4** They promote healthy bones and teeth and help to protect us against heart disease.   **5 a** lead / arsenic / mercury   **b** mines or factories   **6** The water will contain harmful bacteria.

page 191   **1** See the flowchart on page 190.   **2 a** removing solid particles from water by letting them settle to the bottom of the tank   **b** making fine suspended particles stick together to form larger particles that can be skimmed off   **c** the chemical that brings about

flocculation   **3** It kills harmful microbes.   **4** It helps in fighting tooth decay.   **5** It still contains dissolved substances, and some may be at an unsafe level.
6 a It contains impurities, including metal salts.
b distilled water, which has a much higher level of purity
7 a The cobalt(II) chloride paper will turn pink.
b Measure the temperatures at which it boils and freezes; only pure water boils at 100 °C and freezes at 0 °C.

page 193   **1 a** a substance added to the soil to make it more fertile **2** to improve crops yields   **3 a** nitrogen, phosphorus, potassium   **b** Examples are: ammonium nitrate + potassium phosphate or potassium nitrate + ammonium phosphate.   **4** They promote the growth of algae. When these die, bacteria feed on them, using up the oxygen dissolved in the water. So fish and other river life are starved of oxygen.   **5 a** 21.2%   **b** 1 kg of ammonium nitrate   **c** It is very soluble. Rain water could carry it out of the soil, to streams and rivers.

page 195   **1 a** 78   **b** 21   **2** See the pie chart on page 194.
3 It also contains water vapour, waste gases, and particulates.   **4** The atmosphere thins out as altitude increases. At high altitudes there is not enough oxygen, so breathing becomes difficult.   **5** oxygen; many things burn in it   **6** In our cells, glucose reacts with the oxygen we breathe in, and the reaction provides energy.   **7 a** copper(II) oxide, CuO   **b** Pass hydrogen over the hot oxide (see page 72) or heat it with carbon or a more reactive metal such as zinc.

page 197   **1 a** carbon dioxide, carbon monoxide, particulates   **b** nitrogen oxides; the nitrogen and oxygen in air react in hot engines and furnaces   **c** sulfur dioxide, nitrogen oxides   **2** photochemical smog, caused by the effect of sunlight on nitrogen oxides   **3** methane + oxygen → carbon dioxide + water   **4** prevents oxygen being carried around the body   **5 a** calcium oxide + sulfur dioxide → calcium sulfite   **b** using low-sulfur fuels, and installing catalytic converters in cars   **6 a** platinum, palladium, rhodium   **b** $2NO(g) \rightarrow N_2(g) + O_2(g)$ / $2CO(g) + 2NO(g) \rightarrow 2CO_2(g) + N_2(g)$

page 199   **1 a** gases that trap heat in the atmosphere
b Without them the temperature on Earth would be too low to support life.   **c** An increase in the level of greenhouse gases means global air temperatures are rising, causing climate change.   **2 a/b** carbon dioxide through combustion of fossil fuels; methane through rearing grazing animals   **3** the rise in average temperatures around the world   **4** They absorb heat that is reflected from Earth, preventing it from escaping into space.
5 Global warming leads to climate change.   **6 a** In some places farmers will be unable to grow their usual crops / drought will kill crops / new crop pests will arrive. In other places a warmer climate will help crop production.   **b** In some places, lower rainfall / shrinking glaciers will reduce the water supply. But where rainfall increases, there will more water available on and below the ground, for a water supply.   **7** It will affect us all because it affects the natural environment we depend on.

page 201   **1 a** wind   **b** moving water   **2** solar power
3 a source of energy you can keep using, that will not get used up   **4** Examples are energy from sunlight, waves, tides, and fast-flowing rivers. These sources do not get used up: the sun continues to shine, waves roll, tides rise and fall, and rivers flow.   **5** Trees take in carbon dioxide for photosynthesis.   **6** See the equation on page 201.
7 a It can be charged using non-carbon electricity.
b Grazing animals, unlike fish and chicken, add methane to the atmosphere.   **8** You can do something, however small; for example, switch off a light that is not needed, walk or cycle rather than take a petrol- or diesel-fuelled vehicle, adjust your diet.

page 202   **Core**   **1 a** *nitrogen / oxygen / carbon dioxide / noble gases*   **b i** incomplete combustion of fuels in car engines   **ii** They increase the risk of respiratory problems and cancer.   **c i** the combustion of fossil fuels containing sulfur   **ii** causes acid rain   **iii** removing sulfur from fossil fuels before combustion / using calcium oxide to remove sulfur dioxide from flue gases (flue gas desulfurisation)
2 a River water will contain higher quantities of harmful substances, because rivers flow past farmland, industrial areas and built-up areas, collecting substances as they go.
b i filtration   **ii** insoluble impurities   **c i** lead, arsenic, mercury   **ii** Carbon removes substances responsible for bad tastes and smells.   **d i** Some microbes cause illness.
ii chlorine   **e** Calcium is needed for healthy bones and teeth. Calcium and magnesium help to protect us against heart disease.

page 203   **Core**   **3 a i** white → blue   **ii** blue → pink
b i You could measure its boiling point and freezing / melting point. **ii** boils at 100 °C, freezes (melts) at 0 °C
c Distilling removes impurities such as metal salts.
4 a the amount of a crop harvested per unit of land area
b fertilisers   **c i** calcium phosphate   **ii** phosphorus
d i ammonium nitrate   **ii** nitrogen **e i** They may be washed off the land and into rivers by rain.   **ii** phosphate ion, nitrate ion   **iii** It decreases.   **iv** Fish suffocate.
5 a i the rise in average temperatures around the world   **ii** See the bullet points on page 229.   **b** reduce livestock farming of grazing (ruminant) animals
c carbon dioxide   **d** Greenhouse gas emissions from any country affect all countries, so governments need to cooperate to tackle the problem. / New laws may be needed to reduce greenhouse gas emissions, and this requires politicians. / If no action is taken, the consequences may be drastic for future generations.   **6 a i** in the exhaust pipe from the engine   **ii** to convert harmful gases to harmless gases   **iii** transition elements   **b i** by reaction between nitrogen and oxygen, in the hot air in the car engine **ii** They causes respiration (breathing) problems. / They produce a photochemical smog in sunlight.

Extended   **7 a** greenhouse gases   **b** ❶ incoming light energy from the Sun, ❷ thermal energy from Earth's surface that passes through the atmosphere and escapes into space, ❸ thermal energy trapped by greenhouse gases in the atmosphere   **c i** As greenhouse gas levels rise, relatively more heat (thermal energy) will be trapped ❸, and relatively less heat will escape to space ❷.
ii Earth's average temperature will continue to rise.
8 a i It is a poisonous gas.   **ii** It causes breathing problems / forms acid rain.   **b** A catalyst is present, to increase the rate. (Rapid reactions are essential because

the gases flow quickly from the engine, along the exhaust pipe.) **c** $2CO(g) + 2NO(g) \rightarrow 2CO_2(g) + N_2(g)$

Chapter 16

page 205 **1** crude oil **2** It was formed from the remains of sea organisms that lived millions of years ago. **3** coal and natural gas **4** carbon and hydrogen **5** covalent **6 a** It is mainly methane. **b** It is a mixture of organic compounds, mainly hydrocarbons. **7** It shows all the atoms and the bonds in a molecule. See page 204 or Unit 16.3 for examples. **8** It provides most of the world's transport fuels, as well as the starting materials for making plastics, medical drugs, and many other useful products. **9** The fossil fuels produce carbon dioxide when they burn. This is a greenhouse gas, linked to climate change. Switching to solar power, wind power, and other renewable energy sources will help us to tackle climate change, and will reduce the demand for fossil fuels.

page 207 **1** evaporating and condensing **2** fraction **3 a** evaporates easily to form a gas **b** thick and sticky **c** catches fire easily **4** They differ in boiling point, volatility, viscosity, and flammability. **5 a** gasoline **b** refinery gas **c** bitumen **d** fuel oil **6** It is thin and runny, volatile, and highly flammable. Also a liquid is easy to store, transport and pour into tanks. **7 a** It does not burn easily. **b** It softens on heating so is easy to spread, it binds stone chips to form a tough surface, and it is flexible rather than brittle so will not break up under traffic.

page 209 **1** alkanes, alkenes, alcohols, carboxylic acids **2 a 3 b** alcohols **3 a** methane **b** butane **4** methane   butane

5 the part of the molecule that largely dictates how the compound will react; for example the C=C bond / OH group / COOH group **6** The compounds in a homologous series have the same functional group, they fit a general formula, the chain length increases by $-CH_2-$ each time, and there is a gradual change in physical properties as the chain gets longer. **7** molecular: C_6H_{14} structural: $CH_3CH_2CH_2CH_2CH_2CH_3$ displayed:

8 a $C_{20}H_{42}$ **b** $C_{32}H_{66}$

page 211 **1** Each C atom shares electrons with 3 H atoms and another C atom. See the diagram at the top of page 210. **2** They burn readily, providing plenty of thermal energy. They are easy to transport by pipeline, or in cylinders under pressure. **3** butane + oxygen → carbon dioxide + water **4** because chlorine atoms take the place of hydrogen atoms **5** energy from light **6 a** Cl–CH$_2$–CH$_2$–Cl and H–CHCl–CH$_2$Cl **b** $C_2H_4Cl_2$ for both

7 a / b Here are the displayed formulae, and matching boiling points: straight chain b. pt. 36 °C; one branch b. pt. 28 °C; two branches b. pt. 9.5 °C. Boiling points decrease as branching increases. This is because branching prevents the molecules from getting close, so there is less attraction between them. So less heat is needed to overcome the attraction and turn the liquid into gas.

page 213 **1** Long chain hydrocarbons are broken down into smaller molecules. **2 a** A catalyst is used. **b** heat **3** an alkene **4** It provides the hydrocarbons that industry wants, including the reactive alkenes used to make plastics and other materials. **5** It produces ethene and hydrogen. Ethene is used to make poly(ethene), and hydrogen is used in the Haber process to make ammonia. **6 a** pentane displayed formula **b** Your reaction should show products with a total of 5 C atoms and 12 H atoms between them. At least one should have a double bond. Examples are $C_2H_4 + C_3H_8$ (ethene + propane), and $C_2H_6 + C_3H_6$ (ethane + propene).

page 215 **1** ethene; for its displayed formula see page 214. **2 a** It will be higher than –6.5 °C. **b** It will have the C=C bond. **3** Their C=C bonds break easily, allowing other atoms to add on. **4 a** It contains a C=C bond. **b** propane (or cyclopropane) **5** Add bromine water. If the orange colour disappears, the liquid is an alkene. **6** $C_2H_4(g) + Br_2(aq) \rightarrow C_2H_4Br_2(l)$ **7** by reaction with: **a** steam **b** hydrogen; both reactions require a catalyst **8** addition **9 a** Structural isomers are compounds with the same molecular formula, but different structural formulae. **b** $CH_2=CHCH_2CH_3$ and $CH_3CH=CHCH_3$

page 217 **1** See the displayed formula in the table on page 216. **2** ethanol + oxygen → carbon dioxide + water **3 a** fermentation using enzymes in yeast as catalyst **b** Carry out the reaction at 300 °C and 60 – 70 atmospheres pressure, using phosphoric acid as a catalyst. **4 a** Choose advantages and disadvantages from the table on page 217. **b** fermentation **5** $CH_3CH_2CH_2OH$ and $CH_3CHOHCH_3$

page 219 **1** COOH **2** See the displayed formula in the table on page 219. **3** The products are, in order: salt + hydrogen   salt + water   salt + water + carbon dioxide **4** oxidised **5** concentrated sulfuric acid **6 a** ethanol and ethanoic acid **b** methanol and propanoic acid **7 a** methyl ethanoate **b** ethyl methanoate

ANSWERS FOR CHAPTERS 1 – 19

8 a [structural formula of methyl methanoate] **b** [structural formula of ethyl methanoate] **f** [equilibrium equation showing ethanoic acid + ethanol ⇌ ethyl ethanoate + water]

page 220 Core 1 a i fractional distillation **ii** boiling points **b i** The range of boiling points shows that it is a group of compounds. **ii** As the petroleum is heated, the compounds in it evaporate. As they rise through the tower they condense to form a liquid. A mixture of these liquids is drawn off together, as naphtha. **c** See page 206. **d i** longer **ii** more viscous **2 a** breaking down long-chain molecules into smaller ones **b** heat and a catalyst **c** It produces hydrocarbons with shorter chains and some with double bonds (alkenes). Both types of product are very useful. Liquid alkanes with shorter chains are more flammable, so they are suitable for petrol, for example. The alkenes are reactive and are the starting chemicals for many important products. **d i** Orange bromine water will turn colourless. **ii** Bromine reacts with the alkene to form a colourless compound. **e** $C_2H_6(g) \rightarrow C_2H_4(g) + H_2(g)$

page 221 Core 3 a carbon and hydrogen **b** methane **c i** 5 **ii** 6 **d** No, because it has no C=C bond. It is saturated. **e** carbon dioxide and water vapour **f** pentane + oxygen → carbon dioxide + water **g** carbon monoxide **h** substitution **4 a** a family of organic compounds with the same functional group and the same general formula **b** The compound contains a carbon-carbon double bond **c** C=C **d i** C_2H_4, ethene **ii** See page 214 for the displayed formula of ethene. **e i** alcohols **ii** a catalyst (phosphoric acid), high temperature (300 °C), high pressure (60 – 70 atm) **iii** catalytic addition **f i** alcohols **ii** ethanol, C_2H_5OH **iii** See page 216 for the displayed formula of ethanol.

Extended 5 a An addition reaction turns an unsaturated alkene into a saturated compound. There is only one product. **b i** $CH_2BrCHBrCH_2CH_3$ **ii** $CH_3CH_2CH_2CH_3$ **iii** $CH_2OHCH_2CH_2CH_3$ or $CH_3CHOHCH_2CH_3$ **c i** See page 215 for the displayed formula of but-2-ene. **ii** Yes, because it has the same functional group, C=C. **6 a** Members have the same general formula, the same functional group, the chain length increases by $-CH_2-$ each time, and their properties show a trend as chain length increases. **b i** the alcohols **ii** the OH or hydroxyl group **c i** $CH_3CH_2CH_2OH$ **ii** $CH_3CHOHCH_3$, propan-2-ol **d** a carboxylic acid **7 a i** the carboxylic acids **ii** $C_nH_{2n+1}COOH$ **b** the COOH or carboxyl group **c i** oxidation **ii** fermentation **d** Only some of its molecules dissociate to form ions: $CH_3COOH (aq) \xrightarrow{\text{some molecules}} CH_3COO^- (aq) + H^+ (aq)$ **e i** fizzing, as carbon dioxide bubbles off **ii** $2CH_3COOH (aq) + Na_2CO_3 (s) \rightarrow 2CH_3COONa (aq) + H_2O (l) + CO_2 (g)$ **f i** propanoic acid, C_2H_5COOH; its displayed formula is on page 218. **ii** $C_2H_5COOH (aq) + NaOH (s) \rightarrow C_2H_5COONa (aq) + H_2O (l)$ **8 a** ethyl ethanoate **b** an ester **c** water **d** It acts as a catalyst. **e** It can go both ways.

9 a i alkenes **ii** C_6H_{12} **b i** $C_6H_{12} (l) + Br_2 (aq) \rightarrow C_6H_{12}Br_2 (l)$ **ii** addition **iii** a colour change from orange to colourless **c** It contains fewer than 6 carbon atoms.

Chapter 17

page 223 1 a a large molecule built up from many smaller molecules called monomers **b** a reaction in which thousands of small molecules join, to form very large molecules (polymers) **c** the small molecule from which a polymer molecule is made **2** False. Polymers are very common in nature. They form the keratin and collagen in our bodies, cellulose in trees and plants, and starch in the food we eat. Many materials we use and wear are made of synthetic polymers. **3** It means *many*. Many ethene molecules join. **4** natural – made by nature; synthetic – made in a factory **5 a** keratin **b** cellulose **6** Answers will vary. **7** addition polymerisation and condensation polymerisation **8** A carbon dioxide molecule has only one carbon atom.

page 225 1 because small molecules add on to each other, with no other product **2** No, because it does not contain a C=C (double) bond. **3** See the diagram on page 224. **4** because they all have different chain lengths **5** Look at this table:

Monomer	Part of the polymer molecule
a CH₂=CHCl	–CH₂–CHCl–CH₂–CHCl–CH₂–CHCl–
b C₆H₅CH=CHCl	–C(C₆H₅)H–CHCl–C(C₆H₅)H–CH–C(C₆H₅)H–CH–

6
H   H
| |
C = C
| |
H   COCH₃

page 227 1 *Condensation*: two different monomers / join by eliminating a small molecule / two products. *Addition*: one monomer / join by addition at a C=C bond / one product. **2** the parts of the molecules not directly involved in the reaction **3 a** See the first two diagrams on page 226. **A** is a dicarboxylic acid, **B** is a diamine. **b** If there was only one functional group, only two units could join. Having two functional groups means thousands of molecules can join. **4** See the third diagram on page 226. **5** [diagram of polyamide linkage showing –C(=O)–▢–C(=O)–N(H)–▬–N(H)–]

6 a dicarboxylic acid and diol **b** dicarboxylic acid

7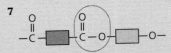

page 229 1 synthetic polymers **2** Choose the most suitable properties from page 228 for each group of items. (*Unreactive* is important for all those uses!)
3 For properties it may have, check the list on page 228.
4 high melting point, unreactive, non-toxic, slippery
5 a H₂C=CHCl (H, H, H, Cl around C=C)
b addition **c** light, unreactive and strong; can be made flexible or rigid depending on the type of pipe needed (see the photo of flexible hoses on page 225)

page 231 1 Waste food decays quite fast (broken down by bacteria). Waste plastic can take hundreds of years.
2 an area of land set aside as a place to dump rubbish
3 See the bullet points on page 230. **4** They are very tiny particles of plastic obtained when plastics break down over time, for example through exposure to sunlight. **5** See the bullet points on page 231. **6** Think of solutions that address points 1 – 6 at the top of page 231. For example provide bins at the beach where people can put their rubbish, or charge a fine for dumping waste in rivers or the ocean. **7** The burning of plastics produces toxic chemicals.

page 233 1 See examples in section 1 and in the first photo on page 232. **2** It reduces the amount of plastics in landfill / rivers / oceans where they degrade only very slowly. It also lowers the demand for plastic so less is made, saving materials and energy. **3** to check that no toxic gases are escaping into the atmosphere
4 *biodegradable*: can be broken down by bacteria (or fungi); *photodegradable*: can be broken down by the action of light **5 a** PET **b** high density poly(ethene) **c** low density poly(ethene) **6** See the answer to question 2.
7 PET bottles for recycling must be collected, sorted and brought to the recycling factory; all this need fuel – for trucks, conveyor belts, etc. The plastic is then melted and shaped into a new product, or broken down to its monomers and re-polymerised. This needs fuel too. If the fuels are fossil fuels, or electricity from fossil fuels, then carbon dioxide and other harmful gases are produced (page 196). So reusing a bottle is much simpler. Just wash it! **8** There is no right or wrong answer. It might help you to consider that 1, 4 and 5 reduce the amount of waste plastic that will be produced, while 2 and 3 deal with the waste plastic.

page 235 1 a H₂N–CHR–COOH (structural formula shown)
b the side chain, which is different in each amino acid
c i NH₂ **ii** COOH **d** proteins and water
2 a (polypeptide chain structure shown)

b four **c** There is no repeat unit because different amino acids can join, in any order. So each block represents a different side chain. **3 a** breaking down a chemical by adding water **b** amino acids (hydrolysis will reverse the polymerisation reaction)

page 236 Core 1 a monomers **b** polymer
c polymerisation **d** (chain of repeating blocks)

2 a a large molecule built up from many smaller molecules called monomers **b** plastics **c** ethene, styrene, chloropropene **d** They all contain a C=C bond.
e poly(ethene), polystyrene, poly(chloropropene)
f polymerisation **3 a i** ethene **ii** poly(ethene) **b i** The monomers add on to each other. **ii** Small molecules join together to form a large molecule. **c i** It can take hundreds of years for them to break down. Animals mistake them for food; the plastic blocks their guts so they starve. **ii** The gases produced when plastics burn are toxic.

page 237 Extended 4 a
b CH₂=CHCONH₂
(–CH₂–CH(CONH₂)–)ₙ repeat unit shown

c acrylamide **d** unsaturated **5 a** the C=C bond
b addition polymerisation **c** –(CF₂–CF₂)– chain shown

d poly(tetrafluroethene) **6 a i** Both are long chains with repeating units, built up from monomers. **ii** An *addition polymer* is built from identical monomers, and there is no other product. A *condensation polymer* is built from two different types of monomer, with water (or other small molecule) as a second product. **b i** group X **ii** group Y **c i** group Y **ii** group X **d i** PET **ii** It allows complete recycling of a plastic item where melting would not work. For example, melting PET bottles to make new bottles works well if the bottles are all the same colour, and contain only PET. But if they are different colours, or contain other plastics too, it may be better to completely recycle the PET by turning it back into its monomers, and re-polymerising. **7 a i** condensation **ii** water
b i HOOC–C₆H₄–COOH and HO–CH₂–CH₂–OH
ii dicarboxylic acid   diol
c i ester linkage **ii** –C(=O)–O–

d PET **e** Break it down to the monomers by hydrolysis, which involves reaction with water under suitable conditions (high temperature and pressure). Then re-polymerise it. **8 a i** B **ii** A **iii** C **b** addition
c

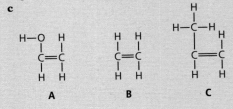

A   B   C

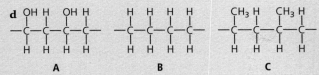

d A, B, C structures shown above

9 a proteins **b i** an amino acid **ii** the side chain (containing carbon atoms), which is different in each amino acid **iii** carboxylic acid

c structure of amino acid shown

Chapter 18

page 239 **1 a** copper(II) oxide + sulfuric acid → copper(II) sulfate + water **b** The copper(II) sulfate is not an impurity because it is the product you want. The water is not an impurity because this is an aqueous solution.
2 step 2, excess copper(II) oxide / step 3, water / step 5, any remaining copper(II) sulfate solution / step 6, water
3 by heating it to evaporate water **4** patted with filter paper **5 A** beaker, **B** conical flask, **C** test-tube, **H** filter funnel, **K** retort stand (with clamp), **G** gas syringe
6 Choose E, the volumetric pipette; titrations depend on accurate volume measurements, and the pipette is by far the most accurate of those three. **7** D, burette **8** Recall experiments you have met already. For example: **a** an experiment requiring a known mass of reactant, or an experiment on rate where a loss of mass occurs
b titrations **c** experiments on rate of reaction
d investigating exothermic and endothermic reactions, or controlling temperature in investigations about rate of reaction **e** measuring gas volumes in experiments on rate of reaction, where only one product is a gas

page 241 **1 a** dissolves in a solvent (the particles separate and spread out among the solvent particles) **b** does not dissolve in a solvent (the particles do not separate and spread) **c** a solution with water as the solvent **d** a solution in which no more of the solute will dissolve at that temperature **2 a** silver nitrate **b** 20 times **3** because so little of it dissolves (only 0.0013 g in 100 g of water)
4 because solubility changes with temperature
5 a around 87.5 g **b** 45 g **6** Crystals of copper(II) sulfate will appear. The solution will remain blue because it still contains dissolved copper(II) sulfate.

page 243 **1 a** filter funnel, filter paper **b** copper(II) oxide **c** water **2 a** Sugar is soluble in water.
b crystallisation **3** Follow the steps outlined on page 238.
4 The difference in solubility is so small that very few crystals will form when the hot solution is cooled. You could keep heating to remove *all* the water by evaporation.

page 245 **1** by simple distillation; see the apparatus at the top of page 244. The water boils off, and the steam is condensed, and then collected, in separate parts of the apparatus. **2** The solvent condenses in them. The cold water keeps the temperature down, to allow condensation to take place. **3** Some water vapour would evaporate with the ethanol, and pass into the condenser. So complete separation would not be possible. **4 a** The glass beads in the fractionating column provide a large surface area on which condensation can take place. **b** To separate the mixture you need to identify each component before it condenses. This is done by measuring the temperature at the top of the column. **5 a** in the tall steel tower
b refining petroleum / separating the gases in air

page 247 **1 a** If the ink line is below the ethanol level, the ink will dissolve in the ethanol and not travel up the paper. **b** Each colour travels up the paper at a different speed. **c** the yellow substance – it has travelled further up the paper **d** The ink colours may be soluble in ethanol, but not in water. **2** The spot where a solution is placed on the paper. The origin is usually on a pencil line drawn on the paper. **3 a** Br, G, O **b** Bl, Br, R, O **c** Bl, Bu
4 Y contains D, but not C. The spot for D is at the same height as one of the Y spots. The spot for C is not.

page 249 **1 a** the leading edge of the solvent, as it travels up the paper **b** You need to measure the distance travelled by the solvent and the spots to calculate R_f. If the paper is not removed the spots will continue to move up, giving false results. **2 a** It shows the location of a colourless substance. **b** because amino acids are colourless **3 a** See the equation on page 249.
b A substance has a fixed R_f value for a given solvent. So we can identify the substance by looking up a table of R_f values for that solvent. **4 a** Only C. C and A have spots at the same height. **b / c** These are the R_f values and amino acid names: **A** 0.14 lysine, 0.26 glycine, 0.73 leucine; **B** 0.38 alanine **C** 0.73 leucine **D** 0.43 proline
E 0.6 valine.

page 251 **1 a** contains particles of only one substance
b another, unwanted, substance present in a substance
2 Y (starts melting at a lower temperature and melts over a bigger temperature range) **3** will be higher than 78 °C.
4 B; has more than one spot **5** water

page 254 Core 1 A iii B v C vi D i E iv F ii
2 a i 100 °C **ii** Seawater has a higher boiling point. (Impurities raise the boiling point.) **b** A flask **c i** It is labelled C. **ii** at the lower end of the condenser

page 255 Core 3 a Bunsen burner, filter funnel, tripod, conical flask, gauze (wire gauze), stirring rod, filter paper, beaker **b** place sample in beaker and add water / place beaker on gauze on tripod / heat to boiling, stirring with glass rod / put filter paper in filter funnel, resting on a conical flask / carefully filter the heated mixture / remove the solid residue and dry in an oven **4 a** mixture
b Simple distillation will not separate the substances properly. A mixture of gases will rise into the condenser, and especially where their boiling points are close. (Nitrogen boils at −196 °C, argon at −186 °C, and oxygen at −183°C.) **c** Nitrogen has the lowest boiling point of the three, and oxygen the highest. **5 a** Add ethanol to the mixture in a beaker and stir. **b** Separate the salt by filtering; see the diagram on page 242. **c** By simple distillation. When most of the ethanol has gone, let sugar crystals form. **d** See page 244 (top diagram).
e Add water to the mixture, stir and heat to dissolve the salt. Filter to remove the sand, and evaporate the filtrate to dryness to obtain the salt. **6 a** The steps are: *1* On the chromatography paper, draw a pencil as shown in the *before* diagram. *2* Add drops of pigment solution as shown

285

in the *before* diagram. **3** Place the paper in a covered beaker containing ethanol. (The solvent level must be below the pencil line.) **4** Allow the ethanol to move up the paper (but not to the top). **5** Remove the paper (the chromatogram) and allow it to dry. **b** It contains two substances. **c** only one spot **7 a** B **b** A **c** red

Extended 8 a proline and valine **b** proline (travels less far) **c** They will not change. (The spots will travel less far too, so the ratio of distances does not change.) **9** See the instructions on pages 248 and 249. Locating agents are needed because the amino acids are colourless. The R_f value is the distance moved by the substance ÷ distance moved by the solvent, and it is unique for each amino acid in a given solvent.

Chapter 19

page 257 1 a reasonable statement that you can test; in chemistry you test it in the lab! **2 a** no **b** yes **c** no **d** yes **3** a variable you change, when you do an experiment, to see how it affects the results; you keep all other conditions unchanged **4** the solvent **5** Repeat the experiment several times, using a different mass of powdered catalyst each time. The volume of hydrogen peroxide, and the temperature, must be kept constant. Do not use too much hydrogen peroxide, to ensure all the oxygen fits in the gas syringe. You could simplify the experiment by measuring how long it takes to collect a given volume of oxygen (so you would take just one time reading in each experiment). **6** Answers will vary.

page 259 1 a Yes, the brand of kitchen cleaner. **b** Yes, the amount of acid needed to neutralise the alkali. **2 a** B conical flask, D burette, E volumetric pipette **b** It delivers the volume you want, more accurately. **c** so that the liquid can be swirled (to mix it) without spilling **d** They allow you to deliver liquids drop by drop, and measure the delivered volume accurately **e** a volumetric pipette; it is easy to fill to the exact volume, because the mark is on its very slim stem **3** to tell you when neutralisation is complete **4 a** There is no sharp colour change. The colour changes continually with pH. **b** thymolphthalein; blue to colourless **5** ... X contains a higher concentration of sodium hydroxide than solution Y does. **6** ... support my hypothesis that X is a better at removing grease than Y, because it has a higher concentration of sodium hydroxide. **7** Make sure to shake the cleaners well before measuring out. Repeat each titration to check the results. **8** NaOH (aq) + HCl (aq) → NaCl (aq) + H$_2$O (l) **9** Titrate the scale remover with an alkaline solution (e.g. sodium hydroxide) of known concentration, using a suitable indicator (e.g. thymolphthalein). **10** F, the thermometer

page 261 1 a See the apparatus on the top left of page 260. **b** It will turn limewater milky. **c** CO$_2$ (g) + Ca(OH)$_2$ (aq) → CaCO$_3$ (s) + H$_2$O (l) **2 a** It is lighter than air. **b** 2H$_2$ (g) + O$_2$ (g) → 2H$_2$O (l) **3 a** ammonium chloride / ammonium sulfate / ammonium nitrate, and calcium hydroxide / sodium hydroxide **b** It is very soluble in water. **c** It turns damp red litmus blue. **4** There may be too little gas present to detect a smell; it may be a harmful gas. **5** A gas syringe is more accurate and easier to use.

page 263 1 cation **2** See the instructions on page 262. **3 a** the sodium ion **b** the copper(II) ion **b** the lithium ion **4** hydroxides **5** Their precipitates are different colours: pale green for Fe^{2+} and red-brown for Fe^{3+}. **6 a** Both cations give a white precipitate. **b** Repeat the test using aqueous ammonia. Only the Zn^{2+} precipitate will redissolve in excess ammonia. **7 a** Potassium hydroxide is soluble, so no precipitate will form. **b** a flame test

page 265 1 negative ions; for example Cl$^-$, Br$^-$, I$^-$, CO$_3^{2-}$, NO$_3^-$ **2** precipitate, insoluble **3** because silver halides are insoluble, with different colours for Cl$^-$, Br$^-$ and I$^-$ **4** See the instructions on page 265. **5** All nitrates are soluble in water. **6** from the sodium hydroxide solution **7 a** because a gas (carbon dioxide) is produced in the reaction, and you can then test for this gas **b i** a solution of calcium hydroxide **ii** A positive result is easy to see: the limewater turns milky.

page 266 Core 1 a Put filter paper in the filter funnel, and place the filter funnel in the conical flask. Pour the mixture into the filter funnel. Add a few drops of universal indicator to the filtrate, using the dropping pipette. The final colour of the indicator gives the pH of the soil sample. **b** by repeating with soil samples from other areas of the garden **2 a** If you do not recognise these pieces of apparatus, look through the textbook to find examples. For example, page 260. (You need to become familiar with apparatus, for answering exam questions.) **b** because sulfur dioxide is soluble in water **3 a** a metal such as magnesium, zinc, or iron, or any carbonate **b** a stopwatch / timer **c** These are the steps:
- Weigh out the required mass of solid Y.
- Using a measuring cylinder, place a known volume of acid in the flask.
- Add Y to the acid, and immediately stopper the flask and start the timer.
- Record the volume of gas collected, at regular time intervals.

d More gas may be produced than the syringe can hold.

page 267 Core 4 a i W pH 1, Y pH 5 **ii** a weak acid **b i** fizzing, and the metal disappears **ii** a magnesium salt, and hydrogen **iii** The reaction will be slower. **c i** white **ii** carbon dioxide **iii** sodium carbonate (or sodium hydrogen carbonate) **d i** barium sulfate **ii** sulfuric acid **5 a i** See the instructions on page 262. **ii** lilac **b i** SO$_2$ **iii** See the test for sulfur dioxide on page 261. **6 a** ammonium ion, NH$_4^+$ **b** See the test for the ammonium ion on page 263. **c** nitrate ion, NO$_3^-$ **d** See the test for the nitrate ion on page 265. **7 a i** a positive ion **ii** the zinc cation, Zn^{2+} **iii** the aluminium cation, Al^{3+} **iv** the calcium cation, Ca^{2+} **b i** the chloride ion, Cl$^-$ **ii** See the test for the chloride ion on page 264. **8 a** Anions: chloride, hydrogen carbonate, nitrate, sulfate. Cations: calcium, magnesium, potassium, sodium. **b i** calcium ion **ii** 2.75 milligrams **c** sulfate ion **d** All the compounds of the other metals are soluble. **e** The concentration of the nitrate ion may be very low. There may be too little to give a positive result in the test.

IGCSE practice questions for Paper 3

Paper 3 covers the Core syllabus content.

1 a Diamond is a giant covalent structure of carbon atoms.
  i Explain what is meant by the terms *giant* and *covalent*. [2]
  ii How many covalent bonds does each carbon atom form, in diamond? [1]
  iii Which property makes diamond useful as a cutting tool? [1]
b Graphite is also a giant covalent structure of carbon atoms:

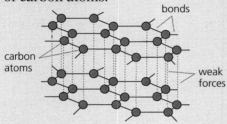

  i How many covalent bonds does each carbon atom form, in graphite? [1]
  ii Explain why graphite is a good lubricant. [1]
  iii Explain why graphite is a good conductor of electricity, while diamond is not. [2]
c Copy and complete this diagram to show the electronic structure of a carbon atom:

[1]

d These symbols represent two naturally occuring isotopes of carbon:

$$^{12}_{6}C \qquad ^{13}_{6}C$$

  i Define the term *isotope*. [1]
  ii Copy and complete this table:

isotope	C-12	C-13
number of protons		
number of neutrons		

[2]

e A third naturally occurring isotope of carbon has 8 neutrons. It is found in very small amounts.
  i How many protons do its atoms contain? [1]
  ii What is the mass (nucleon) number for this isotope? [1]

Total 14 marks

2 The structures of four substances are shown below:

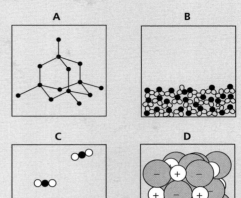

a Which letter, **A**, **B**, **C**, or **D**, represents:
  i a gas? **ii** an element?
  iii a liquid? **iv** an ionic compound? [4]
b The bonding in substance **C** is given below. Only the outer shells are shown:

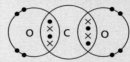

  i Give the name and formula for **C**. [2]
  ii In forming compound **C**, are electrons transferred, or shared? [1]
  iii Name the type of bonding in **C**. [1]
c Chlorine is a simple molecular element.
  i Copy and complete this diagram to show the bonding in chlorine:

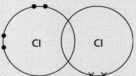
[1]

  ii Give the molecular formula for chlorine. [1]
  iii Is the boiling point of chlorine high, or low? [1]
  iv Which best describes the electrical conductivity of chlorine: poor, or good? [1]
d Substance **D** will conduct electricity when it is melted. Choose the correct word from each pair in brackets below:
Substance **D** conducts electricity when it is melted because the (ions/electrons) in the liquid are (able/unable) to move. [1]

Total 13 marks

3 Diagrams **A** to **F** represent the electronic configuration for six different elements:

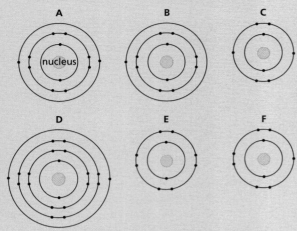

Using the letters **A** to **F**, identify the diagram that shows the electronic configuration for:
- **a** an element with a full shell [1]
- **b** two elements in the same group of the Periodic Table [1]
- **c** two elements in the same period of the table [1]
- **d** a halogen [1]
- **e** an element with atomic number 11 [1]
- **f** the element neon [1]
- **g** an alkali metal [1]
- **h** an element in Period 4 of the Periodic Table [1]
- **i** an element whose atoms lose one electron to to form a positive ion [1]
- **j** an element whose ions have a charge of 2– [1]

Total 10 marks

4 The elements fluorine, lithium, and carbon are all in the same period of the Periodic Table.
- **a**
  - **i** Define the term *element*. [1]
  - **ii** Which period are the three elements in? [1]
  - **iii** Which one has a molecular structure? [1]
  - **iv** How many atoms are in each molecule of the molecular element? [1]
- **b** The structure of lithium allows electrons to move freely.
  - **i** What property does this give to the metal? [1]
  - **ii** Which form of carbon has this same property, diamond or graphite? [1]
  - **iii** Name the type of structure found in both diamond and graphite. [1]
- **c** Name another element with similar structure and properties to:
  - **i** lithium   **ii** carbon   **iii** fluorine [3]

Total 10 marks

5 In winter, some people use disposable hand warmers like the one shown here.

They make use of a chemical reaction which gives out heat.
- **a** Give the name used for reactions which heat their surroundings. [1]
- **b** Three reaction mixtures, X, Y, and Z were tested for use in hand warmers. This shows the results:

reaction mixture	temperature (°C)	
	at the start	after reaction
X	16	21
Y	16	7
Z	16	25

  - **i** Which mixture will not warm hands? Explain why. [1]
  - **ii** Which mixture would be best to use in a hand warmer? Explain your choice. [1]
  - **iii** After some time, hand warmers will no longer keep hands warm. Explain why. [1]
- **c** Many hand warmers, like the one in the photo, use the reaction between iron and oxygen, in the presence of moisture.
  - **i** Copy and complete this word equation:
    iron + oxygen $\longrightarrow$ .... [1]
  - **ii** What is the common name for the oxidation of iron in the presence of moisture? [1]
  - **iii** The iron in the hand warmers is in powder form. Explain why. [1]
  - **iv** Explain why these hand warmers are sold in sealed air-proof sachets. [1]
- **d** The hand warmer above also contains sodium chloride, to speed up the reaction. What term is used for substances that speed up reactions? [1]
- **e** Which of the energy diagrams **i** and **ii** below represents the reaction in the hand warmer?

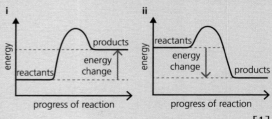

[1]

Total 10 marks

6 Magnesium reacts with dilute hydrochloric acid.
- **a** Complete the word equation for the reaction:
  magnesium + hydrochloric acid $\longrightarrow$
  .... + hydrogen [1]

b A student investigates the rate of the reaction. This diagram shows the apparatus used:

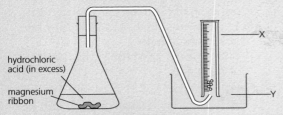

To start the reaction, a strip of magnesium is added to the conical flask. The stopper is replaced immediately. The volume of hydrogen is measured every 15 seconds.

- **i** Why is the stopper replaced immediately? [1]
- **ii** Name the piece of apparatus marked X. [1]
- **iii** Identify the liquid marked Y. [1]
- **iv** Which other piece of equipment is needed, for measuring the reaction rate? [1]

c This table shows the student's results:

time (seconds)	0	15	30	45	60	75	90
volume of gas (cm³)	0	85	135	179	190	200	200

- **i** Plot the results on a grid like this one:

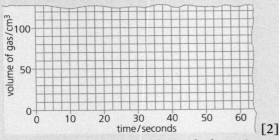
[2]

- **ii** Draw the best curve through the points. [1]
- **iii** Explain why the volume of gas is the same at 75 seconds and 90 seconds. [1]

d The student repeats the experiment using acid at a higher temperature.
- **i** On your grid, sketch the graph you would expect for this higher temperature. [2]
- **ii** Give two other changes that would also increase the rate of this reaction. [2]

Total 13 marks

7 These eight elements are from the same period of the Periodic Table:
lithium   beryllium   boron   carbon
nitrogen   oxygen   fluorine   neon

a From these eight elements, choose **one** that
- **i** is a metal
- **ii** is present in most fertilisers
- **iii** is a gas at room temperature
- **iv** is found in all organic compounds
- **v** is needed for combustion
- **vi** forms an ion with a charge of 1−
- **vii** is found in Group VI of the Periodic Table
- **viii** has 2 electrons in its outer shell [8]

b What is the meaning of the term *element*? [1]

c To which period of the Periodic Table do the eight elements belong? [1]

d All of the listed elements, **except** neon, form compounds with hydrogen.
- **i** Copy and complete this table for three of the elements:

| element | compound formed with hydrogen | |
	name	formula
carbon		
nitrogen		
oxygen		

[6]

- **ii** Which type of bonding is found in all three compounds in your completed table? [1]
- **iii** Explain why neon does not react with hydrogen. [1]

Total 18 marks

8 This table shows some physical properties of the Group I elements.

metal	density (g/cm³)	melting point (°C)	boiling point (°C)
lithium	0.53	181	1342
sodium	0.97	98	883
potassium	0.86	63	759
rubidium	1.53	39	
caesium	1.88	29	669

a Use the information in the table to explain why potassium is a liquid at 100 °C. [1]

b
- **i** What is the trend in melting and boiling points, going *up* the group? [1]
- **ii** The boiling point of rubidium is missing from the table. Suggest a value for it. [1]

c
- **i** What is the general trend in density, going *down* the group? [1]
- **ii** Which element does not follow this trend in density? [1]

d Sodium is soft, and can be cut with a knife.
- **i** Describe the initial appearance of sodium when it is cut. [1]
- **ii** Why does the appearance of sodium change when a cut piece is left in the air? [1]

e Sodium reacts vigorously with water to give a solution of sodium hydroxide.
- **i** Describe two things you would see when a small piece of sodium is placed in a bowl of water. [2]
- **ii** Name the gas released in the reaction. [1]
- **iii** Write a word equation for the reaction. [1]
- **iv** What colour will the final solution be, if it contains universal indicator? [1]

Total 12 marks

9 Concentrated sodium chloride solution can be electrolysed using the apparatus shown below:

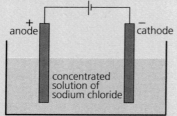

- **a i** What does the term *electrolysis* mean? [1]
  - **ii** What would you see, to indicate that electrolysis was taking place? [1]
- **b i** Name the gas given off at the cathode (negative electrode). [1]
  - **ii** What would be obtained at the cathode if *liquid* sodium chloride was electrolysed? [1]
- **c** Platinum can be used for the electrodes. Give two properties of platinum that make it suitable for this use. [2]
- **d** Chlorine gas is released at the anode.
  - **i** State the formula for chlorine. [1]
  - **ii** How you would identify chlorine gas? Describe the test, and the result. [2]
- **e** Chlorine and iodine are both halogens.
  - **i** In which group of the Periodic Table are the halogens? [1]
  - **ii** Which is less reactive, iodine or chlorine? [1]
- **f i** Copy and complete this word equation for the reaction of chlorine with aqueous potassium iodide:
  chlorine + potassium iodide $\longrightarrow$
  .... + ... [1]
  - **ii** Name another halogen that will react in a similar way, with potassium iodide. [1]

Total 13 marks

10 Chorine and bromine are both halogens.
- **a i** Give the electron configuration for chorine, and explain why the element is in Group VII of the Periodic Table. [2]
  - **ii** What is its state at room temperature? [1]
  - **iii** Describe the appearance of bromine at room temperature. [2]
  - **iv** Which element is more reactive, bromine or chlorine? [1]
- **b** Chlorine reacts with sodium to form X.
  - **i** Give the correct name, and formula, of X. [2]
  - **ii** What type of bonding is in X? [1]
  - **iii** Is X soluble, or insoluble, in water? [1]
- **c i** In which period of the Periodic Table is chlorine? [1]
  - **ii** Name another element, from the same period, that is in the same state as chlorine at room temperature. [1]
  - **iii** Which statement below is correct, A, B, or C, about elements in the same period?
    A All have the same number of outer shell electrons in their atoms.
    B Their atoms have the same number of electron shells.
    C Their atoms have the same number of neutrons. [1]

Total 13 marks

11 Magnesium sulfate is a soluble salt. It can be made by reacting an insoluble metal oxide with an acid.
- **a** Name the metal oxide and acid used to make magnesium sulfate. [2]
- **b** Name the method of purification used to:
  - **i** remove the excess oxide from the magnesium sulfate solution [1]
  - **ii** produce crystals of magnesium sulfate from its solution [1]
- **c** Name another substance that would react with the same acid, to give magnesium sulfate. [1]

Total 5 marks

12 The metals iron, magnesium, zinc, and calcium, all react with dilute hydrochloric acid.
- **a i** Place the four metals in order of reactivity, with the most reactive first. [1]
  - **ii** Copy and complete this word equation for the reaction of zinc:
  zinc + hydrochloric acid $\longrightarrow$
  .... + ... [2]
- **b** The four metals also burn in oxygen.
  - **i** Write out and balance this equation, for magnesium burning in air:
  $Mg + O_2 \longrightarrow MgO$ [1]
  - **ii** What would be seen during the reaction? [1]
  - **iii** Name the product of the reaction. [1]
  - **iv** Explain why the reaction is an oxidation. [1]
- **c** Excess magnesium, as magnesium ribbon, is added to a solution of copper(II) sulfate in a test-tube. A reaction occurs. This shows the test-tube before and after the reaction:

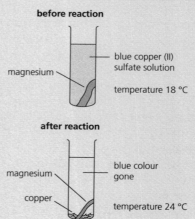

i What does *excess* of a reactant mean? [1]
ii Explain why there is a colour change during the reaction. [2]
iii Write a word equation for the reaction. [2]
iv Give the symbol for the metal ions present in the solution, when the reaction is over. [1]
v Which metal is more reactive, magnesium or copper? Explain your choice. [1]
vi Is the reaction endothermic, or exothermic? Explain your answer. [1]
vii Draw an energy diagram to represent the reaction. It must include the names of the reactants and products. [2]
Total 17 marks

13 Electricity is supplied from power stations in a network called the electricity grid. This uses aluminium cables supported by steel pylons.

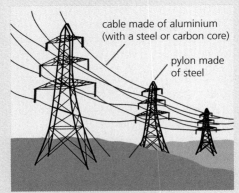

a Steel is an alloy of iron.
  i Name the type of furnace that is used to make iron. [1]
  ii Name one property of steel that makes it a suitable material for the pylons. [1]
b Aluminium and copper are good conductors of electricity. Give one other property that:
  i makes aluminium more suitable than copper, for overhead cables. [1]
  ii is shared by copper and aluminium [1]
  iii belongs to copper compounds, but not to aluminium compounds [1]
c Stainless steel is an important alloy of iron. This bar chart shows the composition of one type of stainless steel:

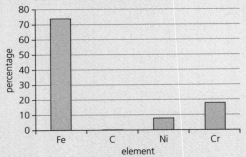

i List the three elements that are added to iron to make this alloy. [3]
ii Is stainless steel a *mixture* of these elements, or a *compound*? Explain your answer. [2]
iii The percentage of carbon in this alloy is about 0.2%. Give approximate values for the percentages of the other three elements. [2]
iv Give two physical properties of stainless steel that make it suitable for use in cutlery. [2]
Total 14 marks

14 Iron is extracted from its ore in the blast furnace.
a i Define the term *ore*. [1]
  ii Which one of these is an ore of iron?
     silicon   hematite   rock salt
     bauxite   granite   rust [1]
b Iron(III) oxide reacts with carbon monoxide in the blast furnace.
  i Balance the equation for the reaction:
     $Fe_2O_3 + 3CO \longrightarrow ...Fe + ...CO_2$ [1]
  ii Which substance is oxidised in this reaction? Explain your answer. [2]
  iii Name the gas produced in the reaction, and explain its link to climate change. [2]
c Iron is a transition element. Give two properties of transition elements that are not shown by sodium. [2]
d The most common isotope of iron, Fe-56, is also shown as ^{56}Fe.
  i How many protons, electrons, and neutrons are there in an atom of this isotope? [2]
  ii The other isotopes of iron differ in atomic structure to Fe-56. In what way? [1]
  iii How many electrons are there in the Fe^{2+} ion? [1]
Total 13 marks

15 Drinks cans are usually made from alloys of iron or aluminium.

a i What is an *alloy*? [1]
  ii Name an alloy of iron. [1]
b This shows the arrangement of atoms in a metal:

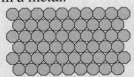

i How does the diagram indicate that this substance is an element? [1]

ii Draw a similar diagram to show the arrangement of atoms in an alloy. [2]
c When manufacturing drinks cans, explain why:
  i pure iron is not used [1]
  ii aluminium alloys are often preferred to iron alloys [1]
d The iron used in the manufacture of drinks cans is extracted from its oxide using carbon.
  i Complete the word equation to represent the overall reaction in the blast furnace:
    iron oxide + carbon ⟶ [2]
  ii Which substance is reduced in the reaction? Explain your answer. [2]
e The aluminium used in drink cans is extracted from its oxide by electrolysis.
  i Explain why aluminium cannot be extracted by reacting its oxide with carbon. [1]
  ii Copy and complete the word equation for the electrolysis:
    aluminium oxide ⟶ .... + ... [1]
  iii At which electrode does the aluminium form? Explain your answer. [2]
    Total 15 marks

16 Ammonia is an important chemical in farming.
a State the formula of ammonia. [1]
b i Name the group of chemicals that are made from ammonia, for use in farming. [1]
  ii Name the essential element that all of these chemicals supply to plants. [1]
  iii Name the two other elements which are also essential for plant growth. [2]
c Ammonia is a gas at room temperature. For transportation, it is turned into a liquid and carried in cylinders, under pressure.
  i Describe how the arrangement of its molecules changes, as the gas is liquefied. [2]
  ii In which state do the molecules move more slowly, gas or liquid? [1]
d Ammonia reacts with acids to produce salts.
  i Copy and complete these word equations:
    Equation 1   ammonia + nitric acid ⟶ ....
    Equation 2   ammonia + ... ⟶ ammonium chloride [2]
  ii Which term describes the reactions in **i**? **oxidation   neutralisation   hydration** [1]
e Ammonia gas can be released from ammonium salts by reacting them with bases.
  i Name a suitable base that could be used. [1]
  ii What else is required for the reaction? [1]
  iii Describe a test to confirm that ammonia has formed, and give its result. [2]
    Total 15 marks

17 The structures show compounds containing sulfur:

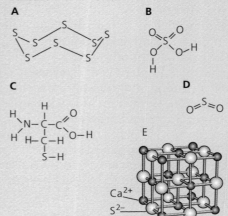

Answer the following questions. You can choose any structure once, more than once, or not at all.

Choose a structure which represents:
a the element sulfur [1]
b a gas that will change the colour of potassium manganate(VII) [1]
c a substance used as a food preservative [1]
d a molecule in organic chemistry [1]
e an ionic solid [1]
f a gas which causes acid rain [1]
g an acid [1]
   Total 7 marks

18 Methane, ethane, and propane are hydrocarbons in the alkane homologous series.
a What is a *hydrocarbon*? [1]
b Give two characteristics of a homologous series. [2]
c Copy and complete for the first three alkanes:

alkane	molecular formula	displayed formula
methane		H–C–H (with H above and below)
ethane	C_2H_6	
propane		H–C–C–C–H (with H's above and below)

[3]

d Which alkane in the table for **c** is the main component of natural gas? [1]
e Give the molecular formula of the alkane that follows propane in the homologous series. [1]
f Ethene belongs to a different homologous series:

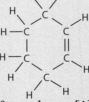

  i What is the molecular formula for ethene? [1]
  ii Name the homologous series it belongs to. [1]
  iii Describe the test to show the difference between ethene and ethane, and its result. [2]
g This is an important reaction of ethene:

  i Which word describes this type of reaction?
  decomposition   **combustion**
  polymerisation   **oxidation** [1]
  ii Name the product formed in the reaction. [1]
  Total 14 marks

19 a Cyclohexene is an **unsaturated** hydrocarbon. This diagram shows the structure of its molecules.

  i Write down its molecular formula. [1]
  ii Explain why it is called *unsaturated*. [1]
b All hydrocarbons burn in oxygen, producing carbon dioxide and water.
  i Copy, complete, and balance this equation:
  $CH_4 + O_2 \rightarrow$ [2]
  ii Describe a test to confirm that carbon dioxide has formed, and give its result. [2]
c Carbon burns in a limited supply of air, to form carbon monoxide gas.
  i Give the formula for carbon monoxide. [1]
  ii Is the structure of carbon monoxide giant, or molecular? Explain your choice. [2]
  iii Which type of bonding is in the compound?
  covalent   **metallic**   **ionic** [1]
  Total 10 marks

20 Vinegar adds flavour to food. It can also be used as a cleaning agent. It removes grease and other stains.

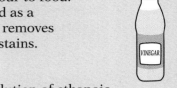

a Vinegar is a solution of ethanoic acid in water.
  i Copy and complete this displayed formula for ethanoic acid, showing all atoms and bonds. [1]
  ii Which ion in vinegar makes it acidic? [1]
  iii What is the pH of vinegar? Choose.
  7   **less than 7**   **more than 7** [1]
b Methyl orange is added to a colourless solution of ethanoic acid.
  i What colour change will be observed? [1]
  ii What name is given to substances like methyl orange, which change colour with pH? [1]
c The apparatus below is used to investigate the amount of ethanoic acid in vinegar.

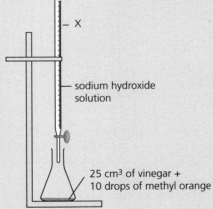

  i State the name of the glass item X. [1]
  ii What is the sodium hydroxide for? [1]
  iii What piece of apparatus would be used to measure out exactly 25 cm³ of vinegar? [1]
  iv How would you know when all the ethanoic acid in the vinegar had reacted? [1]
  v Copy and complete the word equation:
  ethanoic acid + sodium hydroxide
  .... + ... [2]

d Name the gas that forms, and describe the test for it, when ethanoic acid reacts with:
  i magnesium **ii** magnesium carbonate [4]
  Total 15 marks

21 Limonene is a colourless liquid. It is found in the skin of lemons and other citrus fruit. Its molecules have this structure:

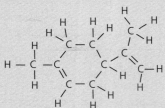

a Is limonene saturated? Give your evidence. [1]
b Give the molecular formula of limonene. [1]
c Name the functional group in the molecule. [1]
d What change will you see if limonene is added to bromine water? [1]
e Limonene can be extracted by this method:

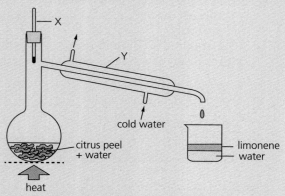

  i Name the items labelled X and Y. [2]
  ii Give evidence that limonene does not form a solution with water, from the diagram. [1]
  iii Which has greater density, limonene or water? Explain your answer. [1]

f Citrus fruits also contain citric acid, $C_6H_8O_7$. A molecule of citric acid has this structure:

$$\underset{HO}{\overset{O}{\diagdown}}C-CH_2-\underset{\underset{HO\diagup \ \diagdown O}{C}}{\overset{OH}{\underset{|}{C}}}-CH_2-\overset{O}{\underset{OH}{\diagup}}C$$

  i The circled functional group appears three times in the structure of citric acid. In which of these compounds is it also found?
  ethanol   **ethanoic acid**   **poly(ethene)** [1]
  ii Identify the other functional group in the citric acid molecule. [1]
  iii Calculate the relative molecular mass of citric acid. (A_r: C =12, H = 1, O =16) [1]
  iv Name the two products that form when citric acid burns in plenty of oxygen. [2]
  Total 13 marks

IGCSE practice questions for Paper 4

Paper 4 covers the Extended syllabus content.

1 These are techniques used in the laboratory:
simple distillation   crystallisation
fractional distillation   chromatography
filtration   electrolysis
   a Which technique depends on ions moving? [1]
   b Which technique allows you to:
      i separate two or more liquids that have different boiling points? [1]
      ii remove an insoluble solid from a liquid? [1]
      iii obtain pure solvent from a solution? [1]
   c Which technique:
      i is used to separate the amino acids obtained from the hydrolysis of a protein? [1]
      ii relies on the change in solubility of a solid with temperature? [1]
      iii brings about a chemical change? [1]
      iv depends on the particles in a mixture having different sizes? [1]

   Total 8 marks

2 Washing soda (hydrated sodium carbonate) is used in homes to remove stubborn stains from laundry, and for other cleaning.

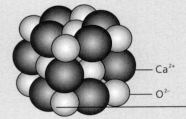

   a Give the formulae of the two ions in sodium carbonate, Na_2CO_3. [2]
   b *Hydrated* sodium carbonate has the formula $Na_2CO_3.xH_2O$, where x is a whole number. Its value can be found using titration with acid.
      i Copy and complete the equation for the reaction: $Na_2CO_3 + 2HCl \rightarrow$ [2]
      ii Which type of reaction is it? [1]
   In an experiment to find the value of x:
   • 4.0 g of hydrated sodium carbonate was dissolved in distilled water to give 500 cm³ of the alkaline solution.
   • 25 cm³ of this solution was then titrated against a 0.1 mol/dm³ solution of hydrochloric acid, with methyl orange as indicator.
   • It was found that an average of 14 cm³ of hydrochloric acid was used in the titration.
   c What colour change will indicate the end point of the titration? [1]
   d You will need these A_r values for the calculations:
      A_r: H = 1, Na = 23, C = 12, O = 16
      i Calculate the number of moles of hydrochloric acid used in the titration. [1]
      ii Calculate the number of moles of sodium carbonate which will react with this many moles of hydrochloric acid. [1]
      iii Using your answer for ii, calculate the number of moles of sodium carbonate in the 500 cm³ of solution. [1]
      iv Calculate the relative molecular mass of sodium carbonate, Na_2CO_3. [1]
      v Using your answers for iii and iv, calculate the mass of sodium carbonate, Na_2CO_3, in the 500 cm³ of solution. [1]
   e i Calculate the mass of water present in 4.0 g of hydrated sodium carbonate. [1]
      ii How many moles of H_2O is present? [1]
   f Calculate the value of x in $Na_2CO_3.xH_2O$, using this formula:
   $$x = \frac{\text{moles of } H_2O \text{ in 4.0 g of the compound}}{\text{moles of } Na_2CO_3 \text{ in 4.0 g of the compound}}$$ [1]

   Total 14 marks

3 This shows the structure of calcium oxide:

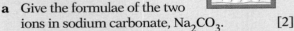

   a Name the type of structure, and type of bonding, in calcium oxide. [2]
   b How can you tell from the diagram above that the formula of calcium oxide is CaO? [1]
   c Calcium oxide forms when the metal calcium burns in oxygen.
      i Give the symbol equation for the reaction. [2]
      ii Explain why the reaction is an *oxidation*. [1]
      iii How do the calcium atoms change during the reaction, in terms of their electrons? [1]
   d The electronic configuration for the calcium particles in calcium oxide is 2,8,8. Give the electronic configuration for the oxygen particles. [1]
   e Calcium oxide is added to water containing universal indicator, and shaken. An exothermic reaction takes place. It produces calcium hydroxide, which is slightly soluble in water.

i Describe three changes you would observe during this reaction. [3]
ii Copy and complete the equation for it:
CaO + H$_2$O $\longrightarrow$ .... [1]
f i Which describes calcium oxide and calcium hydroxide?
acids   bases   salts [1]
ii Write the equation for the reaction between calcium oxide and ammonium nitrate. [2]
g Calcium oxide reacts with dilute hydrochloric acid, giving a colourless solution.
i Which type of chemical reaction is this? [1]
ii Name the two products of the reaction. [2]
iii The oxide of a different metal gives a blue solution, on reaction with hydrochloric acid. Suggest a reason for this. [1]
Total 19 marks

4 Dilute sulfuric acid can be electrolysed using a Hofmann voltameter, as shown below:

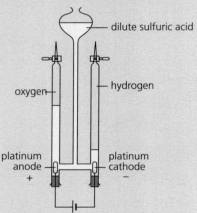

a i State the term used for solutions which can be decomposed by electricity. [1]
ii Why is platinum used for the electrodes? [1]
b Hydroxide ions produce oxygen at the anode.
i Copy and balance the ionic equation:
OH$^-$ $\longrightarrow$ H$_2$O + O$_2$ + e$^-$ [1]
ii Which type of redox reaction is this? Explain your answer. [2]
c Hydrogen is formed at the cathode.
i Which ion reacts to give hydrogen? [1]
ii Write the ionic equation for the reaction. [2]
d Write the overall reaction for the chemical change that takes place in the apparatus. [1]
e The volume of hydrogen produced is twice the volume of oxygen. Explain this. [1]
f i Describe how the design of the apparatus allows the two gases to be kept separate. [1]
ii Samples of the two gases can be collected in test-tubes, by opening the taps. Which gas will burn when a lighted splint is applied? [1]
g Does the concentration of the sulfuric acid in the apparatus change? Explain your answer. [2]
Total 14 marks

5 Marble chips react with dilute hydrochloric acid:
CaCO$_3$ (s) + 2HCl (aq) $\longrightarrow$
CaCl$_2$ (aq) + H$_2$O (l) + CO$_2$ (g)
The rate was investigated using this apparatus:

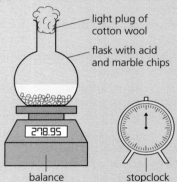

a i What is the purpose of the cotton wool? [1]
ii Explain why a loss of mass occurs. [1]
b This graph shows the results of the experiment:

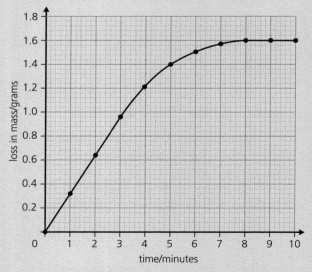

i Describe how the rate of the reaction changed as the reaction proceeded. [3]
ii Why was the rate not constant? [1]
iii How long did the reaction take, in total? [1]
iv In terms of collisions, explain why the size of the marble chips affects the reaction rate. [2]

c Using *collision theory* and *activation energy*, explain why the rate of the reaction will increase with temperature. [3]
Total 12 marks

6 When solid blue copper(II) sulfate ($CuSO_4.5H_2O$) is heated, it turns white in this reaction:
  hydrated copper(II) sulfate $\rightleftharpoons$
    anhydrous copper(II) sulfate + water
  a i Give the chemical equation for the reaction, including state symbols. [2]
  ii What does the symbol $\rightleftharpoons$ mean? [1]
  iii What will you observe when water is added to anhydrous copper(II) sulfate? [1]
  b You will need these A_r values for the calculations:
  A_r: Cu = 64, S = 32, O = 16, H = 1
  i Calculate the mass of one mole of hydrated copper(II) sulfate. [1]
  ii What mass of water is obtained when 5 g of hydrated copper(II) sulfate is converted to anhydrous copper(II) sulfate? [2]
  iii Calculate the percentage of water in hydrated copper(II) sulfate. [1]
  c A solution of copper(II) sulfate is electrolysed:

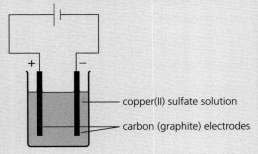

  i Give the formulae of the ions in solid copper(II) sulfate. [2]
  ii What is seen at the positive electrode? [1]
  iii Name the product at each electrode. [2]
  iv Write the ionic equation for the reaction at *one* of the electrodes. [1]
  d Copper(II) sulfate can be made by reacting copper(II) carbonate with sulfuric acid.
  i Copy and complete the chemical equation for this reaction:
  $CuCO_3 + H_2SO_4 \rightarrow$ [2]
  ii Suggest one observation which will indicate that the reaction is taking place. [1]
  Total 17 marks

7 The water supply to many homes contains ions which cause a solid white **scale** to form on shower heads and in kettles. Scale is mainly calcium carbonate.
  a i Give the formulae of calcium carbonate, and each of the two ions in it. [3]
  ii Why do we not *see* this compound in tap water? [1]
  b Vinegar contains ethanoic acid, a weak acid. This reacts with scale to give a gas.
  i Name the gas that is produced. [1]
  ii Suggest why vinegar is used rather than hydrochloric acid, to remove scale. [1]
  iii Write the chemical equation for the reaction between calcium carbonate and ethanoic acid. [2]
  c Calcium carbonate undergoes thermal decomposition, as shown in this equation:
  $CaCO_3\ (s) \longrightarrow CaO\ (s) + CO_2\ (g)$
  i What is needed to enable the decomposition to take place? [1]
  ii Calculate the mass of calcium oxide obtained from the thermal decomposition of 100 kg of calcium carbonate. (A_r: Ca = 40, C = 12, O = 16) [2]
  iii Why is this reaction used in extracting iron from its ore, hematite? [1]
  iv Give the equation showing how calcium oxide reacts in the extraction process. [1]
  d Calcium oxide is converted to calcium hydroxide by adding water. The solution that forms has a pH of 12.
  i Give the formula of calcium hydroxide. [1]
  ii What type of solution has formed? [1]
  iii Give the formula of the ion in the solution, that produces the pH of 12. [1]
  Total 16 marks

8 Phosphorus(V) chloride decomposes on heating, to give phosphorus(III) chloride and chlorine:
  $PCl_5\ (g) \rightleftharpoons PCl_3\ (g) + Cl_2\ (g)$
  a How can you tell that this is an equilibrium reaction? [1]
  b How does the amount of phosphorus(V) chloride that breaks down change, if the chlorine is removed? [1]
  c More phosphorus(V) chloride will decompose, if the temperature is increased. Is the forward reaction exothermic, or endothermic? Explain your answer. [1]
  d What effect, if any, would an increase in pressure have on the amount of phosphorus(V) chloride that decomposes? Explain your answer. [1]

e When phosphorus(V) chloride is added to water, it reacts vigorously in an exothermic reaction.
A mixture of phosphoric acid and hydrochloric acid forms.
  i Copy and balance the equation:
  $PCl_5 (s) + H_2O (l) \longrightarrow H_3PO_4 (aq) + HCl (aq)$ [1]
  ii How will the pH and temperature of the solution change as the phosphorus(V) chloride is added? [2]
f Phosphoric acid is used to make products that protect plants against disease.
  i Find the mass of 1 mole of phosphorus(V) chloride and 1 mole of phosphoric acid. (A_r: P = 31, Cl = 35.5, O = 16, H = 1) [2]
  ii What mass of phosphoric acid can be obtained from 100 g of phosphorus(V) chloride? [1]
g Diammonium phosphate (DAP) is widely used in agriculture. Its formula is $(NH_4)_2HPO_4$.
  i Calculate the percentage of phosphorus in this product, by mass. (A_r: N = 14) [2]
  ii Which other element, essential to plant growth, does DAP provide? [1]
Total 13 marks

9 Ethene is an important chemical, used in making plastics. It is made from ethane.
a The reaction that produces ethene is:
$C_2H_6 \longrightarrow C_2H_4 + H_2$
Which type of chemical reaction is this? [1]
b Ethene is an unsaturated hydrocarbon.
  i What is meant by *unsaturated*? [1]
  ii Name the type of reaction that all unsaturated hydrocarbons can undergo. [1]
  iii Deduce what reacts with ethene to give:
   A $C_2H_4Br_2$ **B** C_2H_5OH [2]
c Ethene burns in oxygen as shown:
$C_2H_4 + 3O_2 \longrightarrow 2CO_2 + 2H_2O$
There is an enthalpy change during the reaction.
  i Which releases energy, *bond making* or *bond breaking*? [1]
  ii Copy and complete this table for the combustion of ethene:

bond	bond energy for this bond (kJ/mol)	number broken, or made	enthalpy change (+ or −) in breaking or making them (kJ/mol)
C=C	612		
C–H	413		
O=O	498		
C=O	749		
H–O	464		

[5]

  iii Calculate the total enthalpy change in breaking the different bonds, in the reaction. [1]
  iv Calculate the total enthalpy change in making all the new bonds, in the reaction. [1]
  v Using the answers from **iii** and **iv**, sketch an energy level diagram for the combustion of ethene (*not to scale*). Calculate and mark in the overall enthalpy change for the reaction. [3]
  vi Ethene does not react with oxygen at room temperature. Explain why. [1]
Total 17 marks

10 Sodium thiosulfate reacts with dilute hydrochloric acid. The equation for the reaction is:
$Na_2S_2O_3 (aq) + 2HCl (aq) \longrightarrow 2NaCl (aq) + 2H_2O (l) + SO_2 (g) + S (s)$
After a time, the solution goes cloudy.
a i Name the four products of this reaction. [2]
  ii Explain why the solution goes cloudy. [2]
b The gas that forms is classified as a pollutant.
  i Explain why it is a pollutant. [1]
  ii Name a reagent used to test for the presence of this gas, and give the result of the test. [2]
c Using test-tubes as below, a student investigates how a change in concentration of the sodium thiosulfate solution affects the reaction rate.

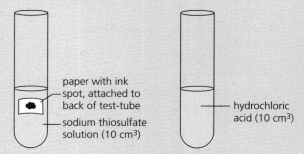

A stopwatch is started when the hydrochloric acid is added to the sodium thiosulfate solution. It is stopped when the ink spot is hidden from view. The experiment is repeated for different concentrations of sodium thiosulfate solution.
  i Give two factors that must be held constant in this investigation, for valid results. [2]
  ii What is the link between time taken for the ink spot to disappear, and reaction rate? [1]

d Predict how an increase in temperature will affect the reaction rate, and explain your answer in terms of colliding particles. [4]
e In a photochemical reaction, which other factor can be used to increase the reaction rate? [1]

Total 15 marks

11 This question is about nickel and its oxide.
a Nickel is a transition element. Sodium is a Group I metal. Give two properties of nickel which are not shared by sodium. [2]
b Nickel oxide is an ionic compound. Its formula is NiO. Give the formulae of the ions in it. [2]
c Nickel oxide has the same lattice structure as sodium chloride. Draw a diagram to show the structure of nickel oxide. [1]
d The atomic number of nickel is 28. Its most stable isotope has a nucleon number of 58. Give the number of protons, neutrons, and electrons, in this isotope of nickel. [3]
e Nickel is extracted from nickel oxide by reduction with carbon. Explain why carbon can be used for this extraction. [2]
f The key reaction in the extraction of nickel is:

NiO ( ) + C ( ) $\longrightarrow$ Ni ( ) + CO ( )

  i Copy the equation, and insert the missing state symbols in the brackets. [2]
  ii Explain why this is a redox reaction. [1]
  iii Calculate the mass of nickel oxide needed to give 100 kg of nickel. (A_r: Ni = 59, O = 16) [2]
g Nickel oxide reacts with hydrochloric acid.
  i Write a balanced chemical equation for the reaction. (Ignore state symbols.) [2]
  ii Describe one observation which would show that the reaction was taking place. [1]

Total 18 marks

12 The elements in Period 3 of the Periodic Table are:

Na	Mg	Al	Si	P	S	Cl	Ar

a Name one element from this period that forms
  i an acidic oxide [1]
  ii an amphoteric oxide [1]
  iii a basic oxide [1]
b Name an element from Period 3 that:
  i forms a chloride with covalent bonding [1]
  ii forms a chloride with ionic bonding [1]
  iii does not react with chlorine [1]
c Explain the difference between ionic and covalent bonding. [2]
d Magnesium oxide melts at 2852 °C, and sulfur dioxide melts at − 72 °C.
  i Give the formulae for the two oxides. [2]
  ii Explain why their melting points are so different. [2]
e Aluminium is extracted from aluminium oxide, Al_2O_3, using electrolysis.
  i Explain the term *electrolysis*. [1]
  ii The aluminium oxide is dissolved in molten cryolite for the electrolysis. Why? [1]
  iii Copy and complete this equation for the electrolysis of aluminium oxide:
$Al_2O_3 \longrightarrow$ [2]
  iv What mass of aluminium will be obtained from 1 kg of aluminium oxide? (A_r: Al = 27, O = 16) [2]

Total 18 marks

13 Ammonia is an important industrial chemical. It is manufactured from hydrogen and nitrogen.
a **i** What is the source of the nitrogen? [1]
  ii How is it obtained from this source? [2]
b The hydrogen for manufacturing ammonia is made by reacting methane with steam.
  i What is the usual source of methane? [1]
  ii Copy and balance the equation for producing hydrogen from methane:
$CH_4(g) + H_2O(g) \longrightarrow CO(g) + H_2(g)$ [1]
  iii Name the second gas that forms. [1]
  iv This gas is reacted with more steam:
$CO(g) + H_2O(g) \longrightarrow CO_2(g) + H_2(g)$
Name the oxidising agent in this redox reaction. [1]
c Ammonia is then made by reacting the nitrogen and hydrogen in the presence of a catalyst.
  i Name the metal commonly used as a catalyst for this reaction. [1]
  ii What is the function of the catalyst? [1]
d **i** Copy, complete, and balance, this equation for the synthesis of ammonia:
.... + $N_2(g) \rightleftharpoons ... NH_3(g)$ [2]
  ii What does the $\rightleftharpoons$ sign tell you? [1]
e During the manufacturing process, the gases are heated to around 450 °C. In terms of the particles present, explain why increasing the temperature speeds up the reaction. [2]
f High pressures are used in the manufacture of ammonia to increase the rate of the reaction, and give a good yield.
  i Why does an increase in pressure increase the rate of reaction? [1]

ii Why does a high pressure increase the yield of ammonia? [1]
Total 16 marks

14 Aluminium is the most abundant metal in Earth's crust. It is extracted from **aluminium oxide**, which is obtained from the aluminium ore bauxite.

a Aluminium has properties which make it very useful. Choose the correct word from each pair in brackets below:
i It has a (high / low) density. [1]
ii It is a (good / poor) conductor of electricity. [1]
iii It forms an unreactive (oxide / hydroxide) layer on its surface. [1]

b i Give the formula for aluminium oxide. [1]
ii It is very difficult to extract aluminium from its oxide. Explain why. [1]
iii Aluminium was not discovered until 1825. Suggest a reason. [1]

c Aluminium is extracted from its oxide by electrolysis:

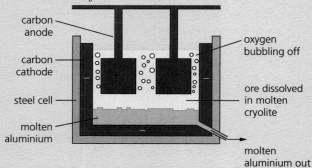

i Why must the aluminium oxide be dissolved before electricity is passed through? [1]
ii If molten cryolite is *not* used, what would the consequence be? [1]
iii Copy and complete the ionic equations for the reaction at the two electrodes:
At the cathode $\longrightarrow$ Al [2]
At the anode $\longrightarrow$ O_2 [2]
iv At which electrode does oxidation occur? [1]

d The carbon anodes need to be replaced at intervals. Explain why. [1]

e Sodium is also extracted by electrolysis. Its ion has a charge of 1+. If the same amount of electricity is used to extract both metals, how will these compare?
i the number of moles of sodium atoms and aluminium atoms produced [1]
ii the masses of sodium and aluminium obtained (A_r: Al = 27, Na = 23) [1]
Total 16 marks

15 Electricity is carried in overhead cables made of aluminium. These are supported on steel pylons. Steel is made from the iron produced in the blast furnace.

a State two properties of aluminium that make it suitable for overhead cables. [2]

b This diagram shows the blast furnace:

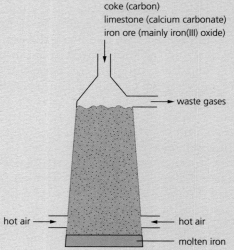

i Name an iron ore used in the blast furnace, and give the formula for the iron ion in it. [2]
ii Outline the role of calcium carbonate in the extraction of iron. [2]
iii The key reaction which produces iron is:
$Fe_2O_3 (s) + 3CO (g) \longrightarrow 2Fe (s) + 3CO_2 (g)$
Write the word equation for the reaction. Include the oxidation number for iron. [2]
iv Name two waste gases from the furnace. [2]
v Calculate the maximum mass of iron that can be produced from 640 tonnes of iron(III) oxide. (A_r: Fe = 56, O = 16) [2]

c i Iron is a typical metal. Draw a diagram to show the structure in a metal. [2]
ii What type of forces hold the metal structure together? [1]

d Iron is used (as steel) to make body panels for cars. Using the information about structure and bonding in metals, explain why the panels:
i conduct electricity [1]
ii can be bent into shape [1]

e The properties of aluminium and iron can be altered by adding other elements to them.
What name is given to a mixture of a metal with another element? [1]
Total 18 marks

IGCSE PRACTICE QUESTIONS FOR PAPER 4

16 Bromine is in Group VII of the Periodic Table.
  a **i** What other name is used for this group? [1]
  ii How many electrons are there in the outer shell of a bromine atom? [1]
  iii Explain how a bromine atom becomes an ion, and give the formula for the ion. [2]
  b Bromine vapour and hydrogen react with a mild explosion when lit, to form hydrogen bromide.
  i Write the chemical equation for the reaction. [2]
  ii Which type of bond is formed between the atoms, in hydrogen bromide? [1]
  iii Draw a diagram to show the bonding in a molecule of hydrogen bromide. Show only the outer shell electrons. [1]
  iv Name another molecule that has bonding similar to hydrogen bromide. [1]
  c A student adds a bromine solution to two test-tubes. One contains a solution of potassium iodide, and the other a solution of potassium chloride. The student observes a reaction in only one test-tube.
  i Explain this observation, indicating which solution reacts with the bromine water, and why. [2]
  ii Write the chemical equation for the reaction. [2]
  d Chlorine is also in Group VII. It is used in the manufacture of titanium.
  In the first stage of the process, chlorine reacts with a mole of titanium oxide (TiO_2) to give a mole of titanium chloride ($TiCl_4$). Calculate the mass of titanium chloride obtained from 800 kg of titanium oxide, if the percentage yield is 60%.
  (A_r: Ti = 48, O = 16, Cl = 35.5) [3]
  Total 16 marks

17 Two useful acids contain the element sulfur. Benzene sulfonic acid, $C_6H_5SO_3H$, is the active ingredient in laundry detergents. Sulfurous acid, H_2SO_3, is used as a disinfectant.
  a Benzene sulfonic acid is a strong acid, while sulfurous acid is a weak acid.
  i What ion causes the acidity in the aqueous solutions of these acids? [1]
  ii Define *strong acid* and *weak acid*. [2]
  iii You are given 0.1 mol/dm³ aqueous solutions of the two acids. Explain how you will distinguish between them. [2]
  b The formula of the benzene sulfonate ion is $C_6H_5SO_3^-$. Write the formula for the similar ion formed from sulfurous acid. [1]
  c Copy and complete these chemical equations:
  i $C_6H_5SO_3H + NaOH \longrightarrow$ [2]
  ii $H_2SO + Na_2CO_3 \longrightarrow$ [2]
  iii $H_2SO_3 + Mg \longrightarrow$ [2]
  d Benzene sulfonic acid reacts with ethanol in a similar way to ethanoic acid:
  $C_6H_5SO_3H + CH_3CH_2OH \longrightarrow C_6H_5SO_3CH_2CH_3 + H_2O$
  i Name this type of reaction. [1]
  ii What else is needed, for the reaction to take place? [1]
  Total 14 marks

18 Titanium is a transition element.
  a Titanium is used in replacement hip joints. Suggest one property it must possess, that makes it suitable for this use. [1]
  b Titanium alloys are used in building aircraft.
  i What is an *alloy*? [1]
  ii The density of titanium is 0.53 g/cm³. The density of iron is 7.87 g/cm³. Explain why titanium alloys are more suitable than steel alloys for use in aircraft. [1]
  c Titanium is extracted from the ore rutile (TiO_2) in a two-stage process.
  Stage 1 $TiO_2 + 2Cl_2 + 2C \longrightarrow TiCl_4 + 2CO$
  i Explain what happens to the titanium in this reaction, in terms of oxidation number. [2]
  ii Suggest one hazard for this reaction. [1]
  d The second stage of the process is carried out in an atmosphere of argon:
  Stage 2 $TiCl_4 + 4Na \longrightarrow Ti + 4NaCl$
  i What does this reaction tell you about the reactivity of titanium relative to sodium? [1]
  ii Why is this reaction carried out in an atmosphere of argon, rather than air? [2]
  iii Write the ionic equation for the reduction of the titanium ion to titanium. [2]
  e In a batch process for stage 2, 80 kg of titanium chloride was added to 40 kg of sodium.
  (A_r: Na = 23, Cl = 35.5, Ti = 48)
  i Which is the limiting reactant in the mixture above? Show your working. [4]
  ii Calculate the maximum mass of titanium that could be obtained from 80 kg of titanium chloride. [2]
  Total 17 marks

19 Propene (C_3H_6) is an alkene.
  a **i** Draw the structure of propene, showing all atoms and bonds. [1]
  ii Why is propene called *unsaturated*? [1]

iii Describe the test to show that propene is unsaturated, and give its result. [2]
b Propene and other alkenes can be made from long-chain hydrocarbons. For example:
C$_8$H$_{18}$ → C$_5$H$_{12}$ + C$_3$H$_6$
octane   heptane   propene
  i Which type of hydrocarbon are octane and heptane? [1]
  ii What is this type of reaction called? [1]
  iii What conditions are needed for this reaction to take place? [1]
c Name the chemical that will react with propene to give these molecules:
  i   ii   iii
  (structures shown) [3]
d These alkenes are structural isomers:
  i Explain why they are structural isomers. [2]
  ii Give their molecular formula. [1]
  iii Draw another isomer with this formula. [1]
e Propene can undergo polymerisation.
  i What type of polymerisation takes place? [1]
  ii Name the polymer made from propene. [1]
  iii Which shows the correct formula for it?
    A   B   C
  [1]
  iv Care should be taken in the disposal of polymers such as this one. Explain why. [1]
  Total 18 marks

20 Below are the structures of the first three members of the homologous series of carboxylic acids:

  a i Give the formula for the functional group of carboxylic acids. [1]
    ii Name the fourth member of the homologous series, and draw its structure. [2]
  b Ethanoic acid is a weak acid.
    i Why is it called *weak*? [1]
    ii Give one example of a strong acid. [1]
  c The concentration of a solution of ethanoic acid can be found by neutralising it with a sodium hydroxide solution, in the presence of an indicator. The reaction is:
  CH$_3$COOH (aq) + NaOH (aq) →
       CH$_3$COONa (aq) + H$_2$O (l)
    i What will show that the reaction is over? [1]
    ii Name the two products of the reaction. [2]
  d 27.5 cm^3 of sodium hydroxide solution, with concentration 0.50 mol/dm^3, neutralised 25 cm^3 of an ethanoic acid solution. Calculate the concentration of the ethanoic acid solution. [3]
  e Ethanoic acid is formed when ethanol reacts with acidified potassium manganate(VII).
    i What type of reaction is this? [1]
    ii Describe an alternative reaction for obtaining ethanoic acid from ethanol. [1]
  f i Ethanol reacts with ethanoic acid. Copy and complete the equation for the reaction:
  C$_2$H$_5$OH + CH$_3$COOH →
       .... + H$_2$O [1]
    ii The reaction requires a catalyst. Why? [1]
    iii Which type of organic compound forms? [1]
    iv Name the organic compound that forms. [1]
  Total 17 marks

21 Methane burns in a plentiful supply of oxygen.
  a Copy and balance this chemical equation for the combustion, and add the state symbols:
  CH$_4$ + O$_2$ → CO$_2$ + H$_2$O [2]
  b Explain why the incomplete combustion of methane is dangerous. [1]
  c This is the energy level diagram for the complete combustion of methane:

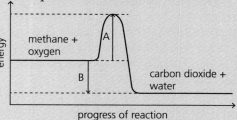

    i What do A and B represent? [2]
    ii Describe a situation in which a mixture of methane and oxygen will not react. [1]
  Total 6 marks

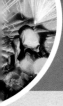

IGCSE practice questions for Paper 6

Paper 6 is the *Alternative to Practical* paper. Use the information on pages 261–265 to help you answer these questions. (It will be given in the exam too.)

1 a Ammonia solution is used in identifying cations.
   i   Give the formula for ammonia. [1]
   ii Is it soluble, or insoluble, in water? [1]
  b Ammonia solution was added to five solutions **A – E**, containing cations. The results are in the table. Identify each cation, and give its formula.

solution	initial effect of adding ammonia	when more (excess) ammonia is added
A	white precipitate	colourless solution
B	white precipitate	no change
C	grey-green precipitate	no change
D	red-brown precipitate	no change
E	pale blue precipitate	dark blue solution

[10]

  c Ammonia reacts with acids to form salts.
   i   What type of chemical reaction is this? [1]
   ii Give the name and formula of the ammonia ion that forms in these reactions. [2]
   iii Name the salt that forms when ammonia reacts with nitric acid. [1]

Total 16 marks

2 A solution contains a mixture of calcium hydroxide and calcium nitrate. It is divided into five portions:

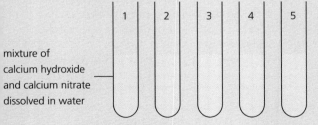

mixture of calcium hydroxide and calcium nitrate dissolved in water

  a **Test-tube 1** Describe:
   i   the appearance of the solution. [1]
   ii what will be observed when universal indicator paper is dipped into it. [1]
  b **Test-tube 2** Predict what will be observed when the following are added:
   i   a few drops of sodium hydroxide solution [2]
   ii excess sodium hydroxide solution [1]
  c **Test-tube 3** Excess ammonia solution is added.
     What will be observed? [1]
  d **Test-tube 4** Sodium hydroxide solution is added, followed by aluminium powder. The mixture is heated, and a gas is released. It is tested with damp red litmus paper.
   i   How will the gas affect the litmus paper? [1]
   ii Name the gas. [1]
  e **Test-tube 5** Carbon dioxide is bubbled into the solution. What will be observed? [1]
  f A second solution contains potassium hydroxide and potassium nitrate. Which of the tests in test-tubes **4** and **5** will give:
   i   the same result as for the calcium compounds? Explain your answer. [2]
   ii a different result than it did for the calcium compounds? Explain the difference. [2]

Total 13 marks

3 The solubility of potassium iodate in water at different temperatures is investigated. 50 cm³ of water is used. The results are shown below.

temperature /°C	mass of potassium iodate dissolved in 50 cm³ of water /g	solubility of potassium iodate g /100 g of water
10	3.1	
20	4.1	
30	5.1	
40	6.2	
50	7.0	

  a Copy the table. Then calculate the solubility of potassium iodate in grams per 100 g of water. Record the values in the third column. (Note: 1 cm³ of water has a mass of 1 g.) [2]
  b i   Now plot a line graph of solubility against temperature, on a grid like this one:

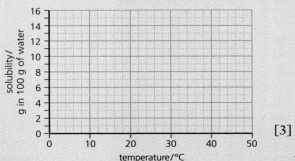

[3]

ii How does the solubility of potassium iodate change with temperature? [1]
iii Predict the solubility of potassium iodate at 5 °C. Show clearly on the graph how you obtained this answer. [2]
iv Calculate the mass of solid potassium iodate obtained, if 50 cm³ of a saturated solution at 50 °C is cooled to 20 °C. [1]

c Here is solubility data for two other compounds:

temperature / °C	solubility (g/100 g of water)	
	sodium chloride	lithium carbonate
10	35.7	1.4
20	35.9	1.3
30	36.1	1.3
40	36.4	1.2
50	36.7	1.1

Compared to potassium iodate, comment on both the solubility, and the effect of temperature on solubility, for:
i sodium chloride [2]
ii lithium carbonate [2]
Total 13 marks

4 A student investigates this reaction:
barium hydroxide + sulfuric acid ⟶ barium sulfate + water
a What type of chemical reaction is it? [1]
b Barium sulfate is an insoluble white compound.
What will you see as sulfuric acid is added to a solution of barium hydroxide? [1]
c The student uses this apparatus:

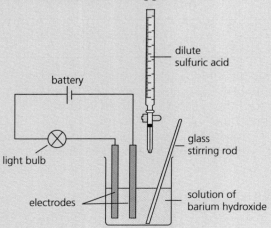

The light will go out when all the barium hydroxide in the beaker has reacted. Explain why, in terms of ions in the beaker. [1]

Two experiments are carried out.

Experiment 1
The burette is filled with dilute sulfuric acid, solution **A**. 50 cm³ of barium hydroxide solution is placed in the beaker, using a measuring cylinder. The electrodes are then placed in the solution, and connected to the bulb.

Solution **A** is slowly added to the beaker, and the contents are stirred with the glass rod.

From this burette diagram, you can tell what volume of solution **A** was required to make the light go out.

Experiment 2
Experiment 1 is repeated using a different solution of sulfuric acid, **B**. The burette diagrams are:

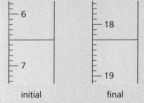

initial final

d i Copy and complete this table of results:

burette readings / cm³	solution A	solution B
final reading		
initial reading	0.0	
volume used		

[4]

ii In which experiment was a greater volume of sulfuric acid used? [1]
iii Suggest an explanation for the difference in the volumes used. [2]
iv Predict the volume of solution **B** needed to make the light go out, if there is 25 cm³ of barium hydroxide solution in the beaker. [1]
v What will you see if more sulfuric acid is added to each beaker at the end of the experiment? Explain your answer. [2]
e To improve the accuracy of the results in the investigation, suggest one change to:
i the apparatus [1]
ii the method [1]
Total 15 marks

5 A student is asked to demonstrate chromatography, using petals from a brightly coloured flower.
a Put these steps for chromatography in the correct order, using their labels **A – G**:

A Put a drop of the filtrate on chromatography paper. Let it dry. Repeat on the same spot.
B Add the petal pieces to the solvent.
C Cut the petals into small pieces.
D Allow the solvent to move up the paper.
E Filter the mixture.
F Heat the mixture, while stirring it.
G Stand the chromatography paper upright in a beaker containing some of the solvent. [3]

b Should the coloured substances in the petals be insoluble in the solvent, or soluble? [1]

c What could the student do, if the spots of filtrate are too pale to show up on the paper? [1]

d This chromatogram shows the student's results:

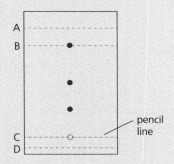

  i On which line A, B, C or D is the origin, where the spot of filtrate was placed? [1]
  ii Which line represents the solvent level at the start of the chromatography? [1]
  iii Deduce one fact from the results. [1]

e The chromatogram below show the results for three colours used in fruit-flavoured sweets:

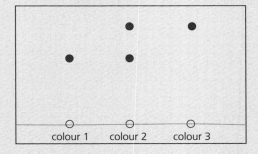

  i What can you deduce about the colours? [3]
  ii The final height of its spot can be used to identify a substance. Suggest why it is important to be able to identify the coloured substances added to sweets. [1]

Total 12 marks

6 When copper(II) carbonate is heated, this chemical change takes place:
copper(II) carbonate ⟶ copper(II) oxide + carbon dioxide

The copper(II) oxide will then react with dilute sulfuric acid to form copper(II) sulfate:
copper(II) oxide + sulfuric acid ⟶ copper(II) sulfate + water

This is how the reactions are carried out:

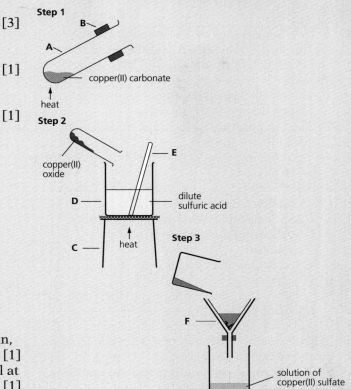

a Name the items of apparatus labelled **A** to **F**. [6]
b What conditions are applied, in turning copper(II) oxide into copper(II) sulfate? [2]
c Which step has a thermal decomposition? [1]
d i In step 2, how could you tell that all the sulfuric acid had reacted? [1]
  ii In step 2, the pH of the liquid changes as the reaction proceeds. Explain how, and why. [2]
e List the further steps you would take to obtain dry copper(II) sulfate crystals from the copper(II) sulfate solution. [4]

Total 16 marks

7 Copy and complete the table below, by describing a chemical test that will distinguish between the substances in each pair. The first row has been filled in for you. (There is more than one correct answer to choose from, in each case.)

IGCSE PRACTICE QUESTIONS FOR PAPER 6

pair of substances	chemical test	result of chemical test
oxygen, hydrogen	glowing splint	splint re-lights in oxygen but not in hydrogen
copper(II) nitrate, sodium nitrate		
chlorine, ammonia		
sodium chloride, sodium carbonate		
potassium sulfate, potassium sulfite		

Total 8 marks

8 A student investigates the temperature changes during displacement reactions between metals and copper(II) sulfate solution. This is the method:
1. 25 cm³ of copper(II) sulfate solution is placed in a polystyrene cup.
2. The temperature of the copper(II) sulfate solution is recorded.
3. 5g of metal power is added and the mixture is stirred.
4. The temperature is recorded, until there is no further change.

a Name a suitable piece of apparatus for:
  i delivering the copper(II) sulfate solution [1]
  ii recording the temperature [1]

b i Give two reasons why the metal is powdered, rather than in another form. [2]
  ii What other variable must be kept constant for this investigation? [1]
  iii Explain the choice of a polystyrene cup for this investigation. [1]

c The thermometer readings are shown below:

Copy and fill in this table, using the thermometer readings displayed at the bottom of the page:

d i Plot the data from **c** as a bar chart. Label the vertical axis as shown below. [2]

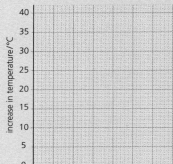

ii Are the four displacement reactions exothermic, or endothermic? Explain your answer. [2]

e The temperature change depends on the reactivity of the metal. The more reactive the metal, the greater the temperature change.
  i Using the results above, place the metals in order of reactivity, least reactive first. [1]
  ii Predict the temperature rise for tin, which is more reactive than copper but less reactive than nickel. [1]
  iii If silver is used there will be no temperature rise. Explain why. [1]
  iv Suggest one reason why the student should not use sodium in this investigation. [1]

Total 18 marks

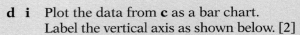

metal used in experiment	temperature/°C		
	initial	final	increase
zinc			
iron			
magnesium			
nickel			

[4]

9 Magnesium reacts with dilute hydrochloric acid to form a gas. The volume of gas given off can be measured using the apparatus below. The values can be used to compare different rates of reaction.

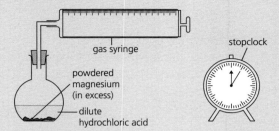

a The gas given off is tested using a lit splint.
  i Name the gas that is given off. [1]
  ii Describe the result of the splint test. [1]
b A student investigating the rate of this reaction chose to use powdered magnesium.
  i Suggest one possible advantage, and one disadvantage, of using powdered magnesium rather than magnesium ribbon. [2]
  ii List three other variables that affect the rate of the reaction, in this investigation. [3]

In one experiment, the volume of gas in the syringe was recorded every 30 seconds. These syringe diagrams show the readings:

time/s	reading on gas syringe/cm³
0	
30	
60	
90	
120	
150	
180	

c Copy and complete this table, using the readings on the gas syringe diagrams:

time/s	volume of gas collected/cm³
0	
30	
60	
90	
120	
150	
180	

d i Plot a graph for the results in your table, using a grid labelled like this:

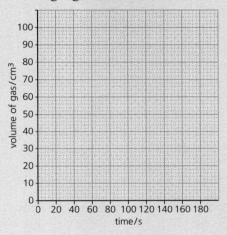
[3]

  ii Describe how the rate changes, as the reaction proceeds. [1]
  iii Explain why the volume of gas collected does not increase after 150 seconds. [1]
  iv On your grid, sketch the shape of the graph line you would expect, if the reaction is carried out at a higher temperature. [2]

Total 17 marks

10 Students are given four colourless solutions. They know they contain potassium iodide, sodium carbonate, potassium carbonate, and potassium chloride, but they do not know which is which.

Task 1: Describe how the students could use dilute hydrochloric acid to place the solutions into the two groups shown below, and suggest suitable apparatus:

group A	group B
potassium chloride	potassium carbonate
potassium iodide	sodium carbonate

[3]

Task 2: Outline a method the students could use to identify the two colourless solutions in Group A. Describe the experiment, and state the results you expect. [5]

Task 3: A flame test can be used to identify the two colourless solutions in Group B. Describe how the students should carry out the test, and state the colours that will be seen. [5]

Total 13 marks

11 Hydrogen peroxide (H_2O_2) decomposes very slowly at room temperature:
$$2H_2O_2\,(aq) \longrightarrow 2H_2O\,(l) + O_2\,(g)$$
Some metal oxides can speed up this reaction. They act as catalysts.

a The apparatus below was used to compare two metal oxides, **A** and **B**, as catalysts. The flask contains 50 cm³ of hydrogen peroxide solution.

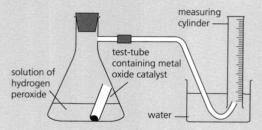

i Name the other piece of equipment needed, in addition to what you see above. [1]
ii Suggest how the reaction was started. [1]
iii What collected in the measuring cylinder? [1]

The volume of gas in the measuring cylinder was recorded every 15 seconds, for each catalyst. The readings are shown on the diagrams below.

b Copy and complete this table, using the readings from the diagrams of the measuring cylinders at the bottom of the page:

time /s	volume of oxygen collected /cm³	
	catalyst A	catalyst B
0		
15		
30		
45		
60		

[3]

c i Plot a graph for both set of results, using a grid like the one below. Label each line.

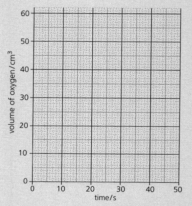

[4]

ii Which metal oxide is the better catalyst for this reaction? Explain your choice. [2]
iii Both catalysts produce the same final volume of oxygen. Explain why. [1]

d Give two other ways of increasing the rate of this decomposition reaction. [2]

Total 15 marks

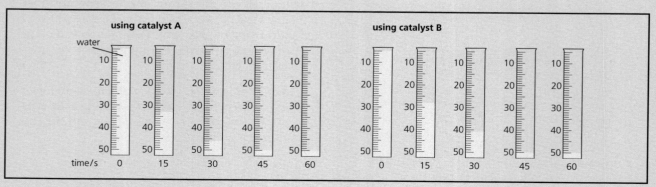

IGCSE PRACTICE QUESTIONS FOR PAPER 6

12 A student investigates the effect of concentration on the rate of the reaction between hydrochloric acid and aqueous sodium thiosulfate.
The equation for the reaction is:
$Na_2S_2O_3 (aq) + 2HCl (aq) \longrightarrow 2NaCl (aq) + 2H_2O (l) + SO_2 (g) + S (s)$
This diagram shows the method used:

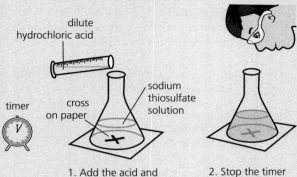

1. Add the acid and immediately start the timer.
2. Stop the timer when the cross is hidden from view.

a i The solution goes cloudy as the reaction proceeds. Explain why. [2]
 ii What is the purpose of the dark cross on the paper under the flask? [1]
 iii The time the student records in this investigation is *not* the total time for the reaction. Explain why. [1]
b Five experiments were carried out. The flask had 50 cm³ of solution each time, but with different concentrations of sodium thiosulfate.
The volume and concentration of the acid were kept constant. All measurements were made at room temperature. This table shows the results:

exp no	volume of liquid in flask / cm³		time for cross to disappear / s
	sodium thiosulfate solution	water	
1	50	0	68
2	40	10	83
3	30	20	113
4	20	30	167
5	10	40	320

 i Name a suitable piece of apparatus for adding the correct volume of sodium thiosulfate solution to the flask. [1]
 ii In which experiment is the concentration of sodium thiosulfate highest? [1]
 iii In which experiment is the rate of reaction fastest? [1]

c i Plot the results obtained on a grid like this one. Join the points smoothly.

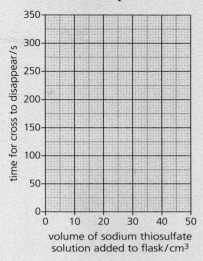

[3]

 ii On your graph, the volume of sodium thiosulfate solution added to the flask is used to represent its concentration. Explain why this is correct to do. [1]
d i Predict how long it will take for the cross to disappear, if 15 cm³ of sodium thiosulfate solution is used. Show clearly on your graph how you worked out your answer. [2]
 ii What volume of water should be added to 15 cm³ of the sodium thiosulfate solution, to ensure a valid result? [1]
e On your grid, sketch the graph line you will expect if the five experiments are repeated at a higher temperature. [2]

Total 16 marks

13 Sodium hydroxide solution in used in the laboratory, to identify cations in solution.
a Define the term *cation*. [1]
b Five solutions were tested using sodium hydroxide solution. This table shows the results:

solution	effect of adding sodium hydroxide solution	
	initial	in excess
A	pale blue precipitate	no change
B	white precipitate	no change
C	ammonia gas released on warming	no change
D	green precipitate	no change
E	green precipitate	green solution

Identify the cation in each solution **A** to **E** and give its formula. [10]

Total 11 marks

14 The electrolysis of lead(II) bromide is carried out:

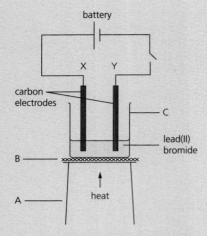

  a Name the items labelled A, B, and C. [3]
b **i** Why is carbon chosen for the electrodes? [1]
    ii Which is the positive electrode, X or Y? [1]
  c Explain why heat must be applied. [1]
  d Name the product, and describe its appearance:
    i at the positive electrode (anode) [2]
    ii at the negative electrode (cathode) [2]
  e This electrolysis is not carried out in the open laboratory. Why not? [1]
      Total 11 marks

15 A mixture of two solids **A** and **B** is analysed. **A** is a white soluble salt. **B** is the insoluble compound copper(II) carbonate. Six tests are carried out.
The tests and some results are given in the table at the bottom of the page.
  a Describe what you expect to observe in tests 1, 2, and 3. (There is one mark for each separate observation.) [6]
  b Name the gas that is given off in test 1. [1]
  c What will be seen if aqueous ammonia is used instead of sodium hydroxide, in test 3? [3]
  d Using the observations for tests 4, 5 and 6, what can you deduce about salt **A**? [3]
      Total 13 marks

16 You are provided with
- standard laboratory equipment
- two powdered metal carbonates, **X** and **Y**
- dilute hydrochloric acid

The two metal carbonates react with dilute hydrochloric acid, to produce carbon dioxide. Plan an investigation to show which compound reacts faster, **X** or **Y**.

Include a diagram in your answer.
      Total 6 marks

tests carried out	observation during tests
test on the solid mixture	
1 Dilute nitric acid is added to the solid mixture in a test-tube. The gas released is tested with lime water.	.... [2] .... [1]
Distilled water is added to the solid mixture in another test-tube. It is shaken, and the contents are filtered, giving a residue on the filter paper, and a filtrate. Tests are carried out as shown below.	
tests on the residue The solid residue is divided into two test-tubes.	
2 The first test-tube is heated. There is a colour change.	.... [1]
3 Dilute hydrochloric acid is added to the second test-tube. A blue solution forms. Excess sodium hydroxide solution is then added.	.... [2]
tests on the filtrate The filtrate is divided into three test-tubes.	
4 Dilute nitric acid is added to the first portion of the filtrate, followed by aqueous silver nitrate.	a pale yellow precipitate forms
5 Aqueous sodium hydroxide is added to the second portion.	a white precipitate forms, and it is soluble in excess sodium hydroxide
6 Aqueous ammonia is added to the third portion.	a white precipitate forms, and it is soluble in excess ammonia

Glossary

A

A_r short for *relative atomic mass of an element* – the average mass of its isotopes, relative to the mass of a carbon-12 atom

acid rain rain that is acidic because gases such as sulfur dioxide are dissolved in it (from burning fossil fuels)

acidic solution has a pH less than 7; an acidic solution contains H^+ ions

acid fermentation the process in which bacteria convert ethanol to ethanoic acid

activation energy the minimum energy that colliding particles must have in order to react; its symbol is E_a

addition reaction where a molecule adds onto an alkene, and the C = C double bond of the alkene changes to a single bond

addition polymerisation where small molecules join to form a very large molecule, by adding on at double bonds

alcohols the family of organic compounds with the OH functional group; ethanol is an example

alkali a soluble base; for example sodium hydroxide

alkali metals Group I of the Periodic Table

alkaline earth metals Group II of the Periodic Table

alkaline solution has a pH above 7; alkaline solutions contain OH^- ions

alkanes a family of saturated hydrocarbons with the general formula C_nH_{2n+2}; *saturated* means they have only single C–C bonds

alkenes a family of unsaturated hydrocarbons with the general formula C_nH_{2n}; their molecules contain a C = C double bond

allotropes different forms of an element; diamond and graphite are allotropes of carbon

alloy a mixture where at least one other substance is added to a metal, to improve its properties; the other substance is often a metal too (but not always)

amphoteric can be both acidic and basic in its reactions; aluminium oxide is amphoteric

anhydrous contains no water molecules in its crystal structure

anion another name for a negative ion

anode the positive electrode of a cell

aquifer a large underground store of water held in rocks; it can be pumped out to give a water supply

atmosphere the layer of gases around the Earth; here at Earth's surface we call it air

atomic number the number of protons in the nucleus of an atom; it is also called the *proton number*

atoms elements are made up of atoms, which contain protons, neutrons, and electrons

Avogadro constant the number of particles in one mole of atoms, molecules or ions; it is a huge number, 6.02×10^{23}

B

bacteria tiny organisms, some of which can cause disease; they can feed on most organic compounds, and some inorganic compounds

balanced equation a chemical equation in which the number of each type of atom is the same on both sides of the arrow

base a metal oxide or hydroxide; a base will neutralise an acid, to form a salt and water

biodegradable can be broken down by bacteria (or fungii)

biopolymer a polymer made by bacteria

blast furnace the chemical plant in which iron is extracted from its ore, iron(III) oxide

boiling increases the rate of a reaction and is unchanged at the end of the reaction

boiling point the temperature at which a substance boils

bond energy the energy needed to break a bond, or released when the bond is formed;
it is given in kilojoules (kJ) per mol

bonding how the atoms are held together in an element or compound; there are three types of bonds: ionic, covalent, and metallic

brittle breaks up easily when struck

burette a piece of lab equipment for delivering a measured volume of liquid

C

carboxylic acids a family of organic acids, which have the COOH functional group

cast iron iron from the blast furnace that is run into molds to harden; it contains a high % of carbon, which makes it brittle

catalyst increases the rate of a reaction and is unchanged at the end of the reaction

catalytic converter a device in a car exhaust, in which harmful gases are converted to harmless ones on catalysts

catalytic cracking where large molecules of hydrocarbons are split up into smaller ones, with the help of a catalyst

cathode the negative electrode of an electrolysis cell

cation another name for a positive ion

ceramic a hard, ureactive material made of baked clay that can withstand high temperatures; ceramics are non-conductors

chalk a rock made of calcium carbonate

change of state a change in the physical state of a substance – for example from solid to liquid, or liquid to gas

chemical change a change in which a new chemical substance forms

chemical equation uses chemical symbols to describe a chemical reaction

chemical reaction a process in which chemical change takes place

chromatogram the paper showing the separated coloured substances, after paper chromatography has been carried out

climate change how Earth's climates are changing, because of global warming

closed system no particles can enter or leave the container where the reaction is taking place

coagulant a substance that makes small particles stick together; it is used in cleaning up water, ready for piping to homes

coke a form of carbon made by heating coal

collision theory the idea that for particles to react, they must collide

311

GLOSSARY

with a certain minimum amount of energy

combustion an exothermic reaction in which the reactant combines with oxygen to form an oxide; also called burning

compound ion contains more than one element; for example the ion NO_3^-

compound a substance in which two or more elements are chemically combined

concentration tells you how much of one substance is dissolved in another; usually given as grams or moles per dm^3

condensation the physical change in which a gas turns into a liquid on cooling

condensation reaction where two molecules join to make a larger molecule, by eliminating a small molecule

condensation polymerisation where many molecules join to make very large molecules, by eliminating small molecules (for example water molecules)

condenser a piece of lab equipment used to cool a gas rapidly, and turn it into a liquid

conductor a substance that allows heat or electricity to pass through it easily

Contact process the industrial process for making sulfuric acid

cooling curve a graph showing how the temperature of a substance changes, while it is being cooled from gas to liquid to solid

corrosion where a substance is attacked by air or water, from the surface inwards; the corrosion of iron is called rusting

corrosive reacts with substances it comes into contact with

covalent bond the chemical bond formed when two atoms share electrons

covalent compound a compound made of atoms joined by covalent bonds

cracking reactions in which long-chain hydrocarbon molecules are broken down to shorter, more useful molecules

crystallisation the process in which crystals form, as a saturated solution cools

D

decomposition reaction where a substance breaks down to give two or more products

denature to destroy the structure of an enzyme by heat, or a change in pH

density the mass of a substance per unit volume; for water it is 1 g/cm^3

diatomic its molecules contain two atoms joined by a covalent bond

diffusion the process in which particles mix by colliding randomly with each other, and bouncing off in all directions

displacement reaction a reaction in which a more reactive element takes the place of a less reactive one, in a compound

displayed formula shows all the atoms and bonds in an organic compound

dissolving the process in which a soluble substance forms a solution

distillation separating a liquid from a mixture by boiling it off, then condensing it

double bond a covalent bond in which two atoms share two pairs of electrons

ductile can easily be drawn out into a wire

dynamic equilibrium where forward and reverse reactions take place at the same rate, so there is no *overall* change

E

effluent liquid waste from factories or sewage works

electrodes the conductors used to carry current into and out of an electrolyte; they may be rods of graphite or platinum (but different metals and alloys can be used)

electrolysis the process of breaking down a compound by passing a current through it

electrolyte a liquid that can conduct electricity (used in electrolysis)

electronic configuration how electrons in an atom are arranged in electron shells; it is shown for example as 2,8,3

electron shells the different energy levels which electrons occupy, around the nucleus

electrons the particles with a charge of 1– and almost no mass, in an atom

electroplating coating one metal with another, using electrolysis

element a substance containing only one type of atom; it cannot be broken down into another substance, in a chemical reaction

empirical found by experiment

empirical formula shows the simplest ratio in which the atoms in a compound are combined

endothermic takes in heat energy from the surroundings

enthalpy change of a reaction the overall energy change in the reaction; its symbol is ΔH. (Say *delta* H.)

enzymes proteins made by living cells, that act as biological catalysts

equation it uses symbols to describe a chemical reaction (but a *word equation* uses just words)

equilibrium when the forward and reverse reactions in a reversible reaction take place at the same rate, so there is no *overall* change

ester a compound formed when an alcohol reacts with a carboxylic acid

evaporation where a liquid turns to a gas at a temperature below its boiling point; this change takes place at the surface of liquid

exothermic gives out heat energy

extract to remove a metal from its ore

F

fermentation the process in which the enzymes in yeast break down sugars, to form ethanol and carbon dioxide

fertiliser substance added to soil to help crops grow well

filtering separating a solid from a liquid by pouring the mixture through filter paper

filtrate the liquid obtained from filtration (after the solid has been removed)

flammable burns easily

flame tests tests to identify metal cations from the colours shown when their compounds are heated in a Bunsen flame

flocculation when particles clump together and can be skimmed off, in water treatment

flue gas desulfurisation the process of removing sulfur dioxide from waste gases in industry, to help prevent acid rain

formula uses symbols and numbers to tell you what elements are in a compound, and the ratio in which they are combined

forward reaction the reaction in which the product is made, in a reversible reaction

GLOSSARY

fossil fuels petroleum (crude oil), natural gas, and coal, all formed over millions of years from the remains of living things

fractional distillation a method used to separate two or more liquids that have different boiling points

fractions the groups of compounds that a mixture is separated into, by fractional distillation; fractions are collected one by one

freezing the change from liquid to solid, that occurs at the freezing point

fuel a substance we use to provide energy; most fuels are burned to release their energy (but nuclear fuels are not)

fuel cell a cell in which a chemical reaction provides electricity

functional group the part of the molecule in an organic compound that largely dictates how the compound reacts; for example the OH group in the alcohol family

G

galvanising coating iron with zinc, to prevent the iron from rusting

giant structure where a very large number of atoms or ions are held in a lattice by strong bonds; metals, diamond, and ionic solids such as sodium chloride are all giant structures

global warming the rise in global temperatures around the world, due to the rising levels of greenhouse gases in the atmosphere

greenhouse gas a gas that traps heat in the atmosphere, preventing its escape into space; carbon dioxide and methane are examples

group a column of the Periodic Table containing elements with similar properties

H

Haber process the industrial process for making ammonia from nitrogen and hydrogen

half-equation shows the reaction taking place at an electrode; it is an ionic equation

halogens the elements of Group VII in the Periodic Table

heating curve a graph showing how the temperature of a substance changes on heating, as it changes state from solid to liquid to gas

homologous series a family of organic compounds, that share the same general formula and have similar properties

hydrated salt has water molecules built into its crystal structure; for example hydrated copper(II) sulfate, $CuSO_4.5H_2O$

hydrocarbon a compound containing *only* carbon and hydrogen

hydrogenation adding hydrogen

hydrogen–oxygen fuel cell uses the reaction between hydrogen and oxygen to give an electric current

hydrolysis the breaking down of a compound by reaction with water

hypothesis a statement you can test by doing an experiment and taking measurements

I

incomplete combustion the burning of fuels in a limited supply of oxygen; it gives carbon monoxide instead of carbon dioxide

indicator a chemical that shows by its colour whether a substance is acidic or alkaline

inert does not react (except under extreme conditions)

inert electrode is not changed during electrolysis; all it does is conduct the current

in excess more than is needed for a reaction; some will be left at the end

insoluble does not dissolve in a solvent

insulator a poor conductor of heat or electricity

intermolecular forces forces of attraction between molecules

ion a charged atom or group of atoms formed by the gain or loss of one or more electrons

ionic bond the bond formed between ions of opposite charge

ionic compound a compound made up of ions, joined by ionic bonds

ionic equation shows only the ions that actually take part in a reaction; the other ions are called *spectator ions*

isomers compounds with the same formula, but a different arrangement of atoms; also called *structural isomers*

isotopes atoms of the same element, that have a different numbers of neutrons

J

joule (J) the unit for measuring energy

L

landfill site an area of land set aside as a place to dump rubbish

lattice a regular arrangement of particles

lime the common name for calcium oxide

limewater a solution of the slightly soluble compound calcium hydroxide

locating agent used in chromatography to show up colourless substances; it reacts with them to give coloured substances

low-grade ore contains only a small % of the desired metal or compound

M

M_r short for *relative molecular mass* or *relative formula mass* for a compound; see the definitions for these

malleable can be bent or hammered into shape

mass number the total number of protons and neutrons in the nucleus of an atom; it is also called the *nucleon number*

mass spectrometer an instrument used to find the masses of atoms and molecules

melting point the temperature at which a solid substance melts

melting the physical change from a solid to a liquid

metal an element that shows metallic properties (for example conducts electricity, and forms positive ions)

metallic bond the bond that holds the atoms together in a metal

metalloid an element that has properties of both a metal and a non-metal

microbes microscopic (very tiny) organisms such as bacterium and viruses

minerals compounds that occur naturally in Earth; rock is made of different minerals

mixture contains two or more substances that are not chemically combined

mole (mol) the amount of a substance that contains 6.02×10^{23} particles; the particles may be atoms, ions, or molecules; this large number is called the Avogadro constant

molecular made up of molecules

molecule a unit of two or more atoms held together by covalent bonds

monatomic made up of single atoms; for example neon is a monatomic element

monomers small molecules that join together to form polymers

GLOSSARY

N

native describes a metal that is found in Earth's crust as the element

negative electrode another name for the cathode, in an electrolysis cell

negative ion an ion with a negative charge

neutral (electrical) has no charge

neutral (solutions) neither acidic nor alkaline; neutral solutions have a pH of 7

neutralisation the chemical reaction between an acid and a base or a carbonate, giving a salt and water

neutron a particle with no charge and a mass of 1 unit, found in the nucleus of an atom

nitrogenous fertiliser it provides nitrogen for plants, in the form of nitrate ions or ammonium ions

noble gases the Group VIII elements of the Periodic Table; they are called *noble* because they are so unreactive

non-metal an element that does not show metallic properties: the non-metals lie to the right of the zig-zag line in the Periodic Table on page 312 (except for hydrogen, which sits alone)

non-renewable resource a resource such as petroleum that we are using up, and which will run out one day

non-toxic not harmful to health

NPK fertiliser a fertiliser that provides nitrogen, phosphorus and potasssium, the three main elements plants need for growth (in addition to the carbon dioxide and water they need for photosynthesis)

nucleon number the total number of protons and neutrons in the nucleus of an atom; it is also called the *mass number*

nuclear fuel contains isotopes with unstable nuclei, such as uranium-235; these break up, giving out a great deal of energy

nucleus the centre part of the atom, made up of protons and neutrons

O

ore rock containing a metal, or metal compound, from which the metal is extracted

organic chemistry the study of organic compounds; there are millions of them!

organic compound a compound containing carbon, and usually hydrogen; carbon forms a great many compounds because of its ability to form chains of different lengths

oxidation a chemical reaction in which a substance gains oxygen, or loses electrons

oxidation number tells you how many electrons the atoms of an element have gained, lost, or shared, in forming a compound; it is given in Roman numerals

oxide a compound formed between oxygen and and other element

oxidising agent a substance that brings about the oxidation of another substance

P

particulates tiny particles of a solid, or tiny droplets of a liquid, suspended in the air; for example tiny particles of soot from diesel engines; particulates can damage health

paper chromatography a way to separate the substances in a mixture, using a solvent and special paper; the dissolved substances separate because they travel over the paper at different speeds

percentage composition what % of each element is present in a compound, by mass

period a horizontal row of elements in the Periodic Table; its number tells you how many electron shells the atoms have

periodicity the pattern of repeating properties that shows up when elements are arranged in order of proton number; you can see it in the groups in the Periodic Table

Periodic Table shows the elements in order of increasing proton number; similar elements are in columns called groups

petroleum a fossil fuel formed over millions of years from the remains of tiny sea plants and animals; it is also called *crude oil*

pH scale a scale that tells you how acidic or alkaline a solution is; it is numbered 0 to 14

photochemical reaction a reaction that depends on light energy

photochemical smog brown haze in the air caused by the effect of sunlight on oxides of nitrogen (from engines and furnaces)

photodegradeable can be broken down by the action of light

photosynthesis the process in which plants convert carbon dioxide and water to glucose and oxygen

physical change a change in which no new chemical substance forms; melting and boiling are physical changes

physical properties properties such as density and melting point (that are not about chemical behaviour)

pipette a piece of laboratory equipment used to deliver a known volume of liquid, accurately

plastics a term used for synthetic polymers (made in factories, rather than in nature)

pollutant a substance that causes harm if it gets into the air or water

pollution when harmful substances are released into the environment

polymer large molecule built up from many smaller molecules called monomers; poly(ethene) is an example

polymerisation a chemical reaction in which many small molecules join to form very large molecules; the product is called a polymer

positive ion an ion with a positive charge

precipitate an insoluble chemical produced during a chemical reaction

precipitation reaction a reaction in which a precipitate forms

product a chemical made in a chemical reaction

protein a polymer made up of many different amino acid units joined together

proton number the number of protons in the nucleus of an atom; it is also called the *atomic number*

proton a particle with a charge of 1+ and a mass of 1 unit, found in the nucleus of an atom

pure there is only one substance in it

PV (photovoltaic) cell converts energy from sunlight into electricity

Q

quicklime another name for calcium oxide

R

radioactive isotope (radioisotope) an atom with an unstable nucleus that breaks down, giving out radiation

random motion the zig-zag path a particle follows as it collides with other particles and bounces away again

rate of reaction how fast a reaction is

reactant a starting chemical for a reaction

reaction pathway diagram shows how the energy levels change as a reaction progresses from reactants to products

reactivity how readily a substance reacts

GLOSSARY

reactivity series the metals listed in order of their reactivity

recycling reusing resources such as scrap metal, glass, paper and plastics

redox reaction any reaction in which electrons are transferred; one substance is oxidised (it loses electrons) and another is reduced (it gains electrons)

reducing agent a substance which brings about the reduction of another substance

reduction when a substance loses oxygen, or gains electrons

refining (petroleum) the process of separating petroleum (crude oil) into groups of compounds with molecules fairly close in size; it is carried out by fractional distillation

relative atomic mass (A_r) the average mass of the isotopes of an element, relative to 1/12 th of the mass of an atom of ^{12}C

relative formula mass (M_r) the mass of one formula unit of an ionic compound; you find it by adding together the relative atomic masses (A_r) of the atoms in the formula

relative molecular mass the average mass of a molecule; you find it by adding the relative atomic masses of the atoms in it

renewable energy energy from a resource such as wind or sunlight, that does not get used up

residue the solid obtained when a solid is separated from a liquid by filtering; the term is also used for solids left after a liquid is removed by evaporation or distillation

respiration the reaction in the cells of living things that provides energy; in humans, the reactants are glucose and oxygen

reversible reaction can go both ways: products can form, then break down again; for example A + B ⇌ C + D; note the arrow symbol used in the equation

reverse reaction in the reversible reaction above C and D react to give A and B again

rusting the special term used for the corrosion of iron; oxygen and water attack the iron, and red-brown rust forms

S

sacrificial protection allowing one metal to corrode, in order to protect another metal

salt an ionic compound formed when an acid reacts with a metal, a base, or a carbonate

saturated compound an organic compound in which all the bonds between carbon atoms are single covalent bonds

saturated solution no more of the solute will dissolve in it, at that temperature

sedimentation when solid particles settle to the bottom of a liquid in a container

sewage waste from toilets and sinks in homes, and waste from farm animals; waste from homes may be piped to sewage works, but in some places it drains into rivers

single bond the bond formed when two atoms share just one pair of electrons

slaked lime calcium hydroxide

solar power converting energy from sunlight into electricity

solubility the amount of a solute that will dissolve in 100 grams of a solvent, at a given temperature

soluble will dissolve in a solvent

solute the substance you dissolve in the solvent, to make a solution

solution a mixture obtained when a solute is dissolved in a solvent

solvent the liquid in which a solute is dissolved, to make a solution

sonorous makes a ringing sound when struck

spectator ions ions in a reaction mixture that do not actually take part in the reaction

state symbols added to a chemical equation to show the physical states of the reactants and products (g = gas, l = liquid, s = solid, aq = aqueous)

structural formula shows how the atoms in a molecule are arranged, but without showing all the bonds; for example CH_3CH_2OH and CH_3COOH

structural isomer see *isomer*

synthetic manufactured, not natural

T

thermal energy another term for *heat*

thermal decomposition the breaking down of a compound using heat

titration a method to find what volume of an acid solution will react with a given volume of alkaline solution, or vice versa

toxic harmful to healths

transition elements make up the wide middle block of the Periodic Table; they are metals, and include iron, copper, and gold

trend a gradual change; the groups in the Periodic Table show trends in their properties

triple bond the bond formed when two atoms share three pairs of electrons; a nitrogen molecule has a triple bond

U

universal indicator a paper or liquid you can use to find the pH of a solution; it changes colour across the whole range of pH

unreactive does not react easily

unsaturated compound an organic compound with at least one double bond between carbon atoms

V

valency a number that tells you how many electrons an atom gains, loses or shares, in forming a compound

variable oxidation number where atoms of an element can form ions with different charges; copper forms Cu^+ and Cu^{2+} ions

viscosity a measure of how runny a liquid is; the more runny it is, the lower its viscosity

viscous thick and sticky

volatile evaporates easily, to form a vapour

W

water of crystallisation water molecules built into the crystal structure of a compound; for example in hydrated copper(II) sulfate, $CuSO_4.5H_2O$

weak acids only some of the molecules are dissociated, to form H^+ ions; ethanoic acid s a weak acid

wind power where the energy of the wind is converted into electricity

Y

yield the actual amount of a product obtained in a reaction; it is often given as a % of the yield calculated from the equation

THE PERIODIC TABLE AND ATOMIC MASSES

Period	I	II												III	IV	V	VI	VII	VIII
1	1 H hydrogen 1																		2 He helium 4
2	3 Li lithium 7	4 Be beryllium 9												5 B boron 11	6 C carbon 12	7 N nitrogen 14	8 O oxygen 16	9 F fluorine 19	10 Ne neon 20
3	11 Na sodium 23	12 Mg magnesium 24					the transition elements							13 Al aluminium 27	14 Si silicon 28	15 P phosphorus 31	16 S sulfur 32	17 Cl chlorine 35.	18 Ar argon 40
4	19 K potassium 39	20 Ca calcium 40	21 Sc scandium 45	22 Ti titanium 48	23 V vanadium 51	24 Cr chromium 52	25 Mn manganese 55	26 Fe iron 56	27 Co cobalt 59	28 Ni nickel 59	29 Cu copper 64	30 Zn zinc 65		31 Ga gallium 70	32 Ge germanium 73	33 As arsenic 75	34 Se selenium 79	35 Br bromine 80	36 Kr krypton 84
5	37 Rb rubidium 85	38 Sr strontium 88	39 Y yttrium 89	40 Zr zirconium 91	41 Nb niobium 93	42 Mo molybdenum 96	43 Tc technetium –	44 Ru ruthenium 101	45 Rh rhodium 103	46 Pd palladium 106	47 Ag silver 108	48 Cd cadmium 112		49 In indium 115	50 Sn tin 117	51 Sb antimony 122	52 Te tellurium 128	53 I iodine 127	54 Xe xenon 131
6	55 Cs caesium 133	56 Ba barium 137	57–71 lanthanoids	72 Hf hafnium 178	73 Ta tantalum 181	74 W tungsten 184	75 Re rhenium 186	76 Os osmium 190	77 Ir iridium 192	78 Pt platinum 195	79 Au gold 197	80 Hg mercury 201		81 Tl thallium 204	82 Pb lead 207	83 Bi bismuth 209	84 Po polonium –	85 At astatine –	86 Rn radon –
7	87 Fr francium –	88 Ra radium –	89–103 actinoids	104 Rf rutherfordium –	105 Db dubnium –	106 Sg seaborgium –	107 Bh bohrium –	108 Hs hassium –	109 Mt meitnerium –	110 Ds darmstadtium –	111 Rg roentgenium –	112 Cn copernicium –		113 Nh nihonium –	114 Fl flerovium –	115 Mc moscovium –	116 Lv livermorium –	117 Ts tennessine –	118 Og oganesson –

lanthanoids

57 La lanthanum 139	58 Ce cerium 140	59 Pr praseodymium 141	60 Nd neodymium 144	61 Pm promethium –	62 Sm samarium 150	63 Eu europium 152	64 Gd gadolinium 157	65 Tb terbium 159	66 Dy dysprosium 163	67 Ho holmium 165	68 Er erbium 167	69 Tm thulium 169	70 Yb ytterbium 173	71 Lu lutetium 175

actinoids

89 Ac	90 Th	91 Pa	92 U	93 Np	94 Pu	95 Am	96 Cm	97 Bk	98 Cf	99 Es	100 Fm	101 Md	102 No	103 Lr

THE PERIODIC TABLE AND ATOMIC MASSES

The symbols and proton numbers of the elements

Element	Symbol	Proton number	Element	Symbol	Proton number	Element	Symbol	Proton number	Element	Symbol	Proton number
actinium	Ac	89	calcium	Ca	20	francium	Fr	87	lawrencium	Lw	103
aluminium	Al	13	californium	Cf	98	gadolinium	Gd	64	lead	Pb	82
americium	Am	95	carbon	C	6	gallium	Ga	31	lithium	Li	3
antimony	Sb	51	cerium	Ce	58	germanium	Ge	32	lutetium	Lu	71
argon	Ar	18	chlorine	Cl	17	gold	Au	79	magnesium	Mg	12
arsenic	As	33	chromium	Cr	24	hafnium	Hf	72	manganese	Mn	25
astatine	At	85	cobalt	Co	27	helium	He	2	mendelevium	Md	101
barium	Ba	56	copper	Cu	29	holmium	Ho	67	mercury	Hg	80
berkelium	Bk	97	curium	Cm	96	hydrogen	H	1	molybdenum	Mo	42
beryllium	Be	4	dysprosium	Dy	66	indium	In	49	neodymium	Nd	60
bismuth	Bi	83	einsteinium	Es	99	iodine	I	53	neon	Ne	10
boron	B	5	erbium	Er	68	iridium	Ir	77	neptunium	Np	93
bromine	Br	35	europium	Eu	63	iron	Fe	26	nickel	Ni	28
cadmium	Cd	48	fermium	Fm	100	krypton	Kr	36	niobium	Nb	41
caesium	Cs	55	fluorine	F	9	lanthanum	La	57	nitrogen	N	7

The symbols and proton numbers of the elements (continued)

Element	Symbol	Proton number	Element	Symbol	Proton number	Element	Symbol	Proton number
nobelium	No	102	rhodium	Rh	45	thallium	Tl	81
osmium	Os	76	rubidium	Rb	37	thorium	Th	90
oxygen	O	8	ruthenium	Ru	44	thulium	Tm	69
palladium	Pd	46	samarium	Sm	62	tin	Sn	50
phosphorus	P	15	scandium	Sc	21	titanium	Ti	22
platinum	Pt	78	selenium	Se	34	tungsten	W	74
plutonium	Pu	94	silicon	Si	14	uranium	U	92
polonium	Po	84	silver	Ag	47	vanadium	V	23
potassium	K	19	sodium	Na	11	xenon	Xe	54
praseodymium	Pr	59	strontium	Sr	38	ytterbium	Yb	70
promethium	Pm	61	sulfur	S	16	yttrium	Y	39
protactinium	Pa	91	tantalum	Ta	73	zinc	Zn	30
radium	Ra	88	technetium	Tc	43	zirconium	Zr	40
radon	Rn	86	tellurium	Te	52			
rhenium	Re	75	terbium	Tb	65			

Relative atomic masses (A_r) for calculations

Element	Symbol	A_r
aluminium	Al	27
bromine	Br	80
calcium	Ca	40
carbon	C	12
chlorine	Cl	35.5
copper	Cu	64
fluorine	F	19
helium	He	4
hydrogen	H	1
iodine	I	127
iron	Fe	56
lead	Pb	207
lithium	Li	7
magnesium	Mg	24
manganese	Mn	55
neon	Ne	20
nitrogen	N	14
oxygen	O	16
phosphorus	P	31
potassium	K	39
silver	Ag	108
sodium	Na	23
sulfur	S	32
zinc	Zn	65

Index

A
accuracy of measurement 139, 239, 257, 260
acid fermentation 218
acidic oxides 137
acid rain 196, 197
acids 128, 130–5
   dissociation 130, 131
   neutralisation reactions 133–5, 139, 258–9
   pH scale 130–1
   proton donors 134, 135
   reactions 132–5
   strong and weak 131
   see also indicators; salt making
   using acids
activation energy (E_a) 95, 113, 211
addition polymerisation 224–5
addition reactions, alkenes 214–16
air, 194–5
air pollution 196–7
alchemists 22, 244
alcohols 208–9, 216–17, 219, 227
   see also ethanol
algae 189, 193
alkali metals (Group I) 29, 148–9, 160
alkaline earth metals (Group II) 31, 160
alkalis 129–31
   see also bases; pH
alkanes 67, 205, 208–13
alkenes 208–9, 212–15, 222, 224
allotropes of carbon 40–1
alloys 177, 180–2
aluminium
   corrosion resistance 167, 179
   discovery 183
   extraction from bauxite ore 176–8
   uses 177–9
aluminium oxide, electrolytic extraction of aluminium 176–7
amide linkage in polymers 226, 234
amino acids 234, 235, 248–9
ammonia
   covalent structure 36
   fertilisers 192, 193
   from cracking ethene 213
   Haber process 122–3
   preparation in the lab 132, 219–21, 260
   reaction with hydrogen chloride 77
   relative formula mass 51
   reversible reaction to make 120–3
   structure 46
   testing for 261, 263, 265
ammonium ions 33, 262, 263
amphoteric oxides 137
anhydrous copper(II) sulfate 118, 191
anhydrous salts 118, 141
anions 29, 264–5
   see also negative ions
anodes 84–7, 176, 177

apparatus in the laboratory 238–9, 260
aqueous solutions
   acids 128, 130, 137
   alkalis 129–31
   electrolysis 85–9
   pH 130
   solubility 240–1
   solubility of ionic versus covalent compounds 39
aquifers 190
A_r (relative atomic mass) 21, 50, 56
argon 28, 147, 157, 194
atmosphere 194–201
   air pollution 99, 187, 196–8
   constituents of air 194
   greenhouse gases 198–9, 200–1
   layers 194
atomic mass units 16, 21, 50
atomic number 16, 146
   see also proton number
atoms 3
   elements 14–21
   forming bonds 28–31, 34–5
   history of understanding 22–3
   isotopes 20–1
   moles 56–7
   nucleus 16–17
   sub-atomic particles 16–19
Avogadro constant 56
Avogadro's Law 60

B
bacteria see microorganisms
balancing
   charges 32, 47
   equations 48, 49
   half equations 74
   valencies 47
bases 129–36
   alkalis 129–31
   insoluble metal oxides 129, 135, 136
   neutralisation reactions 133–5
   pH scale 130–1
   proton acceptors 135
   reactions 132–5
batteries 82–4, 98, 125, 131, 139, 179, 200
biodegradable plastics 233
biofuels and biomass fuels 200, 217
biological catalysts see enzymes
bioplastics 233
bitumen 206, 207
blast furnace 174–5
boiling 4, 7
boiling points 4–5, 8, 9
   covalent compounds 39
   fractional distillation 206–7, 217, 245
   halogens 150
   identifying substances 251
   organic compounds 209, 211
   purity 9, 250–1
bond energy 92, 94, 96–7
bonding/forming bonds 27–31, 34–5, 42–3, 94, 96
boreholes 190
branched chain hydrocarbons 204, 211

brass 180, 181
breaking bonds 94, 96
brittle substances 161, 175
bromine water test 215
bronze 181, 182
burning see combustion; incineration
butane 208, 210–11
but-1-ene 214, 215
but-2-ene 215

C
calcium carbonate 69, 93, 108, 132, 133, 167, 174–5, 189, 240, 260
calcium chloride 49, 132
calcium hydroxide 129, 132, 133, 162, 240, 261
calcium oxide 73, 92, 133, 136, 167, 197
carbon
   coal and coke 174, 196, 197, 204
   competing with metals 164–5
   extracting metals from ores 173–5
   graphite electrodes 84, 88
   isotopes 20–1
   molar mass 56
   reaction with oxygen 48, 52
   reactivity series of metals 166
   structures and properties 140–1, 161
   water treatment 190, 191
carbonates
   ions 33
   making salts 132, 138
   tests 265
   see also calcium carbonate
carbon capture 201
carbon=carbon double bonds 37, 209, 212–13, 224–5
carbon-carbon single bonds 209–11
carbon dioxide
   in atmosphere 196
   burning plastic 232
   carbonic acid 128, 137
   covalent structure 37
   formula 26
   greenhouse gas 198, 200–1, 205
   preparation in the lab 260
   production in blast furnace 175
   testing for 261, 265
carbonic acid 128, 137
carbon monoxide
   air pollution 39, 99, 196, 197
   incomplete combustion 196, 210
   metal extraction in furnace 173–5
   properties 39
carboxylic acids 208–9, 218–19
   see also dicarboxylic acid
car engines see transport
catalysts 112, 113
   enzymes 112–15, 217, 235
   reversible reactions 121, 122, 125
   transition elements 113, 154, 197

catalytic addition 216
catalytic converters 197
catalytic cracking 212–13
cathodes 84–7, 176, 177
cations 29, 32–3, 84, 262–3
   see also positive ions
chain length of polymers 224
chalk 69, 240
changes of state 4, 6–9
charge
   ions 29
   sub-atomic particles 16, 23
chemical changes 26–7
chemical energy 92
chlorine
   ions 29, 30, 31
   isotopes 21
   molecules 34–5
   reaction with hydrocarbons 211
   testing for 261
   water treatment 79, 190, 191
chromatography
   chromatograms 246, 249, 251
   coloured substances 246–7, 251
   colourless substances 248–9
   crime detection 252–3
   gas 252–3
   paper 246–9, 251
   types 252
   uses 247, 251, 253
cleaning products 79, 113, 256, 258–9
climate change 187, 196, 198–201, 205
closed versus open systems 119, 121, 123, 124
coagulants, water treatment 191
coal 174, 196, 197, 204
coke 174–5
collecting gases 260
collision theory, reaction rates 110–11
coloured substances
   halogens 150
   paper chromatography 246–7, 251
   pigments 141
   precipitate tests for ions 141, 263, 264
   transition element compounds 154
combustion
   equations 48–9
   exothermic reactions 93, 94
   hydrocarbons 210, 211
   incomplete combustion 196, 210
   redox reactions 73
   see also fuels
compound ions 33
compounds 15, 26–45
   names and formulae 46–7
   percentage composition 53
   see also bonding; covalent compounds; ionic compounds
concentration
   finding by titration 142–3
   rates of reactions 106, 109–11
   reversible reactions 121
   solutions 62–3

318

INDEX

see also pressure
condensation polymerisation 226–7
condensation reactions 219
condensation of water 4, 7, 9
conductors of electricity 39, 41, 43, 82–3
conductors of heat 43
Contact process 124–5
controls in experiments 112
cooling curves 9
copper 42, 154, 155
    alloys 180–2
    electrodes 88
    extraction from ores 173
    hydrated and anhydrous compounds 118, 141, 191
    ion displacement by iron 165
    uses 179
copper(II) sulfate 59, 62, 88, 118, 132, 138, 163, 238–9, 241–2
corona virus (COVID-19) 234
corrosion resistance 178–81
corrosive chemicals 128, 129
costs/profits, industrial processes 123, 125
covalent bonds 34–5
covalent compounds
    giant structures 40, 41
    molecular 34–9
    names and formulae 46, 47
    properties 39–41, 83
    see also organic compounds
cracking hydrocarbons 122, 211–13
cryolite 176
crystallisation 22, 138, 139, 141, 238, 239, 241, 242
crystal structures 38, 40, 42

D

decane 212
decomposition by heat 97, 167, 175
decomposition of ionic compounds by electricity 83–9
deforestation 201
ΔH, enthalpy change 95–7
denaturation of proteins 235
density 148, 153, 160–1
dependent variables 256
diamines, polymerisation 226
diamond 40, 41, 161
dicarboxylic acid 226, 227
diffusion 3, 11
disinfectants 79, 150, 216
displacement reactions
    of halide ions by halogen 75, 151
    of hydrogen from acids 132, 163, 166, 260
    of metal ions by more reactive metals 165, 166, 183
    preparing gases in the lab 260
dissociation, strong and weak acids 130, 131, 143, 219
dissolving 3
    see also solutions
distillation 22, 206–7, 217, 244–5
distilled water 140, 191, 206, 238, 239, 244, 251
DNA testing 115
double bonds 35, 37, 209, 212–13, 224–5
drinking water 189–91
drive to give up electrons, metals 161–5
ductility of metals 43, 160, 179
dynamic equilibrium 119–21

E

E_a (activation energy) 95, 113, 211
Earth's crust, composition 166, 172
electricity 82–91
    batteries 82–4, 98, 125, 131, 139, 179, 200
    conductivity of materials 39, 41, 82–3, 84–9
    hydrogen–oxygen fuel cells 98–9, 200

power generation 200, 205, 232
electrodes 84–7
electrolysis 84–9
    applications 88–9
    aqueous solutions 85–9
    extracting metals from ores 166, 176–7
    invention 183
    molten ionic compounds 84, 86
    principles 84–5
    producing hydrogen for fuel cells 99
    rules for products from a solution 85
electrolytes 84, 85, 88, 89
electronic configuration of atoms 18–19
    electron shells
electron microscopes 3
electrons 16, 18–19
    atoms becoming ions 29
    covalent bonds 34–5
    discovery 23
    electricity 39, 41, 82, 84
    ionic bonds 30
    metal crystal structure 42
    transfer, half-equations 74–5, 86, 87, 165
electron shells 16, 18–19, 23
    bonding 29–31
    Periodic Table 146, 149, 150, 157
    stability of outer shell 29
electroplating 89
electrostatic attraction 30, 42
elements 14–21
    isotopes 20–1
    mixtures versus compounds 26
    oxidation number zero 76
    relative atomic mass 15, 21
    symbols 14, 15
    see also metals; non-metals; Periodic Table
empirical formulae 64–7
endothermic reactions 93–5, 97
end point of titrations 142, 258
energy
    changes 92–101
    chemical reactions 27
    endothermic reactions 93
    exothermic reactions 92
    forms 92
    reactivity relationship 152
    see also electricity; thermal energy
energy level diagrams 92–3
enthalpy change (ΔH) 95–7
environment 186–203, 230–3
    atmosphere 39, 99, 194–201
    climate change 187, 196, 198–201
    soil 133, 186, 187, 192–3, 230
    water 187–91, 230–1, 233, 253
enzymes 112–15, 217, 235
equations for reactions 48–9
    see also half-equations; moles
equilibrium 119–25
ester linkage 227
    see also polyester
esters 218, 219
ethane 208–11
ethanoic acid 131, 143, 208, 209, 218–19
ethanol 216–17
    distillation 245
    fermentation 114, 217, 245
    as a fuel 217
    production methods 216–17, 245
    structure and formula 26, 208, 209
    uses 216, 217
ethene 208, 209, 212–15
eutrophication 193
evaporation 4, 7, 238, 239, 243
exothermic reactions 92, 94, 95, 97
experiments 256–9
explosions 109
explosive reactions 19, 73, 94, 95, 109
extremophiles 115

F

feedstock chemicals 205, 206, 228
fermentation 114, 217, 218, 245
fertilisers 187, 188, 192–3
filtration 190, 191, 238, 239, 242, 251
flame tests for cations 262
flammability of organic compounds 207, 209
flocculation 190, 191
flue gas desulfurisation 197
forces of attraction, between particles 7, 8
formulae
    balancing valencies or charges 47
    compounds 26, 46–7
    empirical formulae 64–7
    ionic compounds 32
    molecular formulae 66–7
formula mass of ionic compounds 51
fossil fuels
    air pollution 99, 196–7
    coal 174, 196, 197, 204
    global warming 99, 187, 196, 198, 200
    making hydrogen for fuel cells 99
    natural gas 122, 196, 198, 204, 208
    petroleum-based 204–7
fractional distillation 206–7, 217, 245
freezing 4, 7, 9
fruit ripening with ethene 214, 215
fuel cells, hydrogen–oxygen 98–9, 200
fuels *see* biofuels; fossil fuels
functional groups
    organic compounds 209, 214, 216, 218
    polymerisation 226, 227, 234

G

galvanised iron 170, 179
gas chromatography 252–3
gases 4–11
    in atmosphere 186–8
    changes of state 4, 6–9
    collecting 260
    petroleum fractions 206, 210
    preparation 260
    pressure 10
    rates of reactions 103–6, 109, 111
    reversible reactions 121–3, 125
    produced when testing for ions 262, 263, 265
    rate of diffusion 11
    in river water 188, 189, 193
    scrubbing to remove impurities 122
    tests 261, 265
    volume of a mole 60–1
    volume/pressure/temperature relationship 10, 60
gasoline fraction of petroleum 206, 213
    see also petrol
giant structures 30, 38–41, 46, 161
global warming 187, 196, 198–201, 205
gold 88, 163, 166, 172, 173, 178, 183
graphite 41, 82–4, 88, 161
graphs, rates of reactions 104–6, 108
greenhouse gases 198–201, 205
groundwater 190–1
Group I elements (alkali metals) 29, 148–9, 160
Group II elements (alkaline earth metals) 31, 160
Group VI elements 29, 31
Group VII elements (halogens) 29, 75, 150–1
Group VIII elements (noble gases) 19, 28, 147
groups in Periodic Table 15, 19, 47, 146–51

H

Haber process 122–3
half-equations, electron transfer 74–5, 86, 87, 165
halide ions tests 264

INDEX

halides, electrolysis 85–7
halogenated hydrocarbons 211, 215
halogens (Group VII) 29, 75, 150–1
heat energy *see* thermal energy
heating curves 8, 9
helium 19, 28, 57
hematite, iron ore 172, 174, 178
high density poly(ethene) (HDPE) 228, 229
history of chemistry 2, 22–3, 156–7, 182–3, 244
homogenous series of organic chemicals 208–11, 214–19
hydrated salts 118, 138, 141
hydrocarbons 204–21
hydrochloric acid 11, 49, 104, 128, 130–5, 142–3, 163, 260
hydrogen
  from electrolysis 85–7
  ions 85, 130
  molecules 34
  place in Periodic Table 147
  preparation in the lab 260
  production from hydrocarbons 122
  reaction with chlorine 97
  reaction with oxygen 48
  reactivity series of metals 166
  reducing agent 78
  relative molecular mass 51
  sources on Earth 99
  testing for 261
hydrogen chloride 11, 36, 49, 77, 97, 130, 131
hydrogen–oxygen fuel cells 98–9, 200
hydrogen sulfide 47
hydrolysis 227
hydroxide ions (protons) 33, 85, 130, 134
hydroxyl group 212
hypothesis testing 256, 258–9

I
ice 4–7, 38
identification by chromatography 247–9, 252
identification tests 141, 215, 251, 261–5
impurities 122, 239, 247, 250–1
incineration of plastic 231, 232
incomplete combustion 196, 210
independent variables 256
indicators 128, 129, 131, 139, 142, 143, 258
industrial processes 115, 122–5
inert electrodes 84, 88
inert substances *see* unreactive
insoluble bases 129, 135, 136
insoluble salts made using acids 140–1
insoluble substances 240
  see also precipitation; solubility
insulators (electrical) 82, 225
insulators (thermal) 229
intermolecular forces 39
ionic bonds 30–1, 39
ionic compounds 30–3
  alkali metals 149
  balancing charges 32, 47
  decomposition by electricity 83–9
  electrical conduction when liquid/in solution 83
  electrolysis 84–9
  empirical formulae 66
  formulae 32–3
  lattice structures 38
  moles 56, 62
  names 32, 46, 133
  properties 39
  redox reactions 74–9
  relative formula mass 51
ionic equations 75, 134
ions 3
  carrying electric current 39, 83, 84
  charge 29

formation 29, 30
oxidation number 76
iron 154, 155
  alloys 180, 181
  cast iron 175
  early civilisations 182
  extraction in blast furnace 174–5
  galvanised 179
  heating curve 9
  ores 172–6
  reactivity 173
  redox reactions 72
  rusting 167–9
  uses 175, 178
isomerism, hydrocarbons 211, 215
isotopes 20, 50

K
kinetic particle theory 7

L
laboratory apparatus 238–9, 260 landfill sites, plastics 230–3
lattice structures 30, 38–42, 46
laws of chemistry 52
lime *see* calcium oxide
limestone 133, 167, 174, 175
limewater 129, 240, 261, 265
limiting factors 69
linear molecules 37
liquids 4–9
litmus test 128, 129, 134, 136, 137
livestock, methane 198, 201
locating agents for paper chromatography 249
low density poly(ethene) (LDPE) 228, 229
low-sulfur fuels 197

M
magnesium isotopes 21, 50
magnesium oxide 31, 39, 49, 65, 66, 72, 74, 135
malleability of metals 43, 160, 179
mass, *see also* moles; relative atomic mass; relative formula mass
mass number 17, 21
  see also nucleon number
mass of reactants and products of a reactions 52, 58
mass spectrometry 50
mass of sub-atomic particles 16, 17
materials/apparatus/method 238–9
melting 4, 6, 7
melting points 4–5, 8, 9
  alkali metals 148
  comparisons of compounds 39, 42
  giant covalent structures 40
  identifying substances 251
  organic compounds 209, 211
  purity 9, 250–1
membranes
  filtration 191, 242, 251
  hydrogen–oxygen fuel cell 98
Mendeleev, Dmitri 157
metal hydroxides 129
metallic bonds 42–3
metalloids 153
metal oxides
  amphoteric 137
  bases 129, 135, 136
  reduction 72–3, 78
metals
  acid reaction rates 102–6, 110
  alkali metals 29, 148–9, 160
  alkaline earth 31, 160
  alloys 177, 180–2
  cation tests 262, 263
  competing with carbon 164–5
  competing for oxygen 164, 165
  competing to form compounds 164–5
  competing to form ions in solution 165

conductors of electricity 83
electroplating 89
extracting from compounds 166
extraction from ores 173–7
history 182–3
lattice structure 42
Periodic Table 15, 146, 147, 153
properties 42, 160
reactions with dilute hydrochloric acid 163
reactions with oxygen 163
reactions with water 162
reactivity 5, 161–7, 173
refining with electrolysis 88
uses 175, 177–83
see also transition elements
methane 208, 210–11
  covalent structure 36, 37
  greenhouse gas 198, 201
  natural gas 122, 196, 198, 204, 208
methanol 37, 216
methods in practical chemistry 238–9
methyl orange 129, 142, 143, 258
microorganisms (microbes)
  eutrophication 193
  extremophiles 115
  fermentation 114, 217, 218, 245
  making enzymes 114–15
  methane production 198
  oxidising agents/chorine to kill 79, 190
  water pollution 187–90
microparticles, plastic 231
minerals, in river water 188, 189
mixtures 15, 26, 27, 204–7
  see also separation methods
mobile phase, chromatography 252
mobile phones, metals used 179
molar gas volume 60
molar mass 57
molecular compounds 36
  see also covalent compounds; molecules; organic compounds
molecular elements 3, 34–5
molecular formulae, empirical formula comparison 66–7
molecules 3, 34–9
  formulae 46, 66–7
  names and formulae 46, 47
  non-conductors of electricity 83
  properties 39, 40
  relative molecular mass 51
  shapes 37
moles 56–71
  calculations from equations 58–9
  concentration of solutions 62–3
  finding empirical formulae 64–5
  finding molecular formulae 66–7
  percentage yield and % purity 68–9
  titrations 142–3
  volume of gases 60–1
molten ionic compounds, electrolysis 84, 86
monomers 224–7
M_r 51
  see also relative formula mass; relative molecular mass

N
naming compounds 46–7
  ionic compounds 32, 46, 47, 133
  organic compounds 219
naphtha fraction of petroleum 206, 212–13, 228
natural gas 122, 196, 198, 204, 208
natural polymers 223, 234–5
negative ions (anions) 29, 32–3, 264–5
neon 28, 147, 245
neutralisation reactions 133–5, 139, 258–9
neutral oxides 137
neutrons 16–17

INDEX

discovery 23
in isotopes 20
nickel 154, 155
ninhydrin locating agent 249
nitrates
  fertilisers 188, 192–3
  ions 33
  tests 265
  water pollution 188, 189, 191, 193
nitrogen molecules 35
nitrogen oxides, air pollution 196, 197
noble gases (Group VIII) 19, 28, 147
non-carbon power sources 200
non-conductors of electricity 82, 83, 225
non-metal oxides 137
non-metals
  acidic compounds 128
  comparison with metals 160–1
  covalent compounds 34–41
  electrical conduction 83
  from electrolysis 84
  halogens 150–1
  ionic compounds 31, 32
  Periodic Table 15, 146, 147, 153
  properties 161
NPK fertiliser 192
nuclear reactions/power 20, 200
nucleon number (mass number) 17, 21
nucleons 16–17
nucleus of atoms 16–17, 20, 23
nylon 226, 229

O

ocean floor 69, 115, 204
oceans, plastic pollution 231, 233
OILRIG (Oxidation Is Loss of electrons / Reduction Is Gain of electrons) 74
organic compounds 204–21
  families 208–19
  functional groups 209, 214, 216, 218
  petroleum fractions 204–7
  structural isomerism 211, 215, 216
  see also alcohols; alkanes; alkenes; carbonic acids; esters; origin (chromatography) 247
oxidation numbers 76–7, 155
oxidations 72
  ethanol to ethanoic acid 218
  measuring oxygen in air 195
  see also combustion; redox reactions
oxides
  acidic 137, 153
  amphoteric 137, 153
  basic 129, 135, 136, 153
  of metals 31
  neutral 137
oxidising agents 78, 79
oxygen
  in air 186, 194, 195
  from electrolysis 85–7
  importance for life 186, 187, 193, 194, 195
  ionic compounds 31
  isotopes 21
  molecules 35
  preparation in the lab 260
  reactions 48
  testing for 261
  in water 188, 189, 193

P

paper chromatography 246–9
particles 2–3
  collisions in reactions 95
  collision theory of reaction rates 110–11
  in different states of matter 6–7
  history of understanding 22–3
  mass determines rate of diffusion 11
  moles 56
particulates, pollution 187, 196

peer review 256
percentage calculations 52–3
  composition 53
  purity 53, 68–9
  yield 68
periodicity of elements 146
Periodic Table 15, 18–19, 146–59
  comparison of metals with non-metals 153, 160–1
  development/history 156–7
  patterns and trends 19, 146–55
  transition from metals to non-metals 153
period number 19, 147
PET (polyester) 227, 229, 232, 233
petroleum (crude oil)
  constituents 204
  fossil fuels 196–8, 200, 204–5
  fractional distillation/refining 206–7, 245
  uses 99, 196, 197, 205, 207
  see also organic compounds
petrol (gasoline) 67, 99, 197, 206, 207, 211, 213, 217
phosphates 188, 189, 192–3
photochemical reactions 196, 211
photochemical smog 196
photosynthesis 201
pH scale 130–1
physical properties
  alkali metals 148
  halogens 150
  metals 160
  transition elements 155
  trends in Periodic Table 148, 150, 153
physical versus chemical changes 27
physicists, history 23
pigments 141
plant nutrients *see* fertilisers
plastics 228–33
  burning 231, 232
  from petroleum derivatives 205, 213, 215
  pollution 188, 189, 230–1
  properties 225, 228
  recycling 232, 233
  uses 213, 215, 222, 224, 225, 229
  waste disposal 230–4
  see also polymers
pollution
  air pollution 39, 99, 196, 197
  water pollution 187–9, 230–1, 233, 253
polyamide 226
polychloroethene (polyvinyl chloride/PVC) 225, 229
polyester (eg PET) 227, 228, 229, 232, 233
poly(ethene) 213, 215, 222, 224, 228–9, 233
polylactic acid (PLA) 233
polymers 213, 222–37
  addition reactions 224–5
  condensation reactions 226–7
  drawing 225
  from alkenes 213, 215, 222
  identifying monomers 225
  see also plastics; proteins
polyphenylethene (polystyrene) 229
poly(propene) 225
polystyrene (polyphenylethene) 225, 229
polytetrafluoroethene (PTFE/Teflon) 225, 229
polythene 218
  see also poly(ethene)
polyvinyl chloride (PVC/polychloroethene) 225, 229
positive ions (cations) 29, 32–3, 84, 262–3
potassium, fertilisers 192
potassium iodide 46, 79, 151
potassium manganate(VII) 78
power generation 196, 197, 200, 205, 232
practical work in the laboratory 238–55
  gases 260–3
  identification tests 261–5

materials/apparatus/methods 238–41, 260–3
scientific method 256–9
separating substances 242–53
precipitation 140–1, 262–5
pressure 10
  rates of reactions 103–6, 109, 111
  reversible reactions 121–3, 125
  volume of gases 10, 60
products 48, 238
  electrolysis of a solution 85
  energy compared to reactants 92–3
  purity 239, 251
  total mass compared to reactants 52, 58
propane 208, 210–11
propene 214
proteins 223, 234–5
proton number (atomic number) 16, 17, 20, 146, 157
protons 16–17
  discovery 23
  donors and acceptors 134, 135
  see also hydrogen ions
PTFE *see* polytetrafluoroethene
pure water 190–1
  see also distilled water
purity 250
  checking 191, 247, 250–1
  heating and cooling curves 9
  maximising in products 239, 251
  percentage calculations 53
  percentage using moles 68–9
PVC *see* polyvinyl chloride
pyrolysis 232

R

RAC (reduction at cathode) 86
radioactivity 23
radioisotopes 20
rates of reactions 102–17
  catalysts 112–15
  changing concentrations 106, 109–11
  changing pressure of gases 109, 111
  changing surface area 108–9, 111
  changing temperature 107, 111
  collision theory 110–11
  enzymes 112–15
  measuring 103–5
  measuring decrease in reactants 103, 108
  measuring increase in products 103–7
  reversible reactions 120–5
reactants 48, 238
  energy compared to products 92–3
  total mass compared to products 52
reaction pathway diagrams 94–5
reactivity
  alkali metals 149, 152, 162
  halides 151
  halogens 150–2
  metals 5, 161–7, 173
  transition elements 154
  trends in Periodic Table 152
  unsaturated versus saturated hydrocarbons 213
reactivity series (metals plus hydrogen and carbon) 166
recycling 232, 233
redox reactions 72–81
  change in oxidation numbers 77
  electrolysis 86–7
  electron transfer 74–5
  hydrogen–oxygen fuel cells 99
  in the blast furnace 175
  metals in competition 164–5
  oxidising and reducing agents 78–9
  without oxygen 75
reducing agents 78, 79, 164, 167, 175
reduction at cathode, RAC mnemonic 86
reductions 72, 173, 175, 197

see also redox reactions
refinery gas 206, 210
refining metals 88
refining petroleum 206–7, 245
relative atomic mass (Ar) 15, 21, 50, 56
  see also moles
relative formula mass (Mr) 51
relative molecular mass (Mr) 51
  diffusion of gases 11
  moles 56–71
  polymers 224
renewable energy 200
reservoirs 191
respiration 73, 195
retention factor (Rf), chromatography 249
reversible reactions 118–27
  changing conditions 118, 120–5
  closed versus open systems 119, 121, 123, 124
  Contact process 124–5
  dissociation of weak acids 131
  dynamic equilibrium 119–21
  Haber process 122–3
  making ammonia 119–23
  making sulfuric acid 124–5
R_f see retention value
ring hydrocarbons 204
river water pollutants in 187–9, 193, 231
  source of water supply 190
room temperature and pressure, volume of a mole of gas 60
rubidium 19
rusting of iron 72, 102, 167–9, 178

S
sacrificial protection of iron from rusting 169, 179
safety
  clean water supply 190–1
  corrosive chemicals 128, 129
  flammable substances 27, 243
  hydrogen versus petrol 99
  industrial processes 123, 125
  purity of substances 251
salt
  rusting of iron 168
  see also sodium chloride
salts
  making using acids 132, 133, 138, 139
  see also ionic compounds
saturated hydrocarbons 210–11, 214
saturated solutions 138, 239, 241, 242
scientific method 256–9
sedimentation, water treatment 190
separation methods 239, 242–9, 251–3
sewage 187–9
shared electrons 34–5
silicon 153
silicon(IV) oxide (silica) 40, 41
simple distillation 244
single bonds 35, 209–11
slag, creation in blast furnace 175
smog, air pollution 196
sodium 16
sodium chloride 28
  bonding of sodium with chlorine 28–31
  crystal lattice structure 38
  formation by redox reaction 75, 77
  making from sodium hydroxide and hydrochloric acid 133, 139
  molar mass 56
sodium ions 29, 30, 262
soil 186–8, 192–3
solar power 200
solubility 240–1
  ionic versus covalent compounds 39
  of salts 140
  temperature 189, 241
  see also precipitation

soluble bases 129
  see also alkalis, hydroxides
soluble salts, making using acids 138–9
solutes 240–1
solutions 240–1
  concentration in moles 62–3
  see also aqueous solutions
solvent front in paper chromatography 247, 249
solvents
  chromatography 246–9, 252
  cryolite 176
  organic 26, 27, 39, 211, 212, 216, 241, 243
  removing 243–4
  water 62, 188, 240, 243
sonorous metals 160, 181
spectator ions 134, 140
stable compounds 166
stable isotopes 20
stable (unreactive) atoms 19, 28
stainless steel 155, 180, 181
standard atom 12C 21, 50
standard solutions 143
states at room temperature 5
state symbols, equations 48, 49
stationary phase, chromatography 252
steam 4–7, 122, 162, 200, 214, 216, 217, 227
steel 155, 180, 181
straight chain hydrocarbons 204, 208–15
strong acids 131
structural isomers 211, 215, 216
sub-atomic particles 16–19, 23
substitution reactions 211
sulfate/sulfite ions 33, 132, 265
sulfur 124, 197, 207, 212
sulfur dioxide 124–5, 196, 197, 261, 265
sulfuric acid 78, 85, 124–5
sulfur trioxide 124–5
surface area, rates of reactions 108–9, 111
symbol equations 48, 49
symbols of elements 14, 15
synthetic elements 147, 183
synthetic fertilisers 192
synthetic fibres 226–9, 231, 232

T
Teflon (polytetrafluoroethene) 225, 229
temperature
  catalysts speeding up reactions 113
  exothermic reactions 92
  extremophiles 115
  heating and cooling curves 8–9
  rates of reactions 107, 111
  reversible reactions 120, 123, 125
  solubility relationship 189, 241
tests
  for ammonia 261, 263, 265
  for anions 264–5
  for carbon dioxide 261, 265
  for cations 141, 262–3
  for gases 261
  for unsaturated compounds 215
  for water 118, 191
thermal decomposition 97, 167, 175
thermal energy
  activation energy for reactions 95
  changes of state 6–7
  exothermic/endothermic reactions 92–4
  greenhouse effect 198
  pressure and volume of gases 10
thymolphthalein indicator 129, 139, 258
titrations 139, 142–3, 258–9
toxic metals 189
transition elements/metals 146, 147, 154–5, 160
  catalysts 113, 154, 197
  compounds 33
  electron shells after calcium 19
  oxidation numbers 76, 155

transport
  alternative energy 98–9, 200
  catalytic converters 197
  fuelled by alternative energy 98–9, 200
  petroleum-based fuels 67, 99, 196, 205, 206, 207, 211, 213, 217
triple bonds 6, 35

U
universal indicator, pH scale 131
unreactive gases see helium; nitrogen; noble gases
unreactive metals 162, 166, 172, 173, 178
unsaturated hydrocarbons 214–15
unstable isotopes 20

V
valency of elements 47
vehicles see transport
viruses 234
viscosity 206–7, 209
volatility of liquids 39, 207, 216
volume of gas 10, 60–1

W
waste disposal 230
water
  changes of state 4, 6–9
  cooling curve 9
  covalent structure 36
  of crystallisation 141
  decomposing in electrolysis 87
  distilled 191, 244
  formula 26
  from neutralisation reactions 133–5
  heating curve 8
  importance 186
  making hydrogen for fuel cells 99
  molar mass 56
  pollution 187–9, 230–1, 233, 253
  as a solvent 39, 62, 188, 240–1, 243
  supply and treatment 79, 189–91
  tests 118, 191
  see also aqueous solutions
weak acids 131, 143
word equations 48, 49
writing up experiments 257–9

yeast 114, 217
yield of reactions 68, 119–25

zinc
  alloys 180, 181
  extraction from ores 173
  galvanising iron 170, 179
  uses 179